A Textbook of

ASTRONOMY AND ASTROPHYSICS

A Textbook of
ASTRONOMY AND ASTROPHYSICS

Dr. Mohit Kumar Sharma
M.Sc., M.Phil., Ph.D. (Physics)
Young Scientist Awardee (ISCA, 2014)

National Postdoctoral Fellow, SERB, DST, New Delhi
Amity Center for Astronomy & Astrophysics
Amity Institute of Applied Sciences
Amity University Uttar Pradesh
Noida, U.P., India

Prof. (Dr.) Suresh Chandra
AvH Fellow (Germany), FRAS (London), FMAS (Pune)

Deputy Director & Professor
Amity Center for Astronomy & Astrophysics
Amity Institute of Applied Sciences
Amity University Uttar Pradesh
Noida, U.P., India

TECHSAR™
NEW DELHI

Published and Printed by
Techsar Pvt. Ltd.
4435-36/7, Ansari Road, Daryaganj
New Delhi-110 002 (India)
E-mail: tech_info@techsarworld.com
Website: www.techsarworld.com

ISBN 978-93-86768-49-0

Reprint 2024

Dedicated to

Professor Dr. Wilhelm H. Kegel, Germany

Preface

We are aware of various subjects, like Physics, Chemistry, Mathematics, Botany, Zoology, etc. It is, however, obvious that all these subjects emerged out of a common topic, called the Science. Hence, in the early time, all the studies in Physics, Chemistry, Mathematics, Botany, Zoology, etc. were being carried out under one umbrella known as the Science. If we try to go still past, and think, how development of Science would have initiated. A reply may be that when the human-being, after development of sense of recognizing things, looked upon some natural phenomena, such as the rising and setting of Sun each day; appearance of stars and Moon in night (in the absence of Sun); solar as well as lunar eclipses; phases of Moon; etc., they would have intuitively started thinking, why and how these phenomena were taking place. Such events may probably be considered the beginning of the process of understanding of the Science. Hence, the study (the process of thinking) of natural phenomena may be thought as the beginning of Science. The studies of natural phenomena in the present time are carried out under the subject, called the Astronomy & Astrophysics (abbreviated as A&A).

In a clear, dark night, we see sparkling points of light apparently situated on a vast sphere (called, the celestial sphere) in the sky. All these points are not necessarily the stars. They may be stars, star clusters, galaxies, planetary nebulae etc. Because of large distances of these heavenly bodies (also known as the cosmic objects) from us, it is difficult to resolve them with the naked eyes and therefore, each of them appears like a star and we address them as the stars. The A&A is a branch of science which deals with the motions of heavenly bodies and with the science of their interior as well as the atmosphere.

This book deals with the Astronomy and Astrophysics, and can be used as a textbook for a course on Astronomy and Astrophysics. Some courses on A&A are being taught to M.Sc. (Physics or Maths.) students in various universities in India and abroad. In the latest syllabus of B.Sc., one course on the A&A is introduced by the University Grants Commission (UGC), New Delhi. Basically, the text presented here is the print form of the notes, which have been revised from time to time and are used for class room teaching. One of the policies followed in this book is to discuss each topic in detail, in a systematic manner and easily understandable mode as well as language.

After briefing about the beginning of Astronomy, we have discussed about various coordinate system used to specify position of an object on the earth and in the space. Time is an important parameter and can help in having information about the relative occurrences of various events. The limitations of the observations with our eyes can be largely extended with the help of telescopes. Further, the telescopes help us for making observations of those radiations which cannot be seen by human eyes.

The fundamentals used for the study of various phenomena are summarized. Details about the Sun are discussed, as being the closest star, it has has been studied in detail. Information about the solar family is described at length. Thermochemical reactions are responsible for creation of energy in the shining objects. Information about various types of thermochemical reactions is incorporated. Details of the galaxies and the universe are discussed. A brief description about the Cosmology - the science about the formation of universe is mentioned. The text cannot considered as complete in absence of a discussion about the possibility of the existence of life in the universe other than that on the Earth.

While preparing the manuscript, we have consulted various books and got their help. we are grateful to the authors and publishers of those books. Some images are taken from the website of Google. We thankfully acknowledge the Google website. For this book, we have been helped, advised and encouraged by our friends working in various institutions/universities in India as well as abroad, and by our friends in the personal life. We are heartily thankful to all of them. Prof. Chandra would like to thank his students whom he taught the subject and got feed back for the improvement of the manuscript. We would greatly appreciate receiving comments and suggestions from the students, our friend teachers and others, which can be sent also to either of the e-mail addresses:

mksharma4@amity.edu; mohitkumarsharma32@yahoo.in
schandra2@amity.edu; suresh492000@yahoo.co.in

We are grateful to Hon'ble Dr. Ashok K. Chauhan, Founder President, Amity University, Hon'ble Dr. Atul Chauhan, Chancellor Amity University Uttar Pradesh, Prof. Balvinder Shukla, Vice-Chancellor Amity University Uttar Pradesh, Prof. Sunita Rattan, Director, Amity Institute for Applied Sciences, and Prof. S.N. Shukla, Dept of Applied Physics, Amity University, Noida for their moral support and encouragement.

Some part of the text was prepared during the visit of Suresh Chandra to the the Max-Planck-Institut für Radioastronomie, Bonn, Germany, the Technical University, Berlin, Germany and the University of Kassel, Kassel, German. We are grateful to Prof. Dr. D. Breitschwerdt, Prof. Dr. W.H. Kegel, Prof. Dr. T.F. Giesen, and Prof. Dr. K.M. Menten and for their research groups for their warm and memorable hospitality. Thanks are due to the D.F.G., Germany and the Alexander von Humboldt Foundation, Germany for financial support for the visits to Germany.

Suresh Chandra is specially thankful to his wife Mrs. Purnima Sharma for her valuable cooperation in his life and for sharing a major part of the responsibility of family affairs, so that he could spend his time for this book. Mohit Kumar Sharma is thankful to his wife Mrs. Shalini Sharma for her valuable cooperation. We are thankful to Dr. Monika Sharma and Mrs. Shalini Sharma for their help in preparation of the manuscript of this textbook. We are deeply grateful to our family members who always have been a source of inspiration and happiness for us. In the last, but not the least, we are highly thankful to the publisher for bringing out this book nicely.

Mohit Kumar Sharma
Suresh Chandra

CONTENTS

1. Beginning of Astronomy

The eye is one of the senses provided to the human being and the Earth's atmosphere is transparent for the visible part of electromagnetic spectrum, coming from the outer space. Combination of these two factors did not let the human being to remain ignorant about the beautiful phenomena occurring in the nature. For example, from the inception, people noticed the rising and setting of Sun each day; phases of Moon; solar and lunar eclipses; occurrence of seasons; etc. As a natural habit of human being, after noticing such amazing phenomena, one would naturally put questions from oneself or with the others; how and why these phenomena are occurring.

The phases of Moon could be understood by considering spherical shape of the Moon and its revolution around the Earth. The appearance of day and night could be explained with the help of the spin motion of the Earth about its axis. A big breakthrough in this field was to find the depth in the sky, because apparently, all celestial objects look as situated on a common celestial sphere. The difference in the distances of the Sun and Moon from the Earth and with the total solar eclipse provided idea about different sizes of Sun and Earth. Another breakthrough at par was to find the spherical shape of the Earth.

One more milestone may be considered as the idea of heliocentric system, replacing the geo-centric system, which failed to resolve the retrograde motions of Mars, Jupiter, Saturn when observed from the Earth. All these aspects of Astronomy & Astrophysics are discussed in this chapter.

1. Evolution of Astronomy

After noticing a number of phenomena occurring in nature, attempts would have been made to understand them. In the following sections, we have discussed how these phenomena could be understood and how further developments continued.

1.1 Phases of Moon

Apparent shape of the Moon was found to change from day to day, for example, 13 phases of Moon are shown in Figure 1. Sometimes the Moon is full (full Moon night; central position). Then, it decreases in size and the shape changes, and disappears (new Moon night). The Moon again appears (first position on the left), and its size increases and the shape changes in the reverse manner.

For the explanation of the phases of Moon, Anaxagoras (500 BC - 428 BC)[1] formulated the idea that the Moon is spherical in shape and the sunlight is falling

[1]The abbreviation BC stands for the Before Christ. For example, for the year 2019. The 500 BC means 2019 + 500 = 2519 years ago.

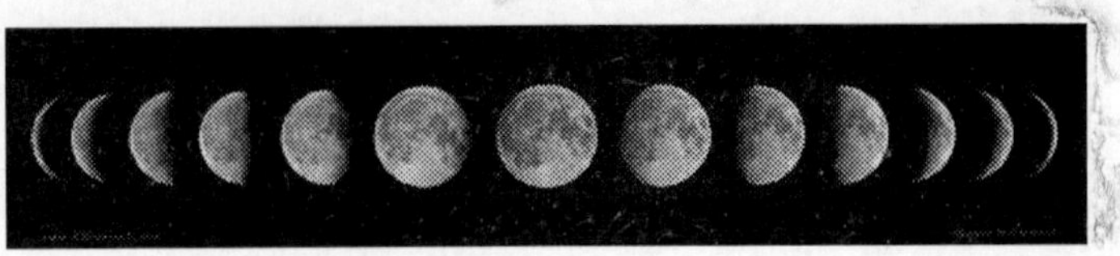

Figure 1: Various phases of Moon. Courtesy to Google website.

on the Moon. Thus, always, half of the surface of Moon is bright and the rest half is dark, as shown in Figure 2, by the inner circle around the central body, the Earth. The Sun (not shown in the figure) is on the left side. As the Moon revolves around the Earth, the area of its surface which reflects the sunlight, in the direction of the Earth, changes with time. Consequently, the phases of Moon are seen, as shown in the outer circle around the Earth. This way, an acceptable explanation for the phases of Moon was provided.

Figure 2: The spherical body at the center is the Earth. In the inner circle around the Earth, eight positions of the Moon are shown. The Sun (not shown in the figure) is on the left side. Half of the surface of Moon is always bright. The outer circle around the Earth shows how the Moon looks like when seen from the Earths surface. Courtesy to Google website.

The shape of the Moon appears to change every day. On the full-moon night, the Moon appears as a full circular plate. After about two weeks, the Moon is not visible, and the night is known as the new-moon night. At the time of full Moon, the Earth is between the Sun and the Moon, and at the time of new Moon, the Moon is between the Sun and the Earth.

1.2 Distances of Sun and Moon

For understanding the phases of Moon, Aristarchus (320 BC - 250 BC) was probably the first person to realize that when the Moon appeared as half in size, the $\angle$ SME is 90°, as shown in Figure 3. Here, S, M, E, respectively, denote the Sun, Moon, and Earth. After measuring the $\angle$ SEM from his position E on the Earth's surface, he could work out the ratio of the distances ES and EM. It is not known how he actually measured the $\angle$ SEM to be 87° and therefore he obtained the distance ES to be 20 times more than the distance EM. (Accurate value of the angle is 89° 51′ and the ratio ES/EM is 391.)

Aristarchus did not realize that he had tackled a very difficult problem as he had no way to know the exact moment when the appearance of Moon was half and possibly he had not obtained an accuracy of 1° in the measurement of angle. This estimation, however, was a remarkable discovery, as he proved that the space has depth and the Sun and Moon do not move in a common surface around the Earth. (In those days all heavenly bodies, including the Sun and Moon, were assumed to revolve around the Earth and the Earth was considered the center of the whole universe. This arrangement is known as the geo-stationary system.)

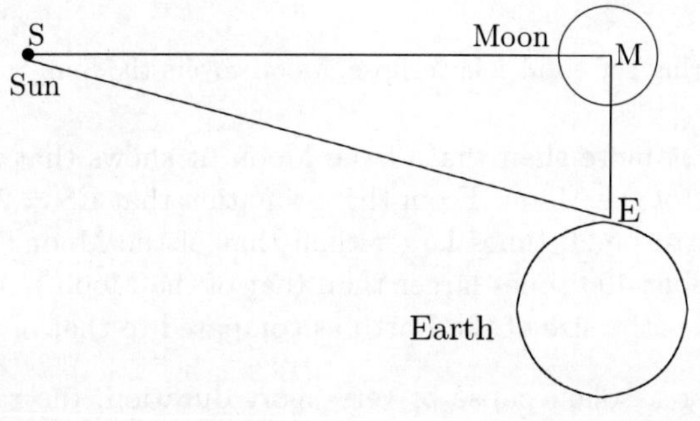

Figure 3: Relative distances of Sun (S) as well as Moon (M) from the Earth (E); not to the scale.

Example 1: By using the idea of the phases of the Moon show that the distance of the Sun from the Earth is more than that of the Moon from the Earth.

Solution: As shown in Figure 3, when the Moon is half, the $\angle$ SME is nearly 90°. In a triangle, we can have only one right angle and other two angles are acute. Using the sin-law, we have

$$\frac{\sin \angle SME}{SE} = \frac{\sin \angle SEM}{SM} = \frac{\sin \angle MSE}{ME}$$

Since $\angle SME$ is larger than the $\angle MSE$, we have SE larger than ME. That is, the distance of the Sun from the Earth is more than that of the Moon from the Earth.

1.3 Sizes of Sun and Moon

After having knowledge that from the Earth, the distance of Sun is larger than that of the Moon, an explanation for the occurrence of total solar eclipse was that at the time of the eclipse, the Moon covers the Sun, as shown in Figure 4. Since the duration of a total solar eclipse is not very large, it was thought that the angular sizes of both the Moon and Sun were nearly equal. Thus, we have

$$\text{angle} = \frac{\text{size of Moon}}{\text{distance of Moon}} = \frac{\text{size of Sun}}{\text{distance of Sun}}$$

Therefore, we have

$$\text{size of Sun} = \frac{\text{distance of Sun}}{\text{distance of Moon}} \times \text{size of Moon}$$

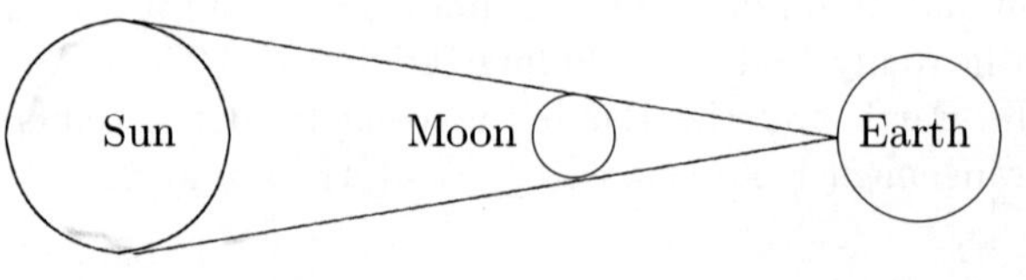

Figure 4: During the total solar eclipse, Moon covers the Sun.

Since the distance of Sun is more than that of the Moon, it shows that the size of the Sun is larger than that of the Moon. From the estimation that ES = 20 EM, the size of Sun was estimated to be 20 times larger than that of the Moon. (Accurate value is that the Sun is about 400 times larger than that of the Moon.) Aristarchus then posed a question about the size of the Earth as compared to that of the Moon.

Example 2: Suppose, for a solar eclipse of very short duration, the ratio of the distances of Sun and Moon from the Earth is 390. In the case of a solar eclipse of large duration, comment for the ratio of the distances of the Sun and Moon from the Earth.

Solution: The sizes of the Sun and Moon are constant. For an eclipse of large duration, the angular size of Moon is larger than that of the Sun. Hence, the Moon is closer to the Earth, as compared to its earlier position. Thus, the ratio of the distances of the Sun and Moon from the Earth is more than 390.

Example 3: Suppose, for a solar eclipse of very short duration, the ratio of the distances of the Sun and Moon from the Earth is 390. Comment for the ratio of the distances of Sun and Moon from the Earth, so that the solar eclipse does not take place.

Solution: The sizes of the Sun and Moon are constant. For no eclipse, the angular size of Moon is smaller than that of the Sun. Hence, the Moon is farther away from the Earth, as compared to its earlier position. Thus, the ratio of the distances of the Sun and Moon from the Earth is smaller than 390.

1.4 Size of Earth

It was clear that a lunar eclipse occurs when the Moon moves into the shadow of the Earth, as shown in Figure 5. During the lunar eclipse, the Earth is between the Moon and Sun. According to Aristarchus, when the arc ABC of the Earth's shadow is completed to describe the circle ABCDA, the diameter of this circle is equal to the Earth's diameter. Once again, he had encountered a difficult problem. He estimated the diameter of the Earth to be twice that of the Moon. (Actual diameter of the Earth is 3.7 times that of the Moon.)

Thus, according to Aristarchus, the size of Sun was 10 times larger than that of the Earth. (Actual diameter of the Sun is 100 times that of the Earth.) The importance of the work of Aristarchus lies in the fact that he proved first time that the Earth, although larger than the Moon, was relatively smaller than the Sun.

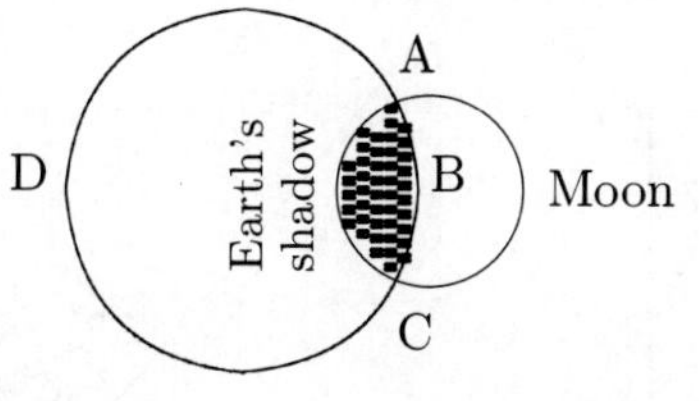

Figure 5: Earth's shadow eclipses the Moon.

Note: It was therefore reasonable to assume that the Moon being smaller than the Earth revolves around the Earth. But, it was not easy to accept that the Earth being smaller than the Sun should revolve around the Sun. Archimedes, who was some younger than Aristarchus, mentioned in his writing that Aristarchus had a model that the Earth revolves around the Sun along with its five bodies (Moon, Venus, Mars, Jupiter and Saturn). The bodies, named as the Venus, Mars, Jupiter, and Saturn were identified by that time. Aristarchus believed also correctly that the stars remain fixed as they are very far away from the Earth.

1.5 Shape of Earth

Some 60 or 70 years later, Eratosthenes (276 BC - 194 BC) made another major step to know the shape of the Earth. He discovered that when the Sun was farthest north, on the day of solstice in mid-summer, it was vertically above the town of Syene and cast no shadow on the water in a well at that place when it was mid-day. At the same time, the pole planted vertically at Alexandria, 4450 stadia (824 km) to the north of Syene, cast a shadow which made an angle of 7.5° with the vertical, as shown in Figure 6.

He tried to understand the observations by assuming that the Earth is flat and the Sun S is vertically above the Syene B, so that the $\angle$ SBA $= 90°$, where A denotes the Alexandria. At the same time, the Sun's rays subtend an angle of 7.5° with the top of a vertical pole at the Alexandria. Thus, $\angle$ SAB $= 82.5°$. The calculated value of SB was found to be 33800 stadia. It was however already known that the Sun is very far away from the Earth.

But, Eratosthenes did not fall into the trap of assuming that the Earth is flat. He had sufficient proof to assume that the Earth is spherical in shape and not flat, and for that he argued that the Sun is so much away that its rays can be taken as parallel when they arrive the Earth. Hence, although the Sun's rays are vertical at B, the angle of 7.5° between the rays and the vertical pole at A occurs because the Earth's surface is curved and not flat. (In absence of curvature, the angle would be zero.)

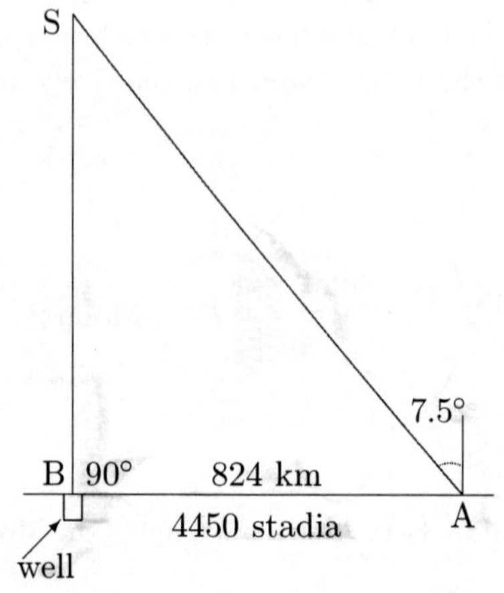

Figure 6: Flat Earth between the Syene B and Alexandria A. S denotes the Sun.

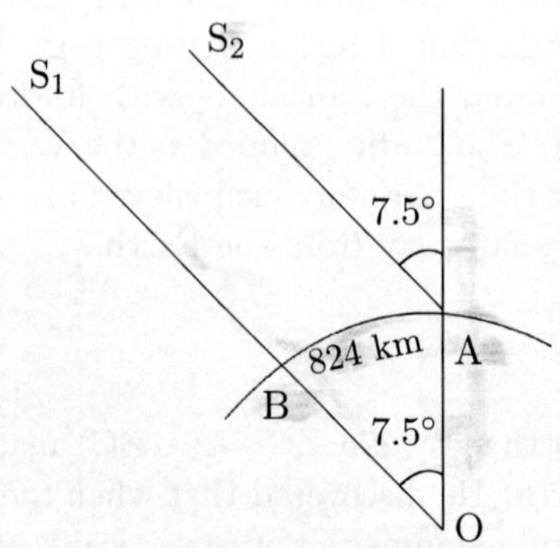

Figure 7: Arc BA subtends an angle 7.5° at the Earth's center.

The angle of 7.5° will thus be equal to the ∠ BOA at the Earth's center, as shown in Figure 7. Hence, the circumference of Earth = (360/7.5) × 824 = 39552 km. (Accurate value of the circumference of Earth through north pole is 40009 km. There is error of only 1.1%. Actually, he was lucky that the angle was 7.5°, which is rather easily measurable, as the successive division of 360° gives 7.5°.)

1.6 Seasons on Earth

Eratosthenes also succeeded in measuring the angle between the Earth's equator and the ecliptic[2]. His value was very close to the correct value 23.5°. It is this inclination of Earth's equator to the ecliptic that is responsible for the seasons on the Earth. In case, the angle of inclination had been zero, there had been no seasons (there had always been the same atmosphere) as the Earth's rotation axis would then be at right angle to the solar rays. There would had been hot strips on both sides of the equator, with subsequent strips becoming progressively cooler as they were further away from the equator.

1.7 Day and night on Earth

The Earth spins about its own axis in a period of 24 hours. The day and night are due to the spin motion of the Earth. A plane tangential to the surface of Earth and passing through the position of the observer, is known as the **horizon** of the observer. When the sun moves from lower side to upper side of the plane, it is called the rising of sun. On the other side, when the sun moves from upper side to lower side of the plane, it is called the setting of sun. The duration from the rising to setting of sun is called the day, and the duration from the setting to rising of sun is called the night.

1.8 Revolution of Moon

The ancient Greeks knew that the Moon takes 29.5 days between one new Moon and the next. This period is known as the Moon's synodic period. To move once around the sky and to come back to the same position with regard to the stars took only 27.5 days. This period is known as the sidereal period, discussed later on.

1.9 Sidereal periods of Mars, Jupiter and Saturn

The Greeks later on found that the sidereal period of Mars was 687 days, of Jupiter 12 years, and of Saturn 30 years. These three planets (of Earth) had to be much away from the Earth as compared to the Moon. At that time, the Earth was considered at the center of the universe, and the Moon, Sun, Mars, Jupiter and Saturn were considered the planets of the Earth (geo-centric system).

1.10 Motions of Mercury and Venus

Venus and Mercury did not show movement right across the sky, but oscillated to and fro about the Sun. Venus never moved more than 47° away from the Sun and Mercury never moved more than 28° away from the Sun. It was very difficult to explain these observations. A reasonable guess was that the Venus and Mercury were nearer to the Sun than to the Earth.

---

[2]Information about the Earth's equator and the ecliptic is discussed later on.

1.11 Positions of stars

With the passage of time, even ages, the positions of the stars remained unchanged. It was understood in terms that the distances of the stars were much larger than those of the planets.

In early days, Greek astronomy reached at its height with Hipparchus in the second century BC. He compiled a star atlas in which he registered 1000 stars. He very painstakingly measured the positions of stars to an accuracy of $(1/15)^\circ$, and noticed that the positions of the stars on his star-charts differed from those on older charts by small amounts in a systematic manner. He deduced a brilliant idea that this was due to the slow westward movement of the point (now called the vernal equinox) where the equator intersects the ecliptic. The movement was estimated to be $40''$ per year. At this rate, one circle is completed in 32400 years. (We know that the precession of equinox is actually $50.3''$ per year and the Earth's axis describes a circle in 25,765 years.)

1.12 Length of day

Hipparchus measured the length of a day to an accuracy of six minutes.

1.13 Magnitudes of stars

Hipparchus classified the stars according to their apparent brightness. The brightest stars were classified as having magnitude +1, and the faintest, that the eye could detect, as having the magnitude +6. The star having magnitude +6 was found about 100 times fainter than that having the magnitude +1. That is, all the visible stars had magnitudes from +1 to +6.

2. Copernicus revolution

At that time, scientists were aware of the Moon, Sun, Mercury, Venus, Mars, Jupiter, Saturn, besides the Earth. The idea about their arrangement has been that the Earth is situated at the the center and the Moon, Sun, Mars, Jupiter, Saturn revolve around the Earth (geo-centric system). Further, the Mercury and Venus have been associated with the Sun. For 1600 years after Hipparchus, no remarkable progress was made in astronomy. The world had to wait until 1543 when Copernicus' book on the solar system was published.

Nicolaus Copernicus (Figure 8) was a priest in the Roman Catholic church. He spent many years on the study of the movement of Sun, Moon and planets against the background stars. One of the main problems faced by him has been the retrograde motion shown by the Mars, Jupiter and Saturn, when observed from the Earth.

Copernicus tried to understand the retrograde motion which can be as shown in Figure 9.[3] Being at larger distance from the Sun as compared to the distance of the Earth from the Sun, the Mars moves with a slower speed as compared to that of the Earth. In Figure 9, at six different increasing times, positions of Earth (E_1, E_2,

[3]Notice that the geocentric system was given Before Christ (BC).

Figure 8: Nicolaus Copernicus (1473 - 1543). Courtesy to Google website.

E_3, E_4, E_5, E_6) and corresponding positions of Mars (M_1, M_2, M_3, M_4, M_5, M_6) are shown. Notice that when the Earth and Mars are observed from the Sun, both are moving properly (along the continuous orbits, without any retrograde motion).

However, when the Mars is seen from the Earth, *i.e.*, E_1 is combined with M_1, and further extension gives position 1 on the celestial sphere (discussed later on). Similarly, E_2, E_3, E_4, E_5, E_6 are combined with M_2, M_3, M_4, M_5, M_6, respectively, and their further extension gives the positions 2, 3, 4, 5, 6, respectively, on the celestial sphere. Thus, when seen from the Earth, the Mars does not always show motion in one direction, but also shows motion in the opposite direction. This explains the retrograde motion of Mars when seen from the Earth.

Similarly, the retrograde motion of Jupiter and Saturn seen from the Earth can be understood.

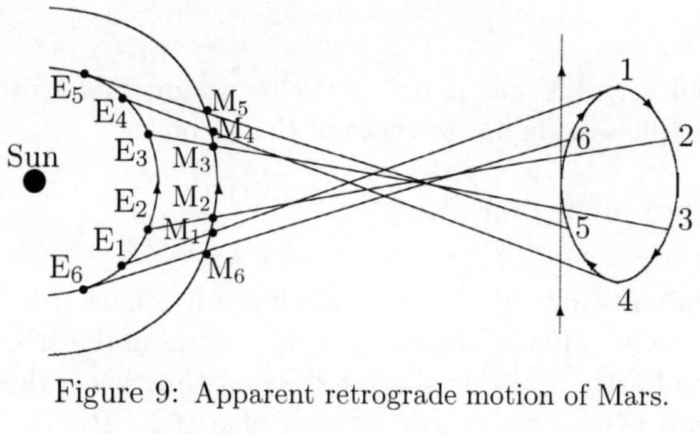

Figure 9: Apparent retrograde motion of Mars.

Based on his observations, Copernicus derived important conclusions. Some of them are the following.

1. The Earth is spherical in shape.
2. The Earth spins from west towards east, and the atmosphere moves along with its surface. In case the atmosphere was not to move, there should have been a continuous wind blowing from east towards west. Because of the spin motion of the Earth, the Sun, Moon, and stars appear to rise in the east and to set in the west; in the opposite direction to that of the spin motion of the Earth.
3. The Earth is not the center of the universe, but the Sun is the center of the universe and the Earth as well as other planets revolve in the circular paths around the Sun.
4. The Earth describes a circle around the Sun, moving from west to east; the stellar background must move in the opposite direction from east to west, and this is exactly what is observed.
5. The planets, Mars, Jupiter and Saturn sometimes show retrograde motion. That is, when their directions of motion are opposite to that of the Sun. Retrograde motions decreases in sequence of Mars, Jupiter and Saturn. This shows that the Mars is nearer to the Sun as compared to the Jupiter, and Jupiter is nearer as compared to the Saturn.
6. The sidereal periods of the planets are indicators of their distances from the Sun. Mars takes 687 days, Jupiter takes 12 years, and Saturn takes 30 years. Hence, the Saturn is farthest from the Sun and the Mars is the nearest.
7. The Earth having a period of revolution of 365 days is nearer to the Sun as compared to the Mars.
8. Mercury and Venus never move completely around the sky, as they move in the circles within the circular orbit of the Earth. Their sidereal periods are 88 and 225 days, respectively. Thus, Mercury must be nearer to the Sun as compared to the Venus.
9. Since the Mercury and Venus move in circles within that of the Earth, they should show the phases, similar to those of the Moon.
10. The Moon revolves around the Earth.

Copernicus system of revolving planets is as shown in Figure 10. This is known as the helio-centric system. That is, the Sun is at the center of the orbits of revolving planets, including the Earth. Thus, it was for the first time when the idea of helio-centric system instead of the geo-centric system emerged. By 1530, Copernicus completed his writing, but he continued to revise and edit the work. In 1541, the first report about the work of Copernicus was published.

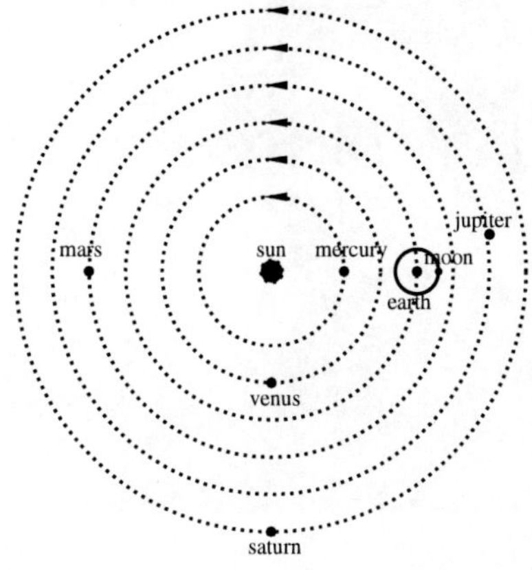

Figure 10: Copernicus system of revolving planets.

2.1 Reaction on the concept of Copernicus

Soon after knowing about the Copernicus system of revolving planets around the Sun, where Earth also was revolving around the Sun, like other planets, attacks started against him by Roman catholic churches as well as by many others. In the meantime, the manuscript of the writing of Copernicus was in the hands of Osiander, who was a churchman. He took it to a publisher. Aware of the attacks which had been made on the first report, Osiander added a preface to the effect that the ideas in the book were not to be taken as the absolute truth, but merely as a mathematical method for calculating the planetary positions.

The book entitled 'On the Revolution of the Celestial Spheres' was published in 1543, just two months before the death of Copernicus (24 May 1543). Copernicus acknowledged the contributions made by Aristarchus and stated that the stars do not show annual motion due to parallax because they are very far away from the Earth. Copernicus was the first to give the correct explanation of the fact that Mercury and Venus do not move around the sky, but oscillate east and west of the Sun through 28° and 47°, respectively. It is because their orbits lie within that of the Earth.

2.2 Giordano Bruno

The opposition of the church was something that had to be taken into account. Giordano Bruno (Figure 11) was a keen supporter of the ideas of Copernicus, and he delivered lectures in France, England and Italy. He also said that the planets were the worlds like the Earth. In 1660, he was charged before the Inquisition and after a summary trial, was burnt alive at the stake.

3. Work of Tycho Brahe

The first accurate determination of planetary positions were made by Tycho Brahe (Figure 12). He did not accept the Copernicus system of planets revolving around

Figure 11: Giordano Bruno (1548 - 1600). Courtesy to Google website.

the Sun. His argument was that if the Earth revolved around the Sun, the nearer stars would describe small circles against the background of more distant stars, as they will show the effects of parallax, as shown in Figure 13.

There X is a nearby star and there are stars, situated at very large distances. The Earth is revolving around the Sun in an elliptical orbit ABCD. The star X is seen from different positions A, B, C, D on the Earth at different times. The projections of the star X on the celestial sphere are respectively denoted by a, b, c, d. Thus, the star X appears to move along a small ellipse denoted by abcd. The size of the ellipse abcd depends on the distance of the star X from the Earth.

Figure 12: Tycho Brahe (1546 - 1601). Courtesy to Google website.

He further argued that if the Mercury and Venus revolve in the orbits within that of the Earth, they should show phases, similar to those shown by the Moon, and the phases were not seen. (The reason for this was found later to be the limitation of instruments available those days.)

According to Brahe's system, the Moon, Sun, Mars, Jupiter and Saturn revolve around the Earth, while the Mercury and Venus revolve around the Sun, as shown in Figure 14. Even if it is the case, Mercury and Venus still should show the phases when they arrive between the Sun and Earth. But, since Brahe had no telescope he could not see the phases of Mercury and Venus. However, Brahe rendered a great service to astronomy by making accurate measurements of planetary positions, especially of Mars. These data enabled Johannes Kepler to formulate the mathematical laws of planetary motion.

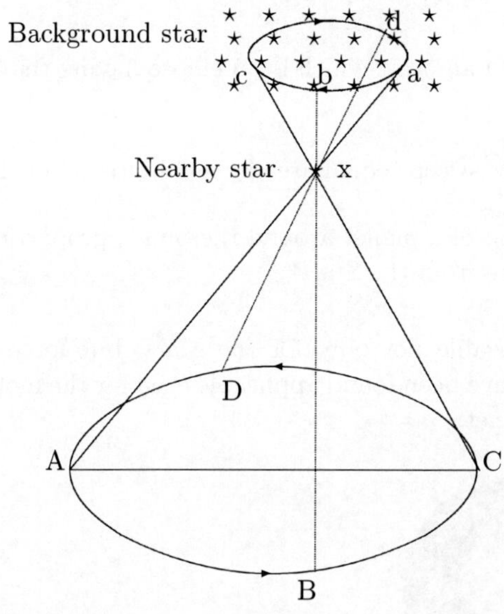

Figure 13: Parallax of a nearby star. The ellipse ABCD denotes the elliptical orbit of Earth around the Sun.

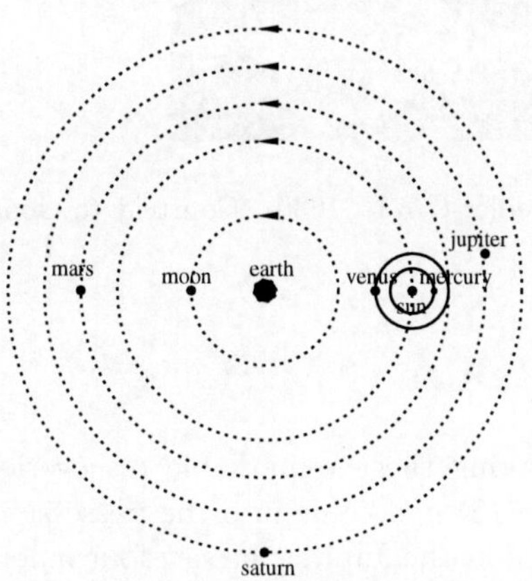

Figure 14: Tycho Brahe system of revolving planets.

4. Work of Johannes Kepler

For one or two years, Johannes Kepler (Figure 15) worked with Tycho Brahe. They had acrimonious arguments about the Copernicus system. Kepler was always in support of the Copernicus system whereas Brahe was not ready to accept it. After the death of Brahe in 1601, Kepler analyzed the Brahe's data for planetary positions and he was able to prove that the orbit of Mars is not a circle but an ellipse. Finally, he formulated the following three laws for planetary motion.

1. A planet revolves around the Sun in an orbit which is an ellipse having the Sun at one of the foci.

2. The line joining the Sun to a planet sweeps equal area in equal interval of time.

3. The square of the time of revolution of a planet around the Sun is proportional to the cube of its average distance from the Sun.

These laws have been found applicable not only for the Mars, but for other planets as well. Moreover, these laws have been found applicable even for the motion of a satellite around the respective planet.

Figure 15: Johannes Kepler (1571 - 1630). Courtesy to Google website.

5. Work of Galileo Galilee

While Kepler was busy in making these remarkable discoveries in planetary dynamics, Galileo Galilee (Figure 16) had come into the possession of one of the first telescopes. When Galileo looked at the Jupiter, he saw four moons revolving around the planet. It was like a solar system in miniature where also the Kepler's laws could be applied. Galileo also observed the phases of Venus. Out of fear of being ridiculed, he stated this finding in the form of a mysterious riddle.

Figure 16: Galileo Galilee (1564 - 1642). Courtesy to Google website.

Galileo noticed that the Saturn appeared to have two ears, which a few years later seemed to have vanished. These were of course, Saturn rings which become invisible every 14.5 years when the Earth is in the same plane of the rings.

The use of telescope, initiated by Galileo, started a revolution in astronomy. In 1633, Galileo was summoned before the Inquisition and was forced to recast his teachings, supporting the Copernicus system. Mindful of Bruno's fate in 1600, Galileo was very wise to recast. He was placed under the house arrest, where he spent the last nine years of his life working out the principles on which Isaac Newton, who was born in the year of Galileo death, was able to formulate his law of gravitation.

6. Work of Newton

Isaac Newton (Figure 17) was born within a year of Galileo death. The work of the principles derived by Galileo was extended by Newton in the form of the law of

Figure 17: Isaac Newton (1643 - 1727). Courtesy to Google website.

gravitation. This law states that the force of attraction F between two point-masses m_1 and m_2, separated by a distance r, is

$$F = -G\,\frac{m_1 m_2}{r^2}$$

where G is universal gravitational constant and the negative sign shows that the force is attractive. This law of gravitation can also be derived from Kepler's laws of planetary motion.

7. Review Questions

1. What are phases of Moon? Provide an explanation for the phases of Moon. When the Moon has been flat in shape, could we see its phases.
2. How the idea was obtained that the distance between the Moon and Earth is smaller than that between the Sun and Earth.
3. How the information about the depth in the sky was obtained.
4. How it was decided that the size of sun is larger than that of the moon.
5. How the size of Earth was estimated for the first time.
6. How it was decided that the Earth's surface is not flat, but spherical in shape?
7. Why the stars are considered to be fixed in position. Define the sidereal period of a celestial body.
8. What is Copernicus' contribution about the planetary system? What was the reaction on it?
9. Though the Mercury and Venus revolve around the Sun, but they are noticed only when they are east or west of the Sun. Comment on it.
10. Define the parallax. How it can be used to distinguish between the nearby and the far distant stars.
11. Write the name of scientist who used the telescope first time for observing outside the Earth. What were the consequences of those observations?
12. When seen from the Earth, the Mars shows its retrograde motion. Explain. How the problem of retrograde motion was sorted out.
13. Write the opinions of Tycho Brahe and Johannes Kepler about the arrangement of planets in our solar system.
14. Write the observations of Copernicus about the helio-centric system. Write the name of the person who propagated this idea of helio-centric system and what have been the consequences of that.
15. Describe the geo-centric system for the arrangement of planets. What is the helio-centric system for the arrangement of planets.

16. Write the sequence of planets in the Tycho Brahe system.
17. Write the sequence of planets in the Copernicus system.
18. Write the names of celestial objects which were known up to the time of Copernicus.
19. Write the names of two celestial objects which were considered planets in the Tycho Brahe system, but are not now.

Ans: Sun, Moon

20. Write the names of one celestial object which was not considered planets in the Tycho Brahe system, but is now.

Ans: Earth

21. Write notes on the following:

(i) Phases of Moon
(ii) Full-Moon night
(iii) New-Moon night
(iv) Seasons on the Earth
(v) Day and night on the Earth
(vi) Magnitudes of stars
(vii) Retrograde motion
(viii) Copernican revolution
(ix) Role of Giordano Bruno
(x) Contribution of Tycho Brahe
(xi) Kepler laws of planetary motion
(xii) Contribution of Galileo Galilee
(xiii) Contribution of Newton

8. Multiple Choice Questions

1. Who proposed the first explanation about the phases of Moon.

A. Einstein B. Galileo C. Anaxagoras D. Newton

Ans. C

2. The idea about the spherical shape of Moon emerged from the

A. solar eclipse B. lunar eclipse

C. phases of Moon D. spherical shape of Earth

Ans. C

3. Out of the Sun, Moon and Earth, which one is in between the others two on the full-moon night.

   A. Sun B. Earth C. Moon D. Any one

   Ans. B

4. Out of the Sun, Moon and Earth, which one is in between the others two on the new-moon night.

   A. Sun B. Earth C. Moon D. Any one

   Ans. C

5. When the Moon is half in shape, the angle between the directions joining the Moon to the Sun and to the Earth is about

   A. 30° B. 45° C. 60° D. 90°

   Ans. D

6. Knowing that the size of Sun is larger than that of the Moon, from the Earth, the distance of Sun is larger than that of the Moon can be decided with the help of

   A. solar eclipse B. lunar eclipse C. phases of Moon D. seasons on Earth

   Ans. A

7. Approximate relation between the angular sizes S and M of the Sun and Moon, respectively, is

   A. $S = 200\ M$ B. $S = 2\ M$ C. $S = M$ D. $2\ S = M$

   Ans. C

8. The size of the Earth was first time estimated with the help of

   A. solar eclipse B. lunar eclipse C. phases of Moon D. seasons on Earth

   Ans. B

9. Who was never convinced with the helio-centric system.

   A. Copernicus B. Tycho Brahe C. Kepler D. Einstein

   Ans. B

10. Who was burnt alive when he supported the helio-centric system.

    A. Copernicus B. Galileo C. Kepler D. Giordano Bruno

    Ans. D

11. Which shows the phases.

    A. Mars B. Jupiter C. Mercury D. Saturn

    Ans. C

12. Which shows the phases.

    A. Mars B. Jupiter C. Venus D. Saturn

    Ans. C

13. Which does not show the phases.

    A. Moon B. Mercury C. Venus D. Mars

    Ans. D

14. Which does not show the phases.

    A. Moon B. Mercury C. Venus D. Jupiter

    Ans. D

15. Which does not show the phases.

    A. Moon B. Mercury C. Venus D. Saturn

    Ans. D

16. The phases are shown by

    A. inner planets only B. both inner and outer planets

    C. outer planets only D. neither inner nor outer planets

    Ans. A

17. Who was the author of the book entitled 'On the Revolution of the Celestial Spheres'?

    A. Galileo B. Copernicus C. Hipparchus D. Tycho Brahe

    Ans. B

18. Who gave the magnitude system of stars.

    A. Galileo B. Copernicus C. Hipparchus D. Tycho Brahe

    Ans. C

19. Who derived the famous laws for the planetary motion around the Sun.

    A. Copernicus B. Tycho Brahe C. Kepler D. Einstein

    Ans. C

20. Whose data were used for the derivation of Kepler laws for planetary motion revolving around the Sun.

    A. Copernicus B. Tycho Brahe C. Kepler D. Einstein

    Ans. B

21. The Sun rises in the east and sets in the west because the

    A. Sun moves from east towards west B. Earth revolves around the Sun

    C. Earth spins about its own axis D. Sun spins about its own axis

    Ans. C

22. The seasons on the Earth are seen because the

A. Sun moves from east towards west B. Earth revolves around the Sun

C. Earth spins about its own axis D. Sun spins about its own axis

Ans. B

23. Which is not a naked-eye planet.

A. Saturn B. Jupiter C. Uranus D. Mars

Ans. C

24. In the geo-centric system, at the center of planetary system is

A. Sun B. Earth C. Moon D. Mars

Ans. B

25. In the helio-centric system, at the center of planetary system is

A. Sun B. Earth C. Moon D. Mars

Ans. A

26. The maximum angular separation of Mercury from the Sun is about

A. 15° B. 28° C. 45° D. 90°

Ans. B

27. The maximum angular separation of Venus from the Sun is about

A. 15° B. 32° C. 47° D. 90°

Ans. C

28. The conclusion that the sidereal periods of planets are indicators of their distances from the Sun, was drawn by

A. Copernicus B. Tycho Brahe C. Kepler D. Einstein

Ans. A

29. Venus appears as a morning star, when situated on which side of the Sun.

A. East B. West C. North D. South

Ans. B

30. Venus appears as a evening star, when situated on which side of the Sun.

A. East B. West C. North D. South

Ans. A

31. Mercury appears as a morning star, when situated on which side of the Sun.

A. East B. West C. North D. South

Ans. B

32. Mercury appears as a evening star, when situated on which side of the Sun.

A. East B. West C. North D. South

Ans. A

33. Who modified the preface in the book entitled 'On the Revolution of the Celestial Spheres' of Nicolaus Copernicus.

A. Osiander B. Galileo C. Kepler D. Bruno

Ans. A

34. Which one does not show the retrograde motion when seen from the Earth.

A. Mercury B. Mars C. Jupiter D. Saturn

Ans. A

35. Which one does not show the retrograde motion when seen from the Earth.

A. Venus B. Mars C. Jupiter D. Saturn

Ans. A

36. Who agreed to recast his teaching given in favour of the helio-centric system.

A. Copernicus B. Tycho Brahe C. Kepler D. Galileo

Ans. D

37. Who was placed under the house arrest as he was in favour of the helio-centric system.

A. Copernicus B. Tycho Brahe C. Kepler D. Galileo

Ans. D

38. Which body was considered a planet in the geo-centric system, but not in the helio-centric system.

A. Mars B. Moon C. Jupiter D. Saturn

Ans. B

39. Which body was considered a planet in the geo-centric system, but not in the helio-centric system.

A. Sun B. Mars C. Jupiter D. Saturn

Ans. A

2. Coordinate Systems

In a clear, dark night, we see sparkling points of light apparently situated on a vast sphere (called, the celestial sphere) in the sky. All these points are not necessarily the stars. They may be stars, star clusters, galaxies, planetary nebulae, etc. Because of large distances of these heavenly bodies (also known as the cosmic objects) from us, it is difficult to resolve them with the naked eyes and therefore, each of them appears like a star. In order to specify the position of an astronomical object on the celestial sphere, one requires a coordinate system. Further, the coordinates of a celestial observed from the Earth depends on the position of the observer on the Earth. Thus, it is also essential to know the position of the observer on the surface of Earth. In this chapter, we have introduced various coordinate systems used to specify the position on the surface of earth and to specify the positions of various celestial bodies.

The Earth moves around the Sun and the seasons on the Earth change regularly. After one year the seasons repeat, and we can get the number of days in one year. How many times the Sun rises (for example) in one year is the number of days in one year. In a period of four years the Sun rises 1461 times. Thus, the number of days in one year is 365.25. Further, it was assumed that the angular displacement of the Earth is 1° per day. In order to make a round figure, it was decided that in one revolution, the angular displacement of the Earth is 360° (instead of 365.25°).[1] In radian, the angular displacement is

$$\text{angle} = \frac{\text{arc}}{\text{radius}} = \frac{2\pi r}{r} = 2\pi$$

Thus, we have

$$2\pi \text{ rad} = 360^\circ$$

The Earth makes a rotation of 360○ in 24 hours. So we have

$$360^\circ \equiv 24 \text{ h} \qquad \text{or} \qquad 15^\circ \equiv 1 \text{ h}$$

Example 1: Convert $5^h\ 16^m\ 27^s$ into angle unit (degrees, arc minutes and arc seconds).

Solution: We have

$$27^s = \frac{27}{60} = 0.45^m$$

[1]The rounding of number could be 370°. The number 370 can be divided by 2, 5, 10, 37 and 370 only whereas the number 360 can be divided by 2, 3, 4, 5, 6, 8, 10 and several other numbers, scientists preferred to consider 360 and not 370.

Therefore,

$$16^m 27^s = 16 + 0.45 = 16.45^m = \frac{16.45}{60} = 0.27417^h$$

and

$$5^h 16^m 27^s = 5 + 0.27417 = 5.27417^h$$

We have $1^h \equiv 15°$. Therefore, we have

$$5.27417^h = 15 \times 5.27417 = 79.11255°$$

Now, we have

$$0.11255° = 60 \times 0.11255 = 6.753'$$

and

$$0.753' = 60 \times 0.753 = 45.18''$$

Thus, we have $5^h\ 16^m\ 27^s \equiv 79°\ 6'\ 45.18''$.

Example 2: Convert 15° 26′ 36″ into time unit (hours, minutes and seconds).

Solution: We have

$$36'' = \frac{36}{60} = 0.6'$$

Therefore,

$$26'36'' = 26 + 0.6 = 26.6' = \frac{26.6}{60} = 0.44333°$$

and

$$15°26'36'' = 15 + 0.44333 = 15.44333°$$

We have $15° \equiv 1^h$. Therefore, we have

$$15.44333° = 15.44333/15 = 1.02956^h$$

Now, we have

$$0.02956^h = 60 \times 0.02956 = 1.7736^m$$

and

$$0.7736^m = 60 \times 0.7736 = 46.42^s$$

Thus, we have $15°\ 26'\ 36'' = 1^h\ 1^m\ 46.42^s$.

1. Point in a plane

In order to specify position of a point in a plane, a coordinate system is required. When a particle is moving, its coordinates change with time. In order to introduce a coordinate system, also called the reference system, in a plane, let us consider a straight line AB (Figure 1) in the plane and a point O on the line. With the help of this reference system comprising the line AB and the point O (Figure 1a), position of a point P in the plane can be specified. Now, it is up to the observer which coordinate system one wants to use.

For the Cartesian coordinate system, we draw a straight line PQ through the point P and perpendicular to the line AB (Figure 1b). When the point O is treated as the origin of the coordinate system, then the distance OQ represents the x-coordinate, and the distance PQ, the y-coordinate of the point P. It is obvious that when we change the position of, either the line AB or the point O, the coordinates (x, y) of the point P change. It shows that the coordinates of a point depend on the reference system.

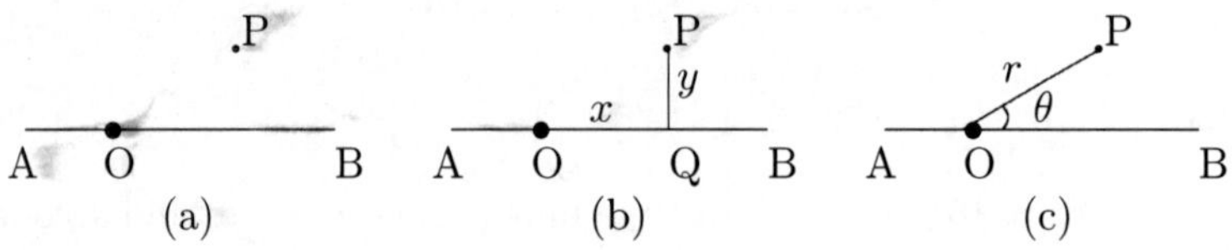

Figure 1: Coordinates in a plane surface.

For the polar coordinate system, we join the points O and P through a straight line (Figure 1c). The distance OP represents the radial coordinate r and the angle POB represents the angular coordinate θ. Thus, the polar coordinates of the point P are (r, θ).

2. Spherical surface

In the Astronomy and Astrophysics,[2] we often deal with the spherical surfaces. For example, the Earth's surface is taken as spherical in shape; the celestial sphere, on which the heavenly bodies are assumed to be situated, is spherical. Before discussing about the coordinate systems for spherical surfaces, let us introduce some definitions pertaining to the spherical surfaces.

2.1 Great-circle and small-circle

We know that intersection of a spherical surface with a plane is a circle. For a given spherical surface, radius of the circle so obtained depends on the position of the

[2]Some persons have opinion that the observational (experimental) aspect is known as the Astronomy whereas the theoretical aspect is known as the Astrophysics. Large number of persons consider Astronomy and Astrophysics as synonyms. Quite often, both the words are used together as Astronomy and Astrophysics (abbreviated as A&A).

plane. Radius of the circle is obviously the largest when the plane passes through the center of the circle. This circle of the largest radius is known as the great-circle. Any of the other circles have radii smaller than that of the great-circle. This circle is known as the small-circle. Parallel to a great-circle, there are obviously a large number of small-circles having various radii. By rotating the position of the intersecting plane, we get another great-circle and corresponding small-circles.

2.2 Poles

For the spherical surface SS, shown in Figure 2, suppose GC is the great-circle obtained by the intersection of a plane passing through the center C of the spherical surface. Now, a straight line perpendicular to the plane of the great-circle and passing through C intersects the spherical surface at two points R and T. These points are known as the poles corresponding to the great-circle. If we consider another great-circle obtained by intersection of another plane passing through C, we get another pair of poles.

On the other side, when we consider a straight line passing through the center C, then a plane whose normal is this line and passes through C intersects the spherical surface in a great-circle.

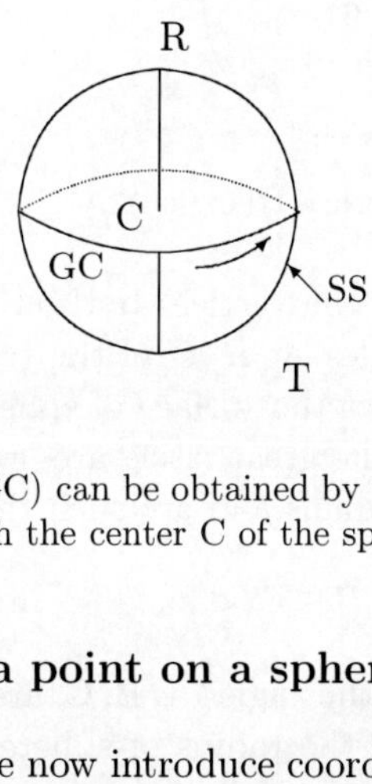

Figure 2: A great circle (GC) can be obtained by the intersection of a spherical surface SS and a plane passing through the center C of the spherical surface.

2.3 Coordinates of a point on a spherical surface

After these definitions, we now introduce coordinate systems for a spherical surface. When we transform from a plane surface (discussed in the preceding section) to a spherical surface, the lines AB and PQ of the plane surface become the arcs of great-circles (Figure 3) and the coordinates of a point P on the spherical surface are expressed in terms of the angles subtended by the arcs OQ and PQ at the center C of the sphere.

For the orthogonal coordinate system, the great-circle of the arc PQ would pass through the poles R and T, corresponding to the great-circle of the arc OQ. Thus, the coordinates of the point P are (ψ, θ), where $\psi = \angle$ OCQ and $\theta = \angle$ PCQ. With the help of these angles ψ and θ, various coordinate systems are introduced, which are used in the Astronomy and Astrophysics.

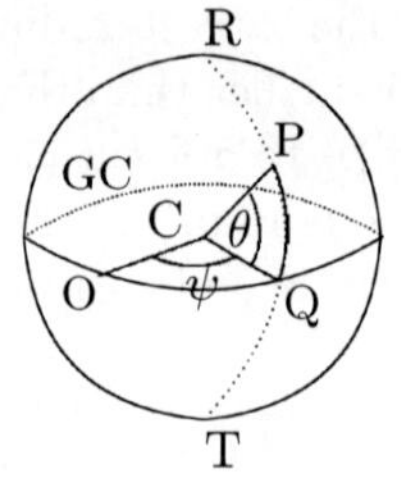

Figure 3: Coordinates (ψ, θ) of the point on a spherical surface.

3. Spherical triangle

Suppose, A, B, C are three points on a spherical surface, as given in Figure 4. Let the arcs AB, BC, CA, lying on the spherical surface, be the parts of great circles. The triangle ABC so formed is known as the spherical triangle.

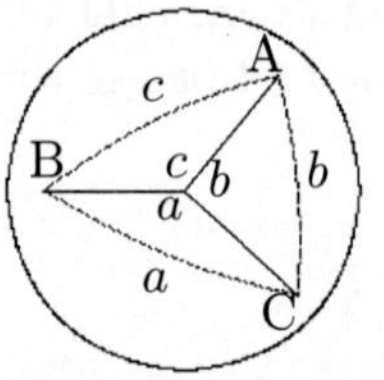

Figure 4: Spherical triangle ABC.

Here, the vertices and the angles at the vertices both are denoted by the same upper case letters A, B and C. The angles A, B, C of the triangle are equal to the angles between the planes which intersect the surface of sphere or, equivalently, the angles between the tangent vectors of the great circle arcs where they meet at the vertices. The angles are measured in radians and are such that[3]

$$\pi < A + B + C < 3\pi$$

The lengths of the sides opposite the the angles A, B, C, are, respectively, denoted by the lower-case letters a, b, c. When the radius of sphere is taken as unity, the lengths of the arcs are numerically equal to the angles (in radian) that the arcs subtend at the center of the sphere.

The radius of the sphere is generally taken as unity. For some specific practical problems on a sphere of radius R, the measured lengths of the sides are divided by R before using the identities given below. Further, after the calculations on the unit sphere the sides a, b, c are multiplied by R.

The spherical triangle is quite different from a triangle in a plane. The trigonometry dealing the spherical triangle is known as the spherical trigonometry, and the cosine and sine rules are as the following.

[3]In case of a triangle ABC in a plane, we have $A + B + C = \pi$.

3.1 Cosine rules

The cosine rules are fundamental identities of spherical trigonometry. They are as the following.

$$\cos a = \cos b \cos c + \sin b \sin c \cos A$$
$$\cos b = \cos c \cos a + \sin c \sin a \cos B$$
$$\cos c = \cos a \cos b + \sin a \sin b \cos C$$

3.2 Sine rules

The sine rules of spherical trigonometry are as the following.

$$\frac{\sin A}{\sin a} = \frac{\sin B}{\sin b} = \frac{\sin C}{\sin c}$$

4. Terrestrial latitude and longitude

This is also known as the geographical coordinate system. For practical purposes, our Earth is taken as a spherical body spinning about its own axis passing through its center C (Figure 5) from west towards east. The axis of rotation intersects the Earth's surface at two points denoted by N and S, called the north pole and south pole, respectively. Corresponding to these poles, the great-circle is known as the equator. This equator obviously divides the Earth into two hemispheres, called the northern hemisphere and southern hemisphere. The equator is used as the reference of the coordinate system. The reference point O on the equator is specified in the following manner.

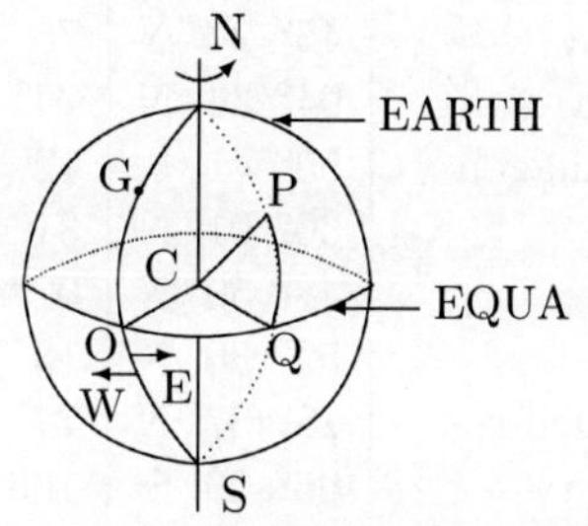

Figure 5: Explanation of latitude and longitude of a point P on the Earth's surface.

The great-circles passing through the poles N and S are known as the meridians. The meridian which passes through the fundamental instrument (the transit circle) installed in the Greenwich observatory (G), in England, is known as the principal or standard meridian. The point of intersection of this standard meridian with the equator is the reference point O. For a point P on the Earth's surface, the angle PCQ is called the latitude of P and the angle OCQ the longitude of P.

When the point P is in between the equator and the north pole N, the latitude is said to be the north latitude. The north latitude can vary from 0° to 90° north. The angle NCP is known as the colatitude of the point P. Therefore, we have

$$\text{colatitude} = 90^\circ - \text{latitude}$$

When the point P is in between the equator and the south pole S, the latitude is said to be the south latitude. The south latitude can vary from 0° to 90° south. The angle SCP is the colatitude of the point P.

Longitudes are measured from 0° to 180° east and from 0° to 180° west from the Greenwich meridian. Thus, the i position of a point on the Earth's surface can be expressed in terms of latitude and longitude for that point. Latitude and longitude of capitals of some countries are given in Table 1. These are the coordinates of a specific point in the city.

5. Celestial sphere

When we look towards the sky, the cosmic objects appear to be situated on a vast spherical surface of which the individual observer is the center. This imaginary sphere is known as the celestial sphere. The eye, of course, fails to give any indication about the distances of these objects. However, it allows us to make the estimates of the angle subtended by a pair of objects at the observer's eye. This angle is known as the angular separation between the two objects. Hence, for each position of an observer, there is a celestial sphere whose center is at the position of the observer.

Table 1. Coordinates of capitals of some countries

Country	Capital	Latitude	Longitude
Argentina	Buenos Aires	36° 30′ S	60° 00′ W
Austria	Vienna	48° 12′ N	16° 22′ E
Brazil	Brasilia	15° 47′ S	47° 55′ W
Cambodia	Phnom Penh	11° 33′ N	104° 55′ E
Canada	Ottawa	45° 27′ N	75° 42′ W
Colombia	Bogota	04° 34′ N	74° 00′ W
Denmark	Copenhagen	55° 41′ N	12° 34′ E
Egypt	Cairo	30° 01′ N	31° 14′ E
France	Paris	48° 50′ N	02° 20′ E
Germany	Berlin	52° 30′ N	13° 25′ E
India	New Delhi	28° 37′ N	77° 13′ E
Indonesia	Jakarta	06° 09′ S	106° 49′ E
Iraq	Baghdad	33° 20′ N	44° 30′ E
Israel	Jerusalem	31° 47′ N	35° 12′ E
Italy	Rome	41° 54′ N	12° 29′ E
Kenya	Nairobi	01° 30′ N	36° 48′ E
Liberia	Monrovia	06° 18′ N	10° 47′ W
Malaysia	Kuala Lumpur	03° 09′ N	101° 41′ E
Mexico	Mexico	19° 20′ N	99° 10′ W
Russia	Moskua	55° 45′ N	37° 35′ E

6. Altitude and azimuth

There are several coordinate systems to specify the positions of an astronomical object (*e.g.*, stars) on the celestial sphere. One of them is the altazimuth coordinate system. This is also known as the horizon coordinate system. In this system (Figure 6), the observer D (on the surface of the Earth) is the center of the celestial sphere. The line joining the observer D and the center of the Earth intersects the celestial sphere (CS) at two points Z and K, called the zenith and the nadir, respectively. The zenith is vertically overhead. The great circle corresponding to the poles Z and K is known as the celestial horizon or simply the horizon. Thus, the horizon is the intersection of the celestial sphere and the plane tangential to the Earth's surface passing through the observer D.

The horizon obviously divides the celestial sphere into two hemispheres, of which the upper hemisphere is visible to the observer, whereas the lower one is hidden from the observer by the Earth. The horizon is used as a great circle for the reference of the coordinate system. Now, we need to specify the position of the reference point on the horizon. This is done in the following manner.

Let the line DP be parallel to the spin-axis of the Earth and passing through the observer. Suppose, the line DP interacts the celestial sphere at the point P. When the latitude of the observer is north, the point P is known as the north celestial pole or simply the north pole. The great circles passing through Z and K are known as the vertical circles. The vertical circle passing through P, that is, ZPNK is defined as the prime vertical. It intersects the horizon at the point N, which is taken as the reference point on the horizon. This point N is also known as the north point on the horizon. The point S on the horizon, exactly opposite to N, is the south point, and the east (E) and west (W) points have the directions at right angles to the directions of N and S. The point N, E, S and W are known as the cardinal points.

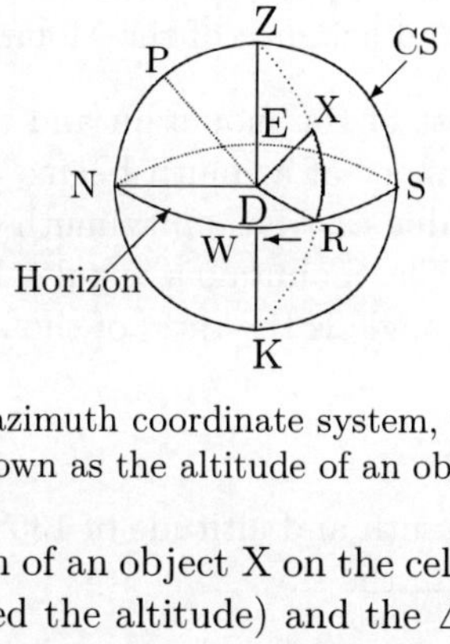

Figure 6: In the altazimuth coordinate system, the angle NESR is known as the azimuth and angle XDR is known as the altitude of an object X.

Now the position of an object X on the celestial sphere can be expressed in terms of the ∠ XDR (called the altitude) and the ∠ NESR, measured along the direction of the arrow with N as the reference point (called the azimuth). The altitude can assume values from 0° to 90°. The angle ZDX is known as the zenith distance (ZD) of the object X. Thus, we have

$$\text{ZD} = 90^\circ - \text{altitude}$$

Since the hemisphere below the horizon is hidden from the observer by the Earth, the altitude is negative for the objects below the horizon. The azimuth can assume the values from 0° to 360°.

It can be noticed that the altitude and azimuth both of an object at a given time depend on the position of the observer on the Earth. Further, as the Earth spins from west towards east about its own axis passing through its center, all the stars appear to travel across the sky along the circular paths from east towards west around the spin-axis. Hence, the altitude and azimuth coordinates of the objects change continuously with time. In the northern hemisphere, there is however a star whose position appears to change very little. This star, named Polaris, is visible to naked eyes.

Example 3: Show that for an observer, the altitude of the north pole is equal to the observer's latitude.

Solution: Suppose, the latitude of the observer on the Earth's surface is ϕ degree north. Since PD is parallel to the spin-axis of the Earth, the angle PDZ is equal to the colatitude of the observer. Thus, $\angle$ PDZ $= 90° - \phi$. As the $\angle$ NDZ $= 90°$, angle NDP is equal to ϕ. It is obviously the altitude of the north pole. Thus, the altitude of the north pole is equal to the observer's latitude.

Example 4: For an observer, the altitude and azimuth of a celestial object are 43° 20′ 13″ and 52° 14′ 38″, respectively. Calculate the zenith distance of the object.

Solution: The zenith distance has no consideration of the azimuth. The zenith distance is

$$\text{ZD} = 90° - 43°20'13'' = 46°39'47''$$

Example 5: An observer is standing on a bench 2.5 km due north-west of a mountain that has a height of 1500 m. The Moon appears to be directly behind the peak of the mountain. Calculate the altitude and azimuth of the Moon.

Solution: The observer is due north-west of the mountain and thus, the Moon is due south-east of the observer. Measurement of azimuth begins at due north and proceeds in an easterly direction so that due east has an azimuth of 90° and south-east a further 45° at (90 + 45 =) 135°. The mountain's angular height is equal to the Moon's altitude. The tangent of this angle is the ratio of the mountain's height to its distance at (1500/2500) = 0.6.

The Moon's altitude $= \tan^{-1} 0.6 = 31°$

At this moment the Moon has an azimuth and altitude of 135° and 31°, respectively, which uniquely define its position in the sky.

7. Standard celestial sphere

It has been mentioned that around each point on the surface of the Earth, there is a celestial sphere. For example, there is a celestial sphere around the observer D. For each celestial sphere, the positions of the celestial poles and celestial equator etc. depend on the position of the observer. Hereafter, we shall refer to a celestial sphere whose center is at the center of the Earth. This celestial sphere is often referred to as the standard celestial sphere or geo-centric celestial sphere. Now, the positions of the celestial poles and celestial equator do not depend on the position of the observer.

The celestial poles, or simply poles, are now defined as the points where the spin-axis of the Earth intersects the standard celestial sphere. The great circle corresponding to these poles is known as the celestial equator. Obviously, the celestial equator is the intersection of the celestial sphere and the plane of the terrestrial equator. With the introduction of the standard celestial sphere, the horizon also would have the center at the Earth's center.

7.1 Celestial poles, celestial equator and celestial meridians

The spin-axis of Earth intersects the celestial sphere at two points, called the celestial north pole and celestial south pole. Thus, the terrestrial and celestial poles both lie on a common line passing through the center of Earth (Figure 7).

The intersection of the plane of terrestrial equator with the celestial sphere is a circle, called the celestial equator. Thus, the terrestrial equator and celestial equator both lie in a common plane.

The planes containing the spin axis (*i.e.*, passing through the center of Earth, north and south celestial poles) intersect the celestial sphere in the form of arcs, called the celestial meridians. The terrestrial meridian and celestial meridian of a place lie in a common plane.

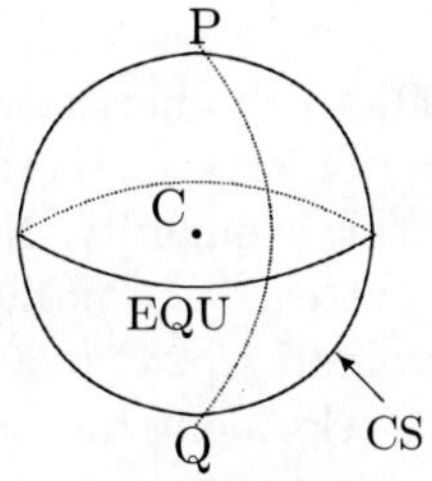

Figure 7: Celestial sphere CS, north celestial pole P, south celestial pole Q, celestial equator EQA; line joining P and Q is celestial meridian.

8. Declination and hour-angle

It is also known as the equatorial coordinate system. The coordinates of an object can also be expressed in terms of the declination and hour-angle. In Figure 8, P and Q are the north and south poles, respectively. These are the points of intersection of the standard celestial sphere (SCS) and the spin-axis of the Earth. The great circle corresponding to the poles is the equator (EQU). For this coordinate system, the equator is used as the great circle for reference. Now, the reference point on the equator can be specified in the following manner.

The great circles passing through the poles P and Q are known as the celestial meridians. Suppose, Z is the zenith for the observer in the northern hemisphere. Then, the celestial meridian passing through Z is known as the observer's meridian. The observer's meridian intersects the equator at the point Y, which is taken as the reference point on the equator. Now, the position of an object X on the celestial sphere can be expressed in terms of the ∠ XCR, called the declination of the star. It is north declination because the object is between the equator and the north pole.

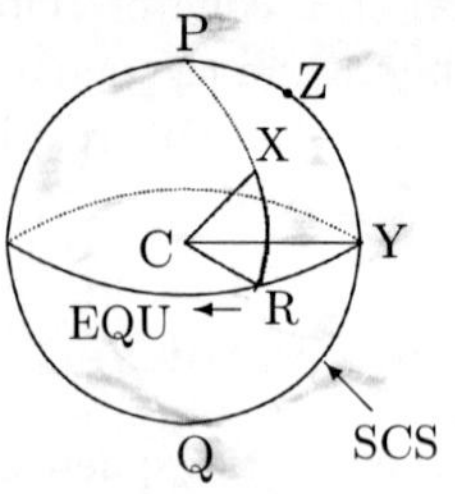

Figure 8: Declination and hour-angle coordinates system. Angle YCR is known as the hour-angle and angle XCR is known as the declination of an object X.

For an object, between the equator and south pole, the declination would be south. The declination is denoted by δ. The ∠ YCR measured along the direction of arrow (*i.e.*, towards the west) with Y as the reference point, is known as the hour-angle of the object and is denoted by H. The angle PCX is known as the north polar distance (NPD) of the star. Thus, we have

$$\text{NPD} = 90^\circ - \delta$$

Declination is measured from 0° to 90° to the north or south of the equator. In order to use the same formulas for the objects in the north and south hemispheres, the declination is treated as an algebraic quantity. North declination carries a positive sign whereas the south declination carries a negative sign. The value of the hour-angle varies from 0° to 360° or from 0^h to 24^h ($360^\circ \equiv 24^h$).

As the objects appear to travel across the sky along the circular paths around the spin-axis of Earth, the declination of an object remains unchanged. However, the hour-angle changes with time due to the rotation of the Earth. It also differs from place to place because of the change in the position of the zenith, corresponding to an observer.

Example 6: The declination and hour angle of an object are 37° 42′ 16″ and 52° 37′ 42″, respectively. Calculate the north polar distance of the object.

Solution: The north polar distance has no consideration of the hour angle. The north polar distance of the object is

$$\text{NPD} = 90^\circ - 37^\circ 42' 16'' = 52^\circ 17' 44''$$

Example 7: From where on the Earth's surface are the declination of all stars equal to their altitudes.

Solution: Declination is measured relative to the celestial equator whereas altitude is measured relative to the observer's horizon. Thus, when the horizon and the celestial equator are equivalent, the declination and altitude are equivalent. Remembering that the celestial equator is a projection of the Earth's equator, then standing on the Earth's equator means that the celestial equator runs in an arc overhead. As the observer moves towards either pole the celestial equator forms an arc that has a progressively lower and lower maximum altitude. At the poles the horizon and celestial equator follow the same line. Therefore, each star has equal altitudes and declinations when observed from the Earth's poles.

9. Rotating sky

The celestial sphere is static and the Earth is spinning inside it. The appearance of sky depends on the time and position of the observer on the Earth. The place in the sky that is invariant when observing from the northern hemisphere is the north celestial pole. In order to understand the path of stars across the sky, it is only necessary to remember that the Earth spins once a day about its axis. Hence, an observer's view of the sky moves through 360° each day. The stars appear to circle the sky (Figure 9). The position of the celestial pole does not change, but the star of interest appears to rotate around the pole.

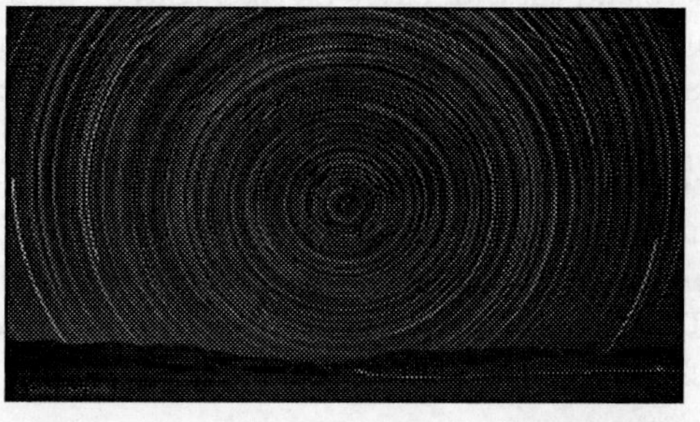

Figure 9: Rotating sky. Courtesy to Google website.

The size and position of circles in the sky described by the stars depend on the position of the star on the celestial sphere and the observer's position on the Earth. It is obvious that the stars close to the celestial pole travel in small circles around the pole. The stars reach their highest position in the sky, known as the upper culmination, when they cross the observer's meridian. At this moment, their altitude is given by the altitude of the celestial pole (equal to the observer's latitude ϕ) plus the difference in the declination between the star and north celestial pole. The north celestial pole has a declination of 90°. Hence, the altitude of a high declination star at upper culmination is expressed as

$$a_{max} = \phi + (90 - \delta) = 90 + \phi - \delta$$

The stars away from the north celestial pole have smaller declination and describe larger circles in the sky. The altitude is measured from the closest point on the horizon. For the southern cardinal point, the maximum altitude is expressed as

$$a_{max} = 180 - (90 + \phi - \delta) = 90 - (\phi - \delta)$$

In both the cases the star's minimum altitude also occurs on the observer's meridian but directly below the north celestial pole. As the motion is circular, the star's minimum latitude is

$$a_{min} = \phi - (90 - \delta) = \phi + \delta - 90$$

Let us see the situation when the observer's latitude plus the star's declination $(\phi + \delta)$ is less than 90° so that the star's minimum latitude is negative. This is the case of the setting of a star which dips below the observer's horizon. The path of such a star across the sky is still circular but part of the circle is hidden by the Earth.

The stars which never leave the sky (even though they are not visible during the day) are said to be circumpolar. The condition for a star to be circumpolar is

$$(\delta + \phi) \leq 90$$

It shows that when the observer is positioned at the north pole ($\phi = 90°$) the same stars remain in the sky at all times. The stars which are visible are those of the north celestial hemisphere and they all circle parallel to the horizon about the north celestial pole which is at the zenith. On movement towards south more stars become visible but some stars are no longer circumpolar and disappear from the view for a period of time.

For an observer on the equator, the north and south celestial poles lie at the north and south cardinal points on the horizon. The celestial equator loops overhead and all the stars describe perfect semi-circle paths across the sky which is parallel to the celestial equator. There are no circumpolar stars but the whole of the celestial sphere appears in the sky through out the course of the day.

Example 8: As observed from a city (latitude 50° N, longitude 8° W), calculate the maximum altitude of a star (right ascension 10°, declination +67°) when it crosses the observer's meridian. Calculate if the star disappears from the sky. If not, calculate the minimum altitude of the star.

Solution: We have $\delta = 67°$ and $\phi = 50°$. Star crosses the observer's meridian at the upper culmination. As $(\delta > \phi)$, the upper culmination takes place north of the zenith. The altitude of star at the upper culmination is

$$a_{max} = 90 + \phi - \delta = 90 + 50 - 67 = 73°$$

As the δ is larger than $(90 - \phi)$, the star is circumpolar when observed from the city. The altitude of star at the lower culmination is

$$a_{min} = \phi + \delta - 90 = 50 + 67 - 90 = 27°$$

Thus, the star remains in the sky at all times with an altitude that varies between 27° and 73°.

Example 9: As observed from York (latitude 54° N, longitude 1° W), calculate the maximum altitude of the star Vega (declination +39°) when it crosses the observer's meridian. Whether the star disappears from the sky. If not, calculate the minimum altitude of the star.

Solution: Star crosses the observer's meridian at the upper culmination. As ($\delta < \phi$), the upper culmination takes place south of the zenith. The altitude of star at the upper culmination is

$$a_{max} = 90 - \phi + \delta = 90 - 54 + 39 = 75°$$

As the δ (39°) is larger than $(90 - \phi)$ (90 - 54 = 36°), the star is circumpolar when observed from York. The altitude of star at lower culmination is

$$a_{min} = \phi + \delta - 90 = 54 + 39 - 90 = 3°$$

Thus, the star remains in the York sky at all times with an altitude that varies between 3° and 75°.

Example 10: As observed from York (latitude 54° N, longitude 1° W), calculate the maximum altitude of the star Spica (declination -11°) when it crosses the observer's meridian. Whether the star disappears from the sky. If not, calculate the minimum altitude of the star.

Solution: Star crosses the observer's meridian at the upper culmination. As ($\delta < \phi$), the upper culmination takes place south of the zenith. The altitude of star at upper culmination is

$$a_{max} = 90 - \phi + \delta = 90 - 54 - 11 = 25°$$

As the δ (-11°) is smaller than $(90-\phi)$ (90 - 54 = 36°), the star is not circumpolar when observed from York. Thus, the star does not remain in the York sky at all times.

10. Ecliptic

Besides the spin motion about its own axis, the Earth revolves around the Sun in one year in an elliptical path, where the Sun, denoted by S, is situated at a focus of the ellipse (Figure 10). The points A, B, D, E, F denote various positions of Earth revolving around the Sun.

Since the observations are generally made from the Earth, the Sun appears to describe an elliptical orbit around the Earth. In Figure 11, C denotes the center of Earth and the ellipse represents the apparent orbit of Sun relative to the Earth. The sequence of positions of the Sun, namely, a, b, c, d, e, f (Figure 11) relative to the Earth corresponds to the sequence of positions A, B, C, D, E, F (Figure 10) of the Earth in its orbit around the Sun, respectively. During one year, the Sun thus appears to make a complete revolution around the Earth.

The plane of the apparent orbit of Sun relative to the Earth (which is also the plane of the orbit of Earth revolving around the Sun) is known as the plane of ecliptic and the great circle obtained with the intersection of this plane and the standard celestial sphere is known as the ecliptic.

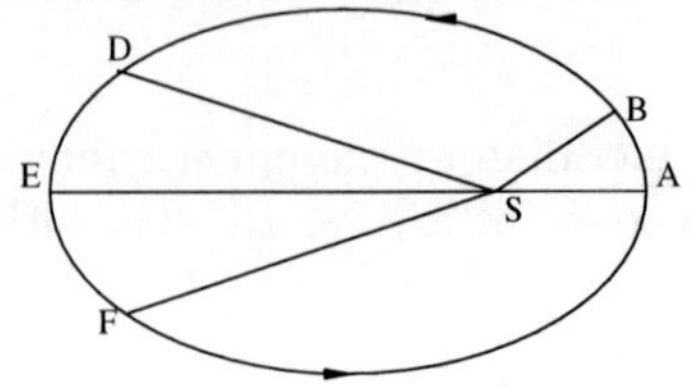

Figure 10: Ellipse shows the path of Earth revolving around the Sun, which is situated at the focus S of the ellipse.

From the observations, it is found that the ecliptic is inclined at an angle of about 23.5° to the celestial equator (Figure 12). This inclination is known as the obliquity of the ecliptic. In Figure 12, P is the pole of the equator (EQU) and K is the pole of the ecliptic (ECL). Relative to the Earth, the Sun appears to move on the celestial sphere along the ecliptic in the direction shown with the help of arrows.

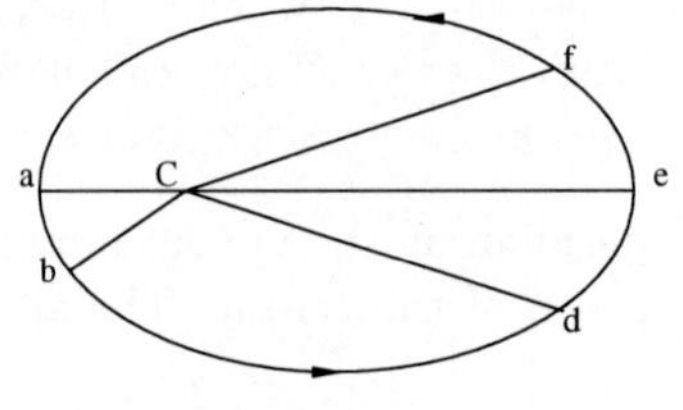

Figure 11: Ellipse shows the apparent path of Sun moving around the Earth, which is situated at the focus C of the ellipse.

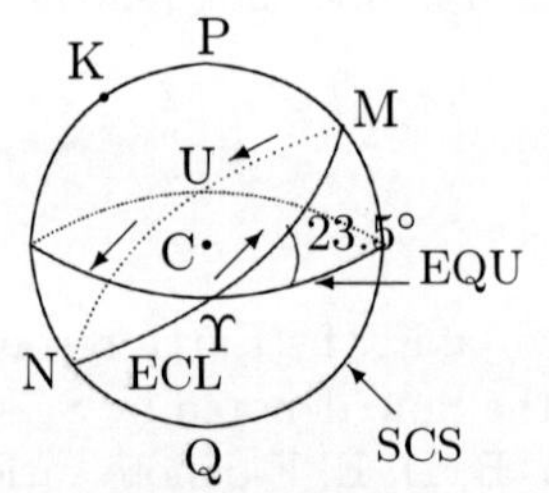

Figure 12: The ecliptic (ECL) is inclined at an angle of about 23.5° to the celestial equator (EQU). The ecliptic intersects the celestial equator at the vernal equinox ϒ and autumnal equinox U.

The ecliptic obviously intersects the equator at two points ϒ and U. At these points, the declination of the Sun is zero degree. During the motion from ϒ to M, and from M to U, the Sun is in the northern hemisphere, and therefore its declination is north. On the other side, during the motion from U to N, and from N to ϒ, the Sun is in the southern hemisphere, and therefore its declination is south. The position ϒ, at which the declination of the Sun changes from south to north is known as the vernal equinox. The Sun is at this position around March 21. The position M at which the north declination of the Sun is the maximum (about 23.5° N) is known as the summer solstice. The Sun is at this position around June 21. The position U at which the declination of the Sun changes from north to south is known as the autumnal equinox. The Sun is at this position around September 23. The position N at which the south declination of the Sun is the maximum (about 23.5° S) is known as the winter solstice. The Sun is at this position around December 21.

11. Right ascension and declination

We have seen that the altitude and azimuth coordinates of a star vary with the rotation of the Earth. Further, the hour-angle coordinate changes with the rotation of the Earth. However, the declination of an object does not depend on the rotation of the Earth. Now, we intend to introduce a coordinate system that does not vary with the rotation of the Earth nor does it vary from place to place. For such requirement, the equator (EQU) can be used as a great circle for reference and the vernal equinox (♈) is taken as the reference point on the equator (Figure 13).

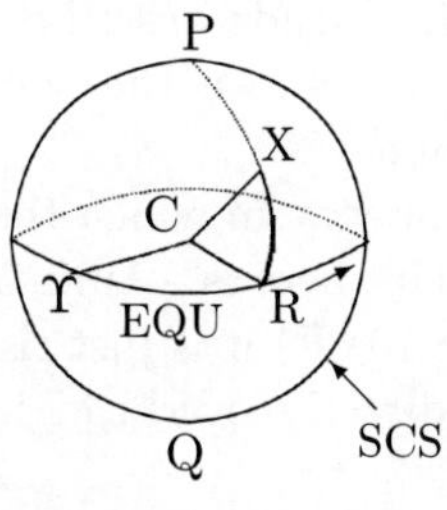

Figure 13: Right ascension and declination coordinate system. Angle ♈CR is known as the right ascension and angle XCR is known as the declination of an object X.

Now, the position of an object X on the celestial sphere can be expressed in terms of the angle XCR (called the declination (Dec) and denoted by δ) and the angle ♈CR, measured along the direction of the arrow (from west to east) with ♈ as the reference point (called the right ascension (RA) and denoted by α). The right ascension also, obviously does not change with the rotation of Earth. The values of right ascension are expressed in terms of hours ($360° \equiv 24^h$). In this coordinate system, four positions of the Sun have the following coordinates.

	δ	α
(i) Vernal equinox	0°	0^h
(ii) Summer solstice	23.5°	6^h
(iii) Autumnal equinox	0°	12^h
(iv) Winter solstice	−23.5°	18^h

Example 11: How does the Sun's highest position in the sky vary throughout the year in Quito which is on the equator in Ecuador, South Africa, and in Libreville which is on the equator in Gabon, Africa?

Solution: On 21 Dec (or close to it, depending on the calendar for that year) the Sun's position on the celestial position reaches its furthest position south, at a declination of -23.5°, when the Earth's southern hemisphere is tilted towards the Sun. The observer's latitude (for both Quito and Libreville) is 0° and so upper culmination takes place south of the zenith ($\delta < \phi$) at this time and the altitude is

$$a_{max} = 90 - \phi + \delta = 90 - 0 - 23.5 = 66.5°$$

As the year progresses, the Sun's declination increases towards zero which it reaches on (or close to) 20 March. At this time the value of a_{max} is 90°. During April and May the Sun continues to increase its declination until it peaks on the summer solstice on (or close to) 21 June. At this time the Sun's declination is 23.5° and the altitude is

$$a_{max} = 90 + \phi - \delta = 90 + 0 - 23.5 = 66.5°$$

The Sun's declination continues to decrease, reaching 0° on (or close to) 22 September. At this time the value of a_{max} is 90°.

Throughout the year, the Sun's peak altitude describes an oscillation between 66.5° above the northern horizon to 66.5° above the southern horizon, passing through the zenith in March and September.

There is essentially no difference in the way in which the Sun travels across the sky in Quito and in Libreville, except that there is a time offset. When the Sun is directly overhead in Libreville (longitude 10° E) it is just rising in Quito (longitude 79° W). Six months later, the Sun is directly overhead in Quito and setting in Libreville.

12. Celestial latitude and longitude

For the study of the motion of planets, a coordinate system is often used in which the ecliptic (ECL) is used as the great circle for the reference and the vernal equinox is taken as the reference point on the ecliptic (Figure 14), where K is the pole of the ecliptic.

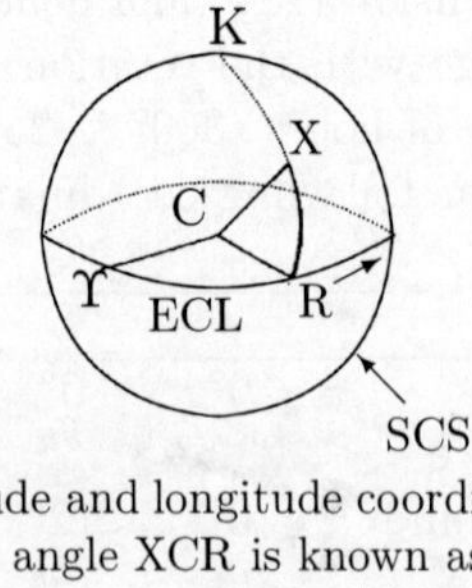

Figure 14: Celestial latitude and longitude coordinate system. Angle ♈CR is known as the celestial longitude and the angle XCR is known as the celestial latitude of a planet X.

Now, the position of a planet X on the celestial sphere can be expressed in terms of the angle XCR (called the celestial latitude and denoted by β) and the angle ♈CR, measured along the direction of the arrow (*i.e.*, from west to east) with ♈ as a reference point (called the celestial longitude and denoted by λ). The latitude above the ecliptic is known as the north latitude and is considered positive, whereas the latitude below the ecliptic is known as the south latitude and is considered negative. The value of λ varies from o° to 360° round the ecliptic.

13. Galactic latitude and longitude

For the study of our galaxy, a coordinate system is used in which the galactic equator (GEQU) (Figure 15) is used as a great circle for the reference. The galactic equator almost coincides the central line of the Milky Way in the sky. GP is the galactic north pole. The reference point O on the galactic equator is a point in the constellation Sagittarius and is located at the center of our galaxy. Now, the position of an astronomical object (*e.g.*, star, cluster) S on the celestial sphere can be specified in terms of the angle SCR (called the galactic latitude and denoted by b) and the angle OCR, measured along the direction of the arrow with O as the reference point (called the galactic longitude and denoted by l). It may be mentioned here that the galactic coordinates cannot be determined by direct observations. They are calculated by using the equatorial coordinates of the star of the galactic pole (GP) and of the point O in the direction of the center of galaxy.

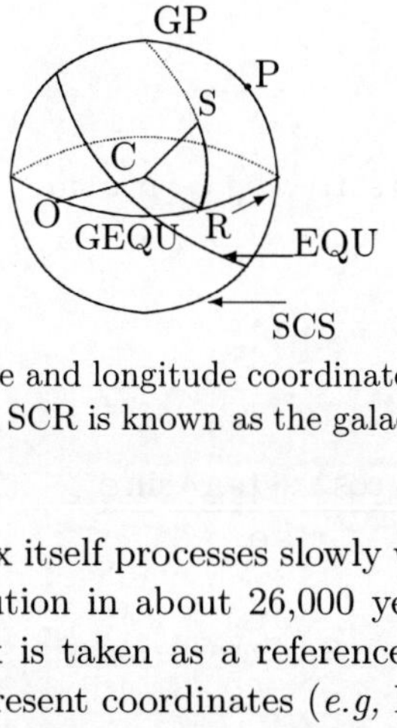

Figure 15: Galactic latitude and longitude coordinate system. Angle OCR is known as the galactic longitude and angle SCR is known as the galactic latitude of an astronomical object S.

Note: The vernal equinox itself processes slowly westward relative to the fixed stars and completes one revolution in about 26,000 years.Therefore, all the coordinates where the vernal equinox is taken as a reference point are modified from time to time. For example, the present coordinates (*e.g,* RA and Dec) are with reference to the position of the vernal equinox on 1 January 2000. Earlier the coordinates were with reference to the position of the vernal equinox on 1 January 1950. This date is known as the epoch.

14. Conversion of coordinates

Let us first summarize the coordinate systems and notations used there.

Horizontal coordinates

A - azimuth; a - altitude

Equatorial coordinates

α - right ascension; δ - declination; h - hour angle

Ecliptic coordinates

λ - ecliptic longitude; β - ecliptic latitude

Galactic coordinates

l - galactic longitude; b - galactic latitude

Miscellaneous

λ_0 - observer's longitude; ϕ_0 - observer's latitude;

ε - obliquity of the ecliptic

θ_{L} - local sidereal time; θ_{G} - Greenwich sidereal time

14.1 Hour angle ⇔ right ascension

$$h = \theta_L - \alpha \qquad \text{or} \qquad h = \theta_G - \lambda_0 - \alpha$$

$$\alpha = \theta_L - h \qquad \text{or} \qquad \alpha = \theta_G - \lambda_0 - h$$

14.2 Equatorial ⇔ ecliptic

We have

$$\cos\beta \sin\lambda = \cos\delta \sin\alpha \cos\varepsilon + \sin\delta \sin\varepsilon$$

$$\cos\beta \cos\lambda = \cos\delta \cos\alpha$$

so that

$$\tan\lambda = \frac{\sin\alpha \cos\varepsilon + \tan\delta \sin\varepsilon}{\cos\alpha}$$

$$\sin\beta = \sin\delta \cos\varepsilon - \cos\delta \sin\varepsilon \sin\alpha$$

We have

$$\cos\delta \sin\alpha = \cos\beta \sin\lambda \cos\varepsilon - \sin\beta \sin\varepsilon$$

$$cos\delta \cos\alpha = \cos\beta \cos\lambda$$

so that

$$\tan\alpha = \frac{\sin\lambda \cos\varepsilon - \tan\beta \sin\varepsilon}{\cos\lambda}$$

$$\sin\delta = \sin\beta \cos\varepsilon + \cos\beta \sin\varepsilon \sin\lambda$$

14.3 Equatorial ⇔ horizontal

Note that the Azimuth (A) is measured from the South point, turning positive to the West. Zenith distance, the angular distance along the great circle from the zenith to a celestial object, is simply the complementary angle of the altitude: $90° - a$. We have

$$\cos a \sin A = \cos \delta \sin h$$

$$cosa \cos A = \cos \delta \cos h \sin \phi_0 - \sin \delta \cos \phi_0$$

so that

$$\tan A = \frac{\sin h}{\cos h \sin \phi_0 - \tan \delta \cos \phi_0}$$

$$\sin a = \sin \phi_0 \sin \delta + \cos \phi_0 \cos \delta \cos h$$

We have

$$\cos \delta \sin h = \cos a \sin A$$

$$\cos \delta \cos h = \sin a \cos \phi_0 + \cos a \cos A \sin \phi_0$$

so that

$$\tan h = \frac{\sin A}{\cos A \sin \phi_0 + \tan a \cos \phi_0}$$

$$\sin \delta = \sin \phi_0 \sin a - \cos \phi_0 \cos a \cos A$$

14.4 Equatorial ⇔ galactic

These equations are for converting equatorial coordinates referred to B1950.0. If the equatorial coordinates are referred to another equinox, they must be precessed to their place at B1950.0 before applying these formulas. We have

$$l = 303° - \arctan \left(\frac{\sin(192.25° - \alpha)}{\cos(192.25° - \alpha) \sin 27.4° - \tan \delta \cos 27.4°} \right)$$

$$\sin b = \sin \delta \sin 27.4° + \cos \delta \cos 27.4° \cos(192.25° - \alpha)$$

These equations convert to the equatorial coordinates referred to B1950.0.

$$\alpha = \arctan \left(\frac{\sin(l - 123°)}{\cos(l - 123°) \sin 27.4° - \tan b \cos 27.4°} \right) + 12.25°$$

$$\sin \delta = \sin b \sin 27.4° + \cos b \cos 27.4° \cos(l - 123°)$$

15. Review Questions

1. Convert $4^h\ 23^m\ 25^s$ into degrees, arc minutes and arc seconds.
2. Convert $45°\ 37'\ 45''$ into hours, minutes and seconds.
3. Describe a reference system for the position of a point in a plane. How it can be modified for a spherical surface.
4. How in general, coordinates of a point on the spherical surface are described.
5. What are the great-circle and small-circle? Is meridian a part of great-circle?
6. On the surface of Earth, define poles, equator, meridian, standard meridian.
7. Describe the terrestrial latitude and longitude systems of coordinates.
8. For a celestial sphere, describe the poles, equator and meridian.
9. Describe the altazimuth coordinate system.
10. For an observer, the altitude and azimuth of a celestial object are $33°\ 25'\ 18''$ and $56°\ 24'\ 48''$, respectively. Calculate the zenith distance of the object.
11. Describe the declination and hour-angle coordinate system.
12. The declination and hour angle of a object are $39°\ 38'\ 17''$ and $62°\ 45'\ 35''$, respectively. Calculate the north polar distance of the object.
13. Discuss about the rotating sky, and describe the circumpolar and non-circumpolar stars. How it is decided if a star is circumpolar or not.
14. Define the ecliptic. Describe the right ascension and declination coordinate system. Write the right ascension and declination of: (i) Vernal equinox, (ii) Summer solstice, (iii) Autumnal equinox, and (iv) Winter solstice.
15. Describe the celestial latitude and longitude coordinate system.
16. Describe the galactic latitude and longitude coordinate system.
17. Write notes on the following:

    (i) Great-circle, small-circle
    (ii) Spherical triangle
    (iii) Equator and poles on the surface of Earth
    (iv) Terrestrial meridians and principal meridian
    (v) Terrestrial latitude and longitude
    (vi) Celestial sphere and standard celestial sphere
    (vii) Altazimuth coordinate system
    (viii) Declination and hour-angle

(ix) Rotating sky

(x) Circumpolar and non-circumpolar stars.

(xi) Ecliptic

(xii) Right ascension and declination

(xiii) Celestial latitude and longitude

(xiv) Galactic latitude and longitude

16. Multiple Choice Questions

1. The number of great circles for a spherical surface is

   A. zero B. one C. two D. infinite

   Ans. D

2. The intersection of a spherical surface with a plane is

   A. ellipse B. hyperbola C. circle D. parabola

   Ans. C

3. For a spherical triangle ABC, the sum of the angles A, B, C satisfy the condition

   A. $A + B + C = \pi$ B. $\pi < A + B + C < 3\pi$

   C. $\pi \leq A + B + C \leq 3\pi$ D. $A + B + C < \pi$

   Ans. B

4. Terrestrial latitude has the range

   A. 0° - 90° N B. $-90°$ - 90° C. 0° - 180° E D. $-180°$ - 180°

   Ans. A

5. The latitude of a place on the Earth is 55°. The colatitude of the place is

   A. 55° B. 35° C. 125° D. 235°

   Ans. B

6. A great circle passing through the poles of the Earth is known as

   A. Equator B. Meridian C. Equinox D. Solstice

   Ans. B

7. For the altitude A, the zenith distance is defined as

   A. $90° + A$ B. $90° - A$ C. $180° - A$ D. $180° + A$

   Ans. B

8. Suppose, the latitude of an observer is ϕ. The colatitude of the observer is

   A. $90° + \phi$ B. $90° - \phi$ C. $180° - \phi$ D. $180° + \phi$

   Ans. B

9. Which coordinate does not change with time.

   A. Declination B. Hour angle C. Altitude D. Azimuth

   Ans. A

10. Which of the following coordinates does not change with time.

    A. Right ascension B. Hour angle C. Altitude D. Azimuth

    Ans. A

11. Through which country, the standard meridian of the Earth passes.

    A. India B. China C. England D. America

    Ans. C

12. The great circles joining the zenith and nadir are known as

    A. Meridians B. Vertical circles C. Equator D. Ecliptic

    Ans. B

13. The points N and S on the observer's horizon are known as the

    A. north and south poles B. meridians C. cardinal points D. equinoxes

    Ans. C

14. The altitude of a celestial object is 40°. The zenith distance of the object is

    A. 40° B. 50° C. 140° D. 220°

    Ans. B

15. The angle between the celestial equator and ecliptic is

    A. 15.5° B. 23.5° C. 45° D. 90°

    Ans. B

16. The azimuth of a celestial objects can assume the values from

    A. 0° to 90° B. 0° to 180° C. 0° to 360° D. -90° to 90°

    Ans. C

17. The spin-axis of Earth intersects the Earth's surface at

    A. north and south poles B. celestial poles C. equinoxes D. cardinal points

    Ans. A

18. For declination 35°, the north polar distance is

    A. 145° B. 215° C. 35° D. 55°

    Ans. D

19. The spin-axis of Earth intersects the celestial sphere at

A. terrestrial poles B. celestial poles C. equinoxes D. Cardinal points

Ans. B

20. The Sun is at the summer solstice around

A. 21 March B. 21 June C. 22 September D. 21 December

Ans. B

21. The Sun is at the vernal equinox around

A. 21 March B. 21 June C. 22 September D. 21 December

Ans. A

22. The Sun is at the autumnal equinox around

A. 21 March B. 21 June C. 22 September D. 21 December

Ans. C

23. The Sun is at the winter solstice around

A. 21 March B. 21 June 6mm C. 22 September D. 21 December

Ans. D

24. For the Sun, the declination $\delta = 0°$ and right ascension $\alpha = 0^h$ at

A. Vernal equinox B. Summer solstice

C. Autumnal equinox D. Winter solstice

Ans. A

25. For the Sun, the declination $\delta = 23.5°$ and right ascension $\alpha = 6^h$ at

A. Vernal equinox B. Summer solstice

C. Autumnal equinox D. Winter solstice

Ans. B

26. For the Sun, the declination $\delta = 0°$ and right ascension $\alpha = 12^h$ at

A. Vernal equinox B. Summer solstice

C. Autumnal equinox D. Winter solstice

Ans. C

27. For he Sun, the declination $\delta = -23.5°$ and right ascension $\alpha = 18^h$ at

A. Vernal equinox B. Summer solstice

C. Autumnal equinox D. Winter solstice

Ans. D

28. Which coordinate system is preferably used for the study of the motion of planets in the solar system.

A. Right ascension and declination B. Celestial latitude and longitude

C. Terrestrial latitude and longitude D. Altitude and azimuth

Ans. B

29. For the study of a distant cosmic object, preferably used coordinate system is

A. Right ascension and declination B. Celestial latitude and longitude

C. Terrestrial latitude and longitude D. Altitude and azimuth

Ans. A

30. For a spherical triangle ABC, the sum of the angles $A + B + C$ is

A. $\pi/2$ B. less than π C. π D. more than π

Ans. D

31. An observer is situated at a place on Earth where latitude is 50° N. The declination of non-circumpolar star is

A. 77° B. 65° C. 42° D. 36°

Ans. D

32. An observer is situated at a place on Earth where latitude is 32° N. The declination of circumpolar star is

A. 77° B. 55° C. 44° D. 18°

Ans. A

3. Time

Time is a very important parameter in Astronomy and Astrophysics (A&A). A periodic motion repeats after an interval of time, called the time-period of the motion. Revolution of a planet around the Sun, spin motion of a planet about its own axis, revolution of a satellite around a planet, spin motion of a satellite, revolution of binary stars about the center-of-mass, are a few examples of periodic motions in the A&A. It is worth mentioning that in spite of the best efforts, the measurement of time could not be transformed to a decimal system, unlike the case of the length and mass. It is because the measurement of time is based on the natural phenomena, rather than on the man-made phenomena. However, the apparent solar day varies regularly and no watch can be fabricated for the apparent solar day. The watches are designed for the motion of a fictitious sun whose motion is parallel to the equator of the earth. The mid-day (12.00 noon) is defined when the fictitious sun is situated at the standard meridian assigned to a country. Thus, local time varies from one country to another. Because of the difference between the sidereal time and the mean solar time, the stars appear early each day, and consequently, the stars in the winter nights are different from those in the summer nights. In this chapter, we have discussed about all these aspects related to time.

1. Apparent solar day

Let us perform a simple experiment. Suppose, we put a tall pole AB vertically in a plane ground PQ (Figure 1). On a sunny day, the shadow of the pole rotates around the pole and its size changes with the apparent motion of the Sun around the Earth. Relative to the Earth, the Sun appears to move from east towards west. In fact, the Earth spins from west towards east and the Sun is stationary.

In the morning hours, the shadow of the pole is quite long and besides the rotation of the shadow, its size decreases with the motion of the Sun. The size becomes the minimum (may be zero) and then again increases. When the Sun is directly overhead the meridian of the pole, the shadow is the minimum. The shadow may disappear (zero size) when the Sun is overhead the pole. That instant is known as the noon at that place. That is, the Sun is at the local meridian. The time interval between two successive returns of the Sun to a given meridian of a place is defined as the apparent solar day. After an apparent solar day, the event repeats and the Sun appears to have covered an angle of 360° around the Earth.

It is however noticed that the time taken by the Sun to complete one circle around the Earth varies from day to day. That is, the difference between two successive appearances of the Sun on a local meridian varies from day to day. The reason for this variation may be assigned to two causes:

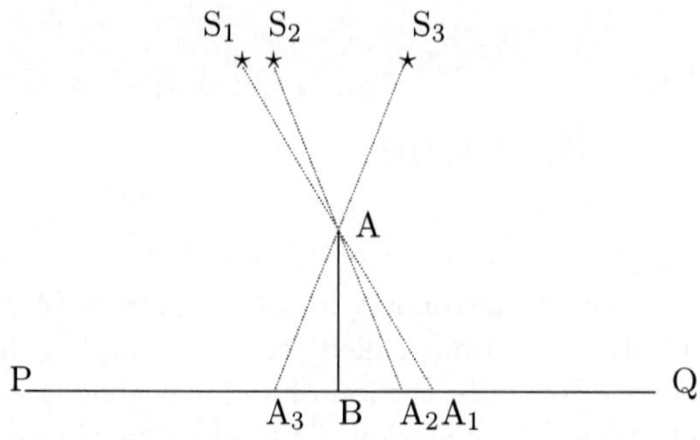

Figure 1: The pole AB is erected vertically upward on a plane ground PQ. For different positions S_1, S_2, S_3 of the Sun, the positions of shadow of the upper most point A of the pole are A_1, A_2, A_3, respectively.

(i) The speed of Earth revolving in the elliptical orbit around the Sun is not constant. It varies with its position. The speed is the largest when the Sun is nearest to the Earth (perihelion) and the smallest when the Sun is farthest from the Earth (aphelion). The variation follows the Kepler's second law of planetary motion, which states that the areal velocity is constant.

(ii) The ecliptic is inclined at an angle of 23.5° with the celestial equator. Owing to the spin motion of the Earth, the apparent motion of the Sun is parallel to the celestial equator whereas its actual motion is along the ecliptic. Therefore, one has to take projection of the motion of Sun, along the ecliptic, on the celestial equator. When the Sun crosses the celestial equator (at both the equinoxes), the projection of the Sun's daily shift (relative to the background stars) on the celestial equator is small. On the other side, when the Sun is farthest from the celestial equator (at both the solstices), the projection of the Sun's daily shift on the celestial equator is large.

Thus, in the months of June and December, the Sun is farthest from the celestial equator and therefore, the Earth's daily shift along the ecliptic corresponds to a large shift on the celestial equator. Further, in the months of March and September, the Sun is closest to the celestial equator and therefore, the Earth's daily shift along the ecliptic corresponds to a small shift on the celestial equator. Consequently, the length of a day is short in the months of March and September as compared to that in the months of June and December. (A day includes the bright part and the night both.)

1.1 Sundial

A sundial (Figure 2) is a device that tells the time of day with the help of the apparent position of the Sun in the sky. It has been in common use in the early days. Thus, a sundial gives the apparent solar time. It can obviously be used when the Sun is visible.

Figure 2: Sundial. Courtesy to Google website.

2. Mean solar day

Since the solar day varies from day to day, it is not possible to design (manufacture) a watch which can work according to the apparent solar day. Therefore, it has been decided by the international scientific community to take average of the solar days over the period of one year and to call it as the mean solar day, or simply a day.

This duration of one day (one mean solar day) is divided into 24 equal parts and each part is known as one hour. Thus, a day has 24 hours. Each hour is divided into 60 equal parts. One such part is known as one minute. Thus, one hour has 60 minutes. One minute is further divided into 60 equal parts. One such part is known as one second. Hence, one minute has 60 seconds.

A watch which runs at a constant rate is designed for the mean solar day.[1] In order to use full bright-time (when the Sun is visible) in a day for the working purposes, the beginning of day was decided to be at the mid-night. Thus, the mid of day, 12 Hours, was named as the noon. Obviously, the mean Sun is at the local meridian at noon (12 hours).

Thus, we have introduced a fictitious body, called the mean Sun, which moves along the celestial equator at a constant speed. This mean Sun may be ahead or behind the real Sun, or at the same position as the real Sun.

It is obvious that there is difference between the apparent solar day and the mean solar day. The difference varies smoothly from day to day. For example, in the year 1998, the length of the apparent solar day on some days has been found as given in Table 1.

[1]Notice that the decimal system could not be introduced in the system of time. It is not possible to introduce the decimal system, as the time depends on the natural phenomena.

Table 1. Length of apparent solar day in 1998

February 11	24 hours
March 26	24 hours − 18.1 seconds
May 14	24 hours
June 19	24 hours + 13.1 seconds
July 26	24 hours
September 16	24 hours − 21.3 seconds
November 3	24 hours
December 22	24 hours + 29.9 seconds

Notice that , for example, the apparent solar time on 26 March 1998 has been shorter than the mean solar day by 18.1 seconds and on 19 June 1998 has been longer than the mean solar day by 13.1 seconds.

The mean solar time is still not perfectly constant from one century to the next. But, it may be considered constant for most purposes.

Note: The length of the mean solar day is slowly increasing due to the tidal acceleration of the Moon by the Earth. In this tidal process, the Earth loses its energy and therefore the Earth's rotational motion is slowing down at very small rate. Consequently, from time to time, the time shown by a watch is decreased by one second. Such second is known as the leap second. This adjustment of time was done first time on 30 June 1972 when the watch was slowed down by 11 seconds. Later on the adjustment is done by 1 second at the beginning of the day of the month. In the night at $23^h\ 59^m\ 59^s$ the time is changed to 24^h. Some dates when the time was modified are given in Table 2.

Table 2. Some dates of modification of time.

31 December 1972	31 December 1973	31 December 1974
31 December 1975	31 December 1976	31 December 1977
31 December 1978	31 December 1979	30 June 1981
30 June 1982	30 June 1983	30 June 1985
31 December 1987	31 December 1989	31 December 1990
30 June 1992	30 June 1993	30 June 1994
31 December 1995	30 June 1997	31 December 1998
31 December 2005	31 December 2008	30 June 2012
31 December 2016		

3. Equation of time

The equation of time describes the difference between the apparent solar time and the mean solar time. It is because the projection of the speed of the real Sun on the equator is not constant, whereas that of the mean Sun (fictitious Sun) is steady. There is no universally accepted definition of the sign of the equation of time. Some scientists show it positive when the sundial is ahead of a clock, and the others when the clock is ahead of the sundial. Thus, we have

$$\text{Equation of time} = \text{Apparent solar time} - \text{Mean solar time}$$

or

$$\text{Equation of time} = \text{Mean solar time} - \text{Apparent solar time}$$

The Equation of time is zero when the mean solar time and the apparent solar time are equal.

4. Local mean time

After the introduction of the mean Sun, the local time is 12 hours (noon) when the mean Sun is overhead the local meridian for that location. Thus, in a country, every place has its own time, as the different places have different meridians. Obviously, it is difficult to manage with the time in a country. Therefore, it has been decided to have one meridian passing over the given country as the standard meridian for that country. Hence, all the places in that country have a common meridian, called the standard meridian for that country.[2] The longitude of the standard meridian for India is 82.5° E.

Table 3. Standard meridian and time-difference for some countries.

Country	Longitude	Difference from GMT
United States (Hawai)	150° W	−10 Hr
United States (Alaska)	135° W	−9 Hr
United States (Colorado)	105° W	−7 Hr
United States (Illinois)	90° W	−6 Hr
Canada (British Columbia)	105° W	− 7 Hr
Canada (Saskatchewan)	90° W	− 6 Hr
Canada (Ontario)	75° W	− 5 Hr
Canada (Nova Scotia)	60° W	− 4 Hr
Argentina	45° W	−3 Hr
Brazil	30° W	−2 Hr
England	0°	0 Hr
Spain	0°	0 Hr
Ireland	0°	0 Hr
Portugal	0°	0 Hr
Germany	15° E	1 Hr
France	15° E	1 Hr
Italy	15° E	1 Hr
Iran	52° 30′ E	3 Hr 30 min
Pakistan	75° E	5 Hr
India	82° 30′ E	5 Hr 30 min
Sri Lanka	82° 30′ E	5 Hr 30 min
Nepal	86° 15′ E	5 Hr 45 min
Bangladesh	90° E	6 Hr
South Korea	135° E	9 Hr
Australia	142° 30′ E	9 Hr 30 min

[2]This standard meridian not necessarily passes over the capital of the country.

Thus, when the mean Sun is over the standard meridian, then it is 12 hours (noon) for that country. The time defined in this manner is known as the local time. The time corresponding to the meridian over the Greenwich is known as the Greenwich Meridian Time (GMT). As the rotation of 360° is equivalent to 24 hours, we have

$$24 \text{ hours} \equiv 360° \qquad \text{and therefore} \qquad 1 \text{ hour} \equiv 15°$$

Obviously, the difference between the local time in India and the GMT is 82.5/15 = 5.5 Hr (5 Hr 30 min). As the longitude of the standard meridian is East, the time in India is large by 5 Hr 30 min as compared to the GMT. That is, if it is 1.00 Hr in England, the corresponding time in India is 6 Hr 30 min. The relation between the local time and the GMT can be expressed as the following.

(i) Suppose, the terrestrial longitude of standard meridian of a country is $a°$ E. Then, the local time in hours is

$$\text{local time} = \text{GMT} + \frac{a}{15}$$

(ii) Suppose, the terrestrial longitude of standard meridian of a country is $a°$ W. Then, the local time in hours is

$$\text{local time} = \text{GMT} - \frac{a}{15}$$

The above two relations can be combined as the following.

$$\text{local time} = \text{GMT} \pm \frac{a}{15}$$

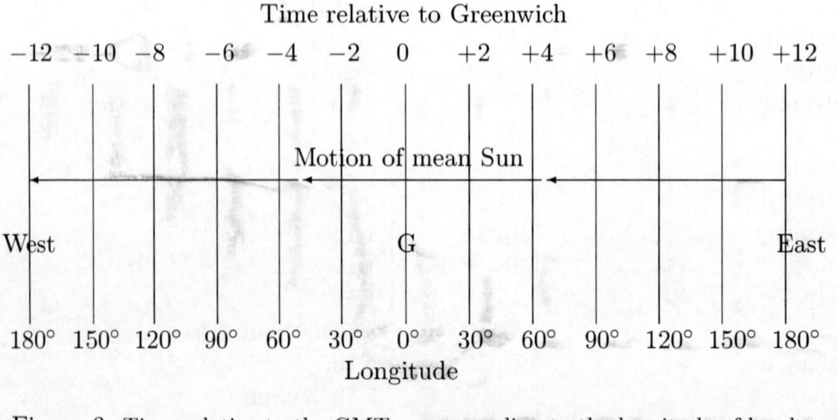

Figure 3: Time relative to the GMT corresponding to the longitude of local meridian.

The plus sign is taken when the longitude is East and the minus sign when the longitude is West. For the given local time and the GMT (both in hours), the longitude (in degrees) of the standard meridian of the country is

$$\text{longitude} = 15(\text{local time} - \text{GMT})$$

When the value of longitude is positive, the longitude is East and when the value is negative, the longitude is West. For some countries, the longitude of the standard meridian and the difference of time with respect to the GMT is given in Table 3. Notice that the countries of large size have two or more standard meridians. The plus sign with time shows that the standard meridian for the country is on the east to the Greenwich meridian and the negative sign shows that the standard meridian for the country is on the west to the Greenwich meridian.
Note: The mean Sun revolves parallel to the celestial equator which is parallel to the terrestrial equator. Thus, the local time depends on the terrestrial longitude and has no dependence on the terrestrial latitude. Longitudes of some meridians and corresponding times relative to the GMT are shown in Figure 3.

Example 1: Two countries A and B have the standard meridians at $37^\circ\ 30'$ E and $22^\circ\ 30'$ W longitudes, respectively. Calculate the local times in the countries when the GMT is 6 Hr 30 min.

Solution: Given, GMT = 6 Hr 30 min = 6.5 Hr. For the country A, the longitude is $37^\circ\ 30'$ E = 37.5° E. As the longitude is East, local time in A is larger than the GMT. Thus, the local time in A is

$$\text{local time} = \text{GMT} + \frac{a}{15} = 6.5 + \frac{37.5}{15} = 9 \text{ Hr}$$

For the country B, the longitude is $22^\circ\ 30'$ W = 22.5° W. As the longitude is West, local time in B is smaller than the GMT. Thus, the local time in B is

$$\text{local time} = \text{GMT} - \frac{b}{15} = 6.5 - \frac{22.5}{15} = 5 \text{ Hr}$$

Example 2: Local times in two countries A and B are 2 Hr 15 min and 8 Hr 45 min, respectively. If the GMT is 5 Hr 15 min, calculate the longitudes of the standard meridians of countries.

Solution: Given, GMT = 5 Hr 15 min = 5.25 Hr. Local time in the country A is 2 Hr 15 min = 2.25 Hr. As the time in A is smaller than that the GMT, the longitude of A is the West. The longitude of standard meridian of A is

$$\text{longitude} = 15(\text{local time} - \text{GMT}) = 15(2.25 - 5.25) = -45^\circ$$

As the negative sign is there, the longitude the standard meridian of A is 45° W.

Local time in the country B is 8 Hr 45 min = 8.75 Hr. As the time in B is larger than GMT, the longitude of B is the East. So, the longitude of the standard meridian of B is

$$\text{longitude} = 15(\text{local time} - \text{GMT}) = 15(8.75 - 5.25) = 52.5^\circ$$

As the positive sign is there, the longitude the standard meridian of B is $52^\circ\ 30'$ E.

Example 3: Longitudes of two countries A and B are at $97^\circ\ 30'$ E and 120° E, respectively. Corresponding to the time $6^h\ 45^m$ in A, calculate the time in B.

Solution: The difference between the longitudes of the countries A and B is

$$120° - 97°\ 30' = 22°\ 30'$$

which is equivalent to $1^h\ 30^m$.

Since the longitude is East and the longitude of B is larger than that of A, the time in B is more than that in A. The time in B is

$$6^h\ 45^m + 1^h\ 30^m = 8^h\ 15^m.$$

Example 4: Local times in two countries A and B are at $5^h\ 45^m$ and $7^h\ 15^m$, respectively. If longitude of A is at 97° 30′ E, calculate the longitude of B.

Solution: The difference between the times in two counties is

$$7^h\ 15^m - 5^h\ 45^m = 1^h\ 30^m$$

which is equivalent to 22° 30″. Since the country A is on the East to the Greenwich and the longitude of A is larger than 22° 30′, the country B also is on the East. As the time in B is larger than that in A, therefore the longitude of B is larger than that of A. So, the longitude of B is

$$97°\ 30' + 22°\ 30' = 120°\ \text{E}$$

5. Sidereal time

In order to understand the sidereal time, let us consider the arrangement, as shown in Figure 4, where the motion (revolution) of the Earth around the Sun S is shown. The Earth is spinning about its own axis. In the figure, E_1 and E_2 are two positions of the Earth on two successive days. Behind the Sun, there are stars situated at very large distances. They are also shown in the figure. When the Earth is at the position E_1, suppose the Sun as well as the distant stars are on the meridian of a person situated at the position P.

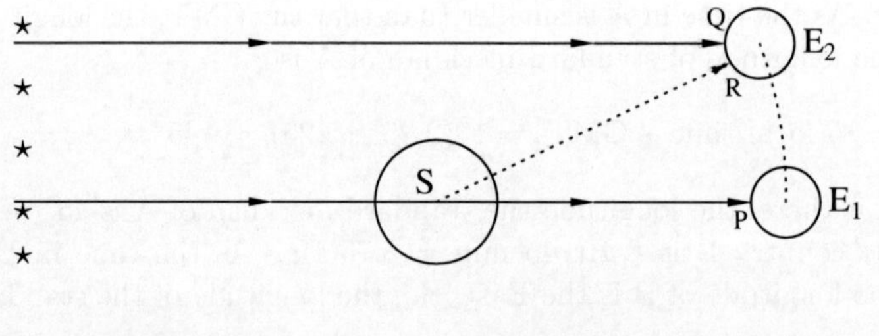

Figure 4: On two successive days, the positions of the Earth, E_1 and E_2. The period of rotation from P to Q is the sidereal day whereas for the rotation from P to R is the solar day.

Now, when the Earth completes one rotation about its axis, the Earth reaches at the position E_2, as it revolves around the Sun. After completion of one rotation, the position of the person on the Earth is Q. In this position, the distant stars are again on the meridian of the person, but the Sun has not yet arrived on the meridian of the person. When the Earth rotates a bit more and the position of the person is at R, the Sun is on the meridian of the person. The time taken for the rotation from P to Q is known as the sidereal day whereas the time taken for the rotation from P to R is known as the solar day. Obviously, the solar day is longer than the sidereal day. Our watch measures the mean solar day. That is, the mean solar day is equal to 24 hours.

Thus, the sidereal time is measured relative to the fixed stars whereas the solar time is measured relative to the Sun.

Let us consider a situation that the Earth has no spin motion about its own axis, what we see on the Earth. Obviously, owing to the revolution around the Sun, we could see only one day relative to the Sun in one revolution. Relative to the stars, there is no change. Hence, in absence of the rotational motion, we have one solar day in one revolution. Thus, the difference between the number of solar days and sidereal days is one. When we account for the spin motion of the Earth, in the time of one year, the Earth makes 365.2425 rotations[3] relative to the Sun. Since the length of sidereal day is smaller than that of solar day, in one year, there are 365.2425 + 1 = 366.2425 rotations relative to the distant stars. The length of the solar day is 24 hours. Suppose, d is the length of the sidereal day. Thus, we have

$$d\ 366.2425 = 24 \times 365.2425$$

Therefore, we have

$$d = \frac{24 \times 365.2425}{366.2425} = 23.9344697 \text{ Hr}$$

$$= 23 \text{ Hr } 56 \text{ min } 4.0907 \text{ s}$$

The difference between a mean solar day and a sidereal day is

$$24 \text{ Hr} - (23 \text{ Hr } 56 \text{ min } 4.0907 \text{ s}) = 3 \text{ m } 55.9093 \text{ s}$$

Therefore, the stars appear $3^m\ 56^s$ early each day. In the time of 6 months, the shift is about 12^h. That is the stars which are visible at the mid-night today appear at noon after six months. Still there are some small corrections to be accounted for. The correct value of the sidereal day is about 23^h, 56^m, 4.0916^s (23.9344699 hours or 0.99726958 mean solar day). It is the time taken by the Earth for making one rotation relative to the vernal equinox. The vernal equinox itself processes slowly westward relative to the fixed stars and completes one revolution in about 26,000

[3]A common person knows that a year (number) which can be divided by 4 is a leap year. For example, the year 1964 is divisible by 4. A leap year has 366 days. Each of the other years, which is not leap year, has 365 days. It is however not the complete story about the leap years. Though each century-year can be divided by 4, but it is not a leap year. In case of the century-years, the year which can be divided by 400 is the leap year. For example, the years 1700, 1800, 1900 are not the leap years, but the years 1600 and 2000 are the leap years. Consequently, in the span of 400 years, there are 97 leap years. Thus, the average number of days in a year is

$$= 365 + \frac{97}{400} = 365.2425$$

years. So, the sidereal day is about 0.0084 seconds shorter than the Earth's period of rotation relative to the fixed stars.[4]

It is a time keeping system used by the astronomers to keep track of the direction to point their telescopes to view a given object (star) in the night sky.

Example 5: Show that the stars which appear at mid-night today appear at noon after six months.

Solution: We know that the daily shift of a star is $3^m\ 56^s = 3.9333^m$. The shift in 6 months (182.5 days) is

$$3.9333^m \times 182.5 = 717.8273^m = 11.96^h$$

Hence, the shift of the distant stars observed in the time period of 6 months is 11.96^h. So, the stars appear at noon.

Example 6: Calculate the angle of rotation by the Earth in one solar day.

Solution: It is obvious that the Earth rotates about its axis through an angle of 360° in one sidereal day. It shows that in one mean solar day (or in one solar day) the Earth rotates through an angle larger than 360°. As the Earth makes one additional rotation in the time period of one year. Therefore, the additional rotation of Earth per day is

$$\frac{360}{366.2425} = 58'59'' \approx 59'$$

Thus, in the time-period of one mean solar day, the Earth rotates about its axis through an angle of 360° 59′.

5.1 Diurnal motion of stars

Diurnal motion (Figure 5) refers to the apparent daily motion of stars around the Earth, or more precisely around the two celestial poles. It is caused by the Earth's rotation about its own axis. So, every star apparently moves on a circle, that is called the diurnal circle. The time for one complete rotation is 23 hours, 56 minutes and 4.09 seconds (1 sidereal day).

Example 7: Suppose, the sidereal and solar clocks show the same time at midnight on 31 December then what time the sidereal clock shows at midnight on 18 August.

Solution: The time period of interest is 230 solar days. If there are 366.26 sidereal days in 365.24 solar days then the number of sidereal days in 230 solar days is

$$\frac{366.26}{365.24}\ 230 = 230.624 \text{ sidereal days}$$

It is equivalent to 230 sidereal days, 15 sidereal hours, 25 sidereal minutes. The sidereal clock therefore reads 15:25 at midnight on 18 August.

[4]The difference of 1 day in 26,000 years corresponds to

$$\frac{24 \times 60 \times 60}{26000 \times 365.2425} = 0.0091 \text{ sec/day}$$

Figure 5: Diurnal motion of stars. Courtesy to Google website.

6. Calendar

Generally, 365 days are taken equal to one year, called the civil year. The interval between two successive passages of the Sun through the vernal equinox is defined as the solar year or the tropical year. In the time of the Roman Emperor Julius Caesar, astronomers regarded 365.25 days as equal to one tropical year. Hence, one day accumulated in every four (civil) years $[(365.25 - 365) \times 4 = 1]$. In order to accommodate this extra day, in the year 46 BC, the Emperor made it a rule that every fourth year, which can be completely divided by 4, should be regarded as consist of 366 days instead of 365 days, so that the civil and tropical years could correspond with each other. Thus, the leap year was introduced and the extra day was added to the month of February.

The astronomers of 16th century, however, found that the actual tropical year was equal to about 365^d, 5^h, 48^m and 46^s. That is, 1 solar year = 365.242199 days. Thus, according to the above convention, 365.25 - 365.242199 = 0.007801 day accumulates each year. Hence, in 400 years, this excess amounts to $0.007801 \times 400 = 3.1204$ days. Therefore, in the year 1582, Pope Gregory of Rome introduced a correction that the century years, which are not completely divisible by 400, has 365 days instead of 366 days. For example, the years 2000 and 2400 are the leap years (as 2000 and 2400 are divisible by 400), but the years 2100, 2200, 2300 are not the leap years (as 2100, 2200, 2300 are not divisible by 400). Thus, a correction of 3 days in 400 years was introduced in the Julian calendar. This correction was, however, adopted in England much later, in 1752. The present calendar is based on this correction and is known as the Gregorian calendar (or Julian calendar with Gregorian correction). Obviously, 0.1204 day per four hundred years is still left unaccounted in the present calendar. This is, however, not troublesome for a long time to come. It amounts to 1 day in $400/0.1204 = 3322.26$ years.

6.1 Systematics in the Gregorian calendar

Some systematics which can be observed in the present calendar are as the following.

1. After a normal year of 365 days, the next year's week-days are shifted by one day, since $365 = (7 \times 52) + 1$.

2. After a leap year of 366 days, the next year's week-days are shifted by two days, since $366 = (7 \times 52) + 2$.

3. If we define the odd days as the days remained after dividing the days by 7, then the normal year has 1 odd day and the leap year has 2 odd days. A century has 24 leap years and 76 ordinary years. So, the total number of odd days is $2 \times 24 + 76 = 124 = (7 \times 17) + 5 = 5$.

4. Once in 400 years one century has 25 leap years and 75 ordinary years, resulting in 6 odd days. Hence, in 400 years, we have $(5 \times 3) + 6 = 21 = (7 \times 3) = 0$ odd day. That is, the Gregorian calendar repeats in every 400 years.

5. Within a century, in 28 years, we have 7 leap years and 21 ordinary years. Thus, we have $21 + (7 \times 2) = 35 = (7 \times 5) = 0$ odd day. That is, within a century, the Gregorian calendar repeats in every 28 years.

6. The number of odd days and shift M' for various months are given in Table 4. Te shift M' for a month is obtained as the sum of the shift of the previous year plus the number of odd days in the previous year. When the shift is larger than 6, we subtract 7 from the shift. In case of a leap year, an asterisk is placed on the number.

Using the above systematics, a day D falling on the particular date, month and year may be expressed as

$$D = \mathrm{R}_7[5\{\mathrm{Rem}(X/4) + \mathrm{Int}(Y/4)\} + \mathrm{Rem}(Y/4) + M' + \delta + \mathrm{DATE}]$$

where DATE is the date, M' the parameter (Table 4) corresponding to the month. X and Y are the parts of the year. For example, for the year 1985, $X = 19$ and $Y = 85$. Rem$(X/4)$ is the remainder after dividing X by 4, Int$(Y/4)$ is the integral part of $Y/4$, and Rem $(Y/4)$ is the remainder after dividing Y by 4. R_7 represents the remainder after dividing by 7. δ is the parameter to be determined according to the days assigned to the value of D. When the values of D are assigned to the days as given in Table 5, then δ comes out to be zero for non-leap years and 1 for the leap years.

Table 4. Number of odd days and shift M'

Month	Days	Odd days	Shift M'
January	31	3	0
February	28 29*	0 1*	3
March	31	3	3 4*
April	30	2	6 0*
May	31	3	1 2*
June	30	2	4 5*
July	31	3	6 0*
August	31	3	2 3*
September	30	2	5 6*
October	31	3	0 1*
November	30	2	3 4*
December	31	3	5 6*

*For the leap year.

Table 5. D versus Day

D	Day
1	Sunday
2	Monday
3	Tuesday
4	Wednesday
5	Thursday
6	Friday
0	Saturday

Table 6. Values of M

Month	Shift M
January	0 6*
February	3 2*
March	3
April	6
May	1
June	4
July	6
August	2
September	5
October	0
November	3
December	5

*For the leap year.

Defining $M = M' + \delta$, we get the values of M as given in Table 6 and the present calendar may be represented by the relation

$$D = \mathrm{R}_7[5\{\mathrm{Rem}(X/4) + \mathrm{Int}(Y/4)\} + \mathrm{Rem}(Y/4) + M + \mathrm{DATE}]$$

However, 0.1204 day per 400 years is left unaccounted. it accumulates and gives a difference of one day after about 3322 years, *i.e.*, around the year (1752 + 3322) = 5074.

Example 8: What day of the week was it on 2 October 1869.

Solution: For 1869, we have $X = 18$, $Y = 69$.

$\mathrm{Rem}(X/4) = 2, \qquad \mathrm{Int}(Y/4) = 17, \qquad \mathrm{Rem}(Y/4) = 1$

For the month October $M = 0$ (Table 6), $\qquad \mathrm{DATE} = 2$

$$D = \mathrm{R}_7[5\{2 + 17\} + 1 + 0 + 2] = \mathrm{R}_7[98] = 0$$

$D = 0$ corresponds to Saturday (Table 5). That is, 2 October 1869 was on Saturday.

Table 7. A chart for the calendar

Y				n			
				0	1	2	3
00				G*	E	C	A
01	29	57	85	A	F	D	B
02	30	58	86	B	G	E	C
03	31	59	87	C	A	F	D
04	32	60	88	E*	C*	A*	F*
05	33	61	89	F	D	B	G
06	34	62	90	G	E	C	A
07	35	63	91	A	F	D	B
08	36	64	92	C*	A*	F*	D*
09	37	65	93	D	B	G	E
10	38	66	94	E	C	A	F
11	39	67	95	F	D	B	G
12	40	68	96	A*	F*	D*	B*
13	41	69	97	B	G	E	C
14	42	70	98	C	A	F	D
15	43	71	99	D	B	G	E
16	44	72		F*	D*	B*	G*
17	45	73		G	E	C	A
18	46	74		A	F	D	B
19	47	75		B	G	E	C
20	48	76		D*	B*	G*	E*
21	49	77		E	C	A	F
22	50	78		F	D	B	G
23	51	79		G	E	C	A
24	52	80		B*	G*	E*	C*
25	53	81		C	A	F	D
26	54	82		D	B	G	E
27	55	83		E	C	A	F
28	56	84		G*	E*	C*	A*

6.2 Finding day without calculations

Using above systematics along with Tables 3 and 4, a chart shown in Table 7 may be obtained. Rem$(X/4)$ is denoted by n. For the given year, find X and Y. Now find the letter (A, B, C, etc.) in the row Y and column n (Table 7). Now look for that letter opposite in the given month (if the letter is written with an asterisk, then see the month with asterisk, *e.g.*, Jan* and Jan are different; Feb* and Feb are different, and move down to a column of days. This column of days along with the numbers written on the left to it is the calendar for the given month and year.

For example, for the year 1869, n = Rem$(18/4) = 2$ and $Y = 69$. The letter below $n = 2$ and opposite to $Y = 69$ is E. For the month of October, the letter E is in the sixth column, and hence, the calendar for October 18969 is as the following.

1	8	15	22	29	F
2	9	16	23	30	St
3	10	17	24	31	S
4	11	18	25		M
5	12	19	26		T
6	13	20	27		W
7	14	21	28		Th

Table 8. Week-days for the letter D and month

Jan*		Apr		Jul	A	B	C	D	E	F	G
Jan		-		Oct	G	A	B	C	D	E	F
-		May		-	F	G	A	B	C	D	E
Feb*		-		Aug	E	F	G	A	B	C	D
Feb		Mar		Nov	D	E	F	G	A	B	C
-		Jun		-	C	D	E	F	G	A	B
Sep		-		Dec	B	C	D	E	F	G	A
1	8	15	22	29	S	M	T	W	Th	F	St
2	9	16	23	30	M	T	W	Th	F	St	S
3	10	17	24	31	T	W	Th	F	St	S	M
4	11	18	25		W	Th	F	St	S	M	T
5	12	19	26		Th	F	St	S	M	T	W
6	13	20	27		F	St	S	M	T	W	Th
7	14	21	28		St	S	M	T	W	Th	F

7. Review Questions

1. Describe the apparent solar day and mean solar day. What is the equation of time? Which one of these is followed in our watches.

2. What is difference between the real Sun and the mean Sun.

3. Two countries A and B have standard meridians at 35° 27″ E and 21° 43″ W longitudes, respectively. Calculate the local times in the countries when the GMT is 6 Hr 43 min.

4. Local times in two countries A and B are 2 Hr 18 min and 9 Hr 37 min, respectively. If the GMT is 4 Hr 25 min, calculate the longitudes of standard meridians of the countries.

5. What is sidereal day for Earth? Explain the reason that the mean solar day is bigger than the sidereal day.

6. Discuss about the sidereal time. Estimate the length of a sidereal day.

7. Show that the Gregorian calendar repeats in every 400 years.

8. Show that the Gregorian calendar repeats in every 28 years.

9. Define a leap year. Which of the following years are the leap years.

   1960, 2100, 2123, 2200, 2275, 2300, 2356, 2400, 2496, 2500

   Ans. 1960, 2356, 2400, 2496

10. Write notes on the following:

    (i) Apparent solar day
    (ii) Sundial
    (iii) Mean Sun
    (iv) Mean solar day
    (v) Equation of time
    (vi) Local mean time
    (vii) GMT
    (viii) Sidereal time
    (ix) Diurnal motion of stars
    (x) Leap year

8. Multiple Choice Questions

1. Solar time varies from day to day because

   A. the Earth is not spherical in shape
   B. the Sun is moving in our galaxy Milky Way
   C. the Earth revolves around the Sun in an elliptical orbit
   D. the Earth spins about its own axis

   Ans. C

2. The mean solar day is equal to

   A. 24 hr   B. 23 hr 56 m   C. 24 hr 3 m   D. 12 hr

   Ans. A

3. The time based on the apparent motion of actual Sun is

   A. sidereal time   B. mean solar time   C. solar time   D. None of them

   Ans. C

4. The apparent motion of real Sun is along the

   A. celestial equator   B. ecliptic   C. local meridian   D. spin axis of Earth

   Ans. B

5. The motion of the mean Sun is along the

   A. celestial equator B. ecliptic C. local meridian D. spin axis of Earth

   Ans. A

6. The longitude of a country is East. The local time in the country is

   A. larger than GMT B. equal to GMT
   C. smaller than GMT D. none of them

   Ans. A

7. The longitude of a given country is West. The local time in the country is

   A. larger than GMT B. equal to GMT
   C. smaller than GMT D. none of them

   Ans. C

8. The local time depends on

   A. latitude B. both latitude and longitude
   C. longitude D. none of latitude and longitude

   Ans. C

9. The local time is larger than the GMT. The longitude of the country is

   A. East B. West C. North D. South

   Ans. A

10. The local time is smaller than the GMT. The longitude of the country is

    A. East B. West C. North D. South

    Ans. B

11. The time measured relative to the distant stars is known as

    A. solar time B. mean solar time C. local time D. sidereal time

    Ans. D

12. The time measured relative to the Sun is known as

    A. solar time B. mean solar time C. local time D. sidereal time

    Ans. A

13. Standard meridians of two countries A and B correspond to 5^h and $6^h\ 30^m$, respectively. Which of the followings is correct.

    A. B is east to A B. B is west to A C. B is north to A D. B is south to A

    Ans. A

14. The Earth rotates 360° in one

    A. solar day B. mean solar day C. sidereal day D. year

    Ans. C

15. In one solar day, the Earth rotates about its own axis through an angle approximately

A. 359° 29′ B. 360° C. 360° 59′ D. 180°

Ans. C

16. The orbit of the fictitious Sun relative to the Earth is

A. ellipse B. circle C. parabola D. hyperbola

Ans. B

17. The angle between the orbit of the fictitious Sun and the celestial equator is

A. 0° B. 23.5° C. 45° D. 60°

Ans. A

18. A sundial gives

A. mean solar time B. apparent solar time C. sidereal time D. local time

Ans. B

4. Telescopes and Detectors

We know that our unaided eyes can see the objects having apparent magnitude up to 6. When we want to study the objects having apparent magnitude larger than 6, we have to take help of some instrument. An instrument used to detect electromagnetic radiations is known as the telescope. The main purpose of a telescope is to collect light from the cosmic object of interest.

Design of a telescope depends on the wavelength of radiation to be detected by the telescope. On the surface of Earth, we have either optical telescopes or radio telescopes, as the Earth's atmosphere is transparent in two bands, known as the optical-window and radio-window. For the study of radiations, outside these two windows, we have to send a telescope to a height where the required radiations are available. For example, for the study of γ-rays and x-rays, one has to send the instrument above the ionosphere.

An optical telescope using lens cannot be of very large size because of its weight and aberrations. Therefore, optical telescopes of very large size use mirror in place of a lens. Since the scattering of radiations in the optical-window is about 10^{20} times larger than of those in the radio-window, therefore, the role of radio telescopes becomes very important, in particular in the context of the study of very distant cosmic objects. However, resolving power of a single dish radio-telescope becomes very poor, which can be improved a lot using the concept of VLBI. Quite often, the instruments are put on a geo-stationary satellite orbiting around the Earth or on a spacecraft going through the interstellar medium. All these concepts about the telescopes and detection of radiations are discussed in this chapter.

1. Astronomical units of distance

Cosmic objects are, in general, situated at very large distances. Hence, in the Astronomy & Astrophysics (A&A), we need to use the units for measuring the distances which are rather large in magnitude. The units of distances used in the A&A are: (i) Light year, (ii) Astronomical unit, and (iii) Parsec.

1.1 Light year (ly)

In the electromagnetic spectrum, all kinds of radiations in vacuum, travel with the same speed c, having the value 3×10^8 m/s. Since the speed of light is finite, it requires finite time to reach the surface of the Earth from a cosmic object. For example, it takes 8 minutes and 20 seconds to reach the Earth from the Sun; 4.2 years to reach the Earth from the second nearest star, Proxima Centauri; 2.5 million years from the nearest large galaxy, Andromeda (also called M 31) situated in the Andromeda constellation. Thus, the distances of the astronomical objects may be expressed in term of the time required for reaching the light from the object to the surface of Earth.

One light year, denoted by ly, is defined as the distance traveled by light in vacuum in one year. Because of the leap years, the average number of days in one year may be taken as 365.25. Thus, we have

1 year = 365.25 days = $365.25 \times 24 \times 60 \times 60 = 3.156 \times 10^7$ s

The speed of light is 3×10^8 m/s

Thus, 1 light year = $(3.156 \times 10^7)\,(3 \times 10^8) = 9.47 \times 10^{15}$ m

1 ly = 9.47×10^{15} m = 9.47×10^{12} km

Hence, the distance traveled by light (photon) in one year in vacuum is 9.47×10^{12} km, which is known as one light year.

Similarly, we can define light-day, light-hour, light-minute, light-second. Equivalences of them are given in Table 1.

Table 1. Equivalences

1 light sec	3×10^5 km
1 light min	1.8×10^7 km
1 light hour	1.08×10^9 km
1 light day	2.59×10^{10} km

1.2 Astronomical unit (AU)

Besides the spin motion about its own axis, the Earth moves around the Sun in an elliptical orbit in the time-period of one year. The distance between the Earth and Sun changes with time; its minimum value is 1.47×10^{11} m, around 3 January each year and the maximum value is 1.53×10^{11} m, around 3 July each year.[1] The average distance between the Sun and Earth is defined as one astronomical unit, denoted by AU. It is equal to 1.50×10^{11} m = 1.50×10^8 km. Distances of planets from the Sun in kilometers and astronomical units are given in Table 2.

Example 1: Calculate the time taken by light to reach the Earth from the Sun. Given, 1 AU = 1.50×10^{11} m.

Solution: The average distance between the Earth and Sun is 1.50×10^{11} m and velocity of light is 3×10^8 m/s. Thus, the time t taken by light for traveling from the Sun to Earth is

$$t = \frac{1.50 \times 10^{11}}{3 \times 10^8} = 500 \text{ sec}$$

Hence, the light takes 8 min and 20 sec to reach the Earth from the Sun.

[1]One should not get confused with the fact that we have winter season in January and summer season in July. We are in the northern hemisphere. In the southern hemisphere, the seasons are reverse.

1.3 Parsec (pc)

One parsec is equal to the length so that the two points, separated by one astronomical unit, subtends an angle of one arc-sec (1″), as shown in Figure 1.

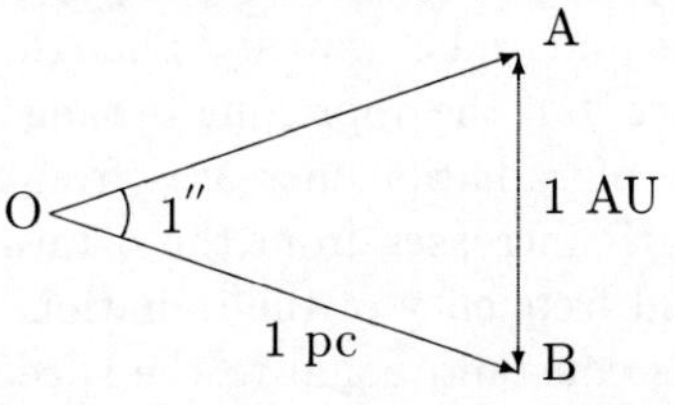

Figure 1: Two objects A and B, separated by a distance of 1 AU, subtend an angle of 1 arc-sec at O. The distance between O and A, or O and B is defined as 1 parsec.

Now, we have

$$1'' = \frac{\pi}{180} \times \frac{1}{60} \times \frac{1}{60} = 4.848 \times 10^{-6} \text{ rad}$$

Thus, we have

$$\text{angle} = \frac{\text{arc}}{\text{radius}} \qquad \text{and therefore} \qquad 4.848 \times 10^{-6} = \frac{1\text{AU}}{1\text{pc}}$$

Therefore, we have

$$1\text{pc} = \frac{1.5 \times 10^{11} \text{ m}}{4.848 \times 10^{-6}} = 3.09 \times 10^{16} \text{ m} = 3.09 \times 10^{13} \text{ km}$$

Therefore, 1 parsec is equal to 3.09 $\times 10^{13}$ km.

Table 2. Distances of planets from the Sun

Planet	Distance	
	km	AU
Mercury	5.81×10^7	0.39
Venus	1.08×10^8	0.72
Earth	1.50×10^8	1.00
Mars	2.29×10^8	1.52
Jupiter	7.80×10^8	5.20
Saturn	1.43×10^9	9.54
Uranus	2.88×10^9	19.18
Neptune	4.51×10^9	30.06

1.4 Relation between pc and ly

We have

$$\frac{1\text{pc}}{1\text{ly}} = \frac{3.094 \times 10^{18} \text{ cm}}{9.467 \times 10^{17} \text{ cm}} = 3.26$$

Therefore, 1 pc = 3.26 ly.

2. Electromagnetic spectrum

The radiations coming from the space towards the Earth have full electromagnetic spectrum, consists of γ-rays, x-rays, ultraviolet (UV) radiation, visible light, infrared (IR) radiation, microwaves and radio waves. Though there are no well defined demarcations for their ranges, but the approximate ranges of these regions are given in Table 3. The frequency of radiation increases from the radio waves to the γ-rays, whereas the wavelength increases from the γ-rays to the radio waves. The variations of wavelength and frequency of the radiations are such that the product of their values in vacuum is constant, equal to the speed of light $c = 3 \times 10^8$ m/s. The frequency ν and wavelength λ are related as $c = \nu\lambda$.

Table 3. Electromagnetic spectrum

Radiation	Frequency
γ-rays	$> 2 \times 10^{19}$ Hz
x-rays	$2 \times 10^{16} - 2 \times 10^{19}$ Hz
EUV	$3 \times 10^{15} - 2 \times 10^{16}$ Hz
UV radiations	$7 \times 10^{14} - 3 \times 10^{15}$ Hz
Visible light	$4 \times 10^{14} - 7 \times 10^{14}$ Hz
IR radiations	$3 \times 10^{12} - 4 \times 10^{14}$ Hz
FIR	$3 \times 10^{11} - 3 \times 10^{12}$ Hz
Microwave	30 – 300 GHz
Radio	0.5 – 30000 MHz

2.1 Optical window and radio window

All of us know that γ-rays, x-rays, EUV and UV radiations are very energetic radiations, and if all of them could reach the Earth's surface, human life obviously could not survive. It is the Earth's atmosphere which plays a magnificent role by shielding us from the merciless, intense radiations coming from the interstellar medium (ISM) towards the Earth.

It is a matter of great pleasure to notice that the γ-rays and x-rays from the ISM are absorbed in the upper part of the Earth's atmosphere, and the molecules dissociate into its constituent atoms; the atoms are stripped of their electrons. As a result of that, the ions (electrons and positively charged nuclei) are produced there. This region of ions at an altitude of about 500 km is known as the **ionosphere**. It contains plasma comprising of electrons, protons, and positively charged ions.

Below the ionosphere, there is a layer of ozone (O_3) gas, which absorbs most of the UV radiations, resulting in the depletion of the ozone layer. Thus, it is the Earth's upper atmosphere that plays a magnificent role in our life, by shielding us from the γ-rays, x-rays and UV radiations coming from the cosmic objects (ISM).[2]

The atmosphere close to the Earth's surface (called the troposphere) contains mainly water (H_2O) and carbon dioxide (CO_2), and thus, it absorbs most of the IR radiations coming from the ISM.

[2]Depletion of the ozone layer has been an important issue for discussion at the national as well as international forums, as it affects the human life on the Earth's surface. The chlorofluorocarbon (CFC) produced in the air conditioners, etc., destroys the ozone layer.

For a plasma, the characteristic frequency (also called the plasma frequency), in Hz, is given by $\nu_p = 9.0\sqrt{N_e}$, where N_e is the density of electrons in m^{-3} in the plasma. In the ionosphere, the electron density ranges from 10^{10} to 10^{12} m^{-3}. As a characteristic of plasma, the radiations having frequency less than the plasma frequency are reflected back by the plasma. Hence, the radiations from the interstellar medium whose frequency is less than the characteristic frequency of plasma in the ionosphere are returned back towards the ISM.

Most of the parts of the electromagnetic spectrum coming from the cosmic objects (ISM) are obviously incapable to transmit through the Earth's atmosphere and to reach the surface of Earth. There are, however, two bands in which the Earth's atmosphere is transparent. These two bands are known as the optical-window and radio-window.

The optical-window ranges from 3000 Å to 10,000 Å, whereas the radio-window ranges from a few mm to several decameters. The radiations which our eyes can see lie in the optical-window. It is the optical-window which did not allow human beings to remain ignorant about the beautiful phenomena, such as rising and setting of Sun, phases of Moon, solar and lunar eclipses, etc., occurring in nature. Therefore, on the surface of the Earth, we have only two types of telescopes, optical telescopes, and radio telescopes.

2.2 Advantage of radio-window

The Earth's atmosphere is transparent in the optical-window and radio-window. The radio-window ranges from mm to decameter in wavelength. In the context of the distant objects, the radiations lying in the radio-window have remarkable advantages over those in the optical-window. The ratio of the average wavelength in the radio-window (λ_R) and the average wavelength in the optical-window (λ_O) is[3]

$$\frac{\lambda_R}{\lambda_O} = 10^5$$

We know that the scattering (S) of a radiation is inversely proportional to the fourth power of its wavelength (λ).

$$S \propto \frac{1}{\lambda^4} \qquad \text{or} \qquad S = \frac{k}{\lambda^4}$$

where k is constant. Therefore, we have

$$S_R = \frac{k}{\lambda_R^4} \qquad \text{and} \qquad S_O = \frac{k}{\lambda_O^4}$$

[3]The mean wavelength in the optical-window is about 5000 Å and in the radio-window is about 5 cm. Thus, we have

$$\frac{\lambda_R}{\lambda_O} = \frac{5 \text{ cm}}{5000\text{Å}} = 10^5$$

Thus, the ratio of scatterings of radiations in the two windows is

$$\frac{S_R}{S_O} = \frac{\lambda_O^4}{\lambda_R^4} = 10^{-20}$$

It shows that the probability of reaching the optical radiations to the Earth's surface is about 10^{20} times less than that of the radio radiations. Hence, in the context of distant objects, the radio-window plays important role as compared to the optical-window. Hence, it is the radio radiations which can bring valuable information about the cosmic objects, situated at very large distances, *e.g.*, at a distance of several hundred light years. That is, the radiations, which have reached the Earth's surface today, would have started several hundred years ago from that cosmic object.

The radio window has importance in the context of communication of information from one place to other. In the past, the communication has been through the reflection with the ionosphere. In that case, the frequency used in the communication had to be smaller than the plasma-frequency.

Nowadays, we communicate with the help of a geo-stationary satellite. The geo-stationary satellite is situated at a distance larger than that of the ionosphere. Obviously, the frequency used in the communication should be larger than that of the plasma, so that the wave could transmit through the ionosphere.

3. Luminosity of a star

The amount of energy emitted per unit time by a cosmic object (star) is defined as its luminosity, and is denoted by L. It is an intrinsic property of the object and it depends on the temperature and composition of the object. For example, the Sun, on an average emits energy at the rate of 3.826 $\times 10^{26}$ J/s. Thus, the luminosity of the Sun, denoted by $L_\odot$, is 3.826×10^{26} J/s.

Further, the luminosity of an object depends on the frequency-band in which the energy is calculated/measured. When no frequency band is given, it is the luminosity of full electromagnetic spectrum emitted by the object.

3.1 Radiant flux of a star

The amount of energy emitted by an object (star) per unit time is spread over a spherical surface in space whose center is at the center of the object.[4] If we assume that the atmosphere around the star is isotropic and homogeneous, and no energy is absorbed during its journey from the object to an observer, the amount of energy from the object received by the observer per unit time per unit surface area is defined as the radiant flux from the object. The radiant flux F measured at a distance r from the object is

$$F = \frac{L}{4\pi r^2}$$

where L is the luminosity of the object. Thus, the radiant flux depends on both the intrinsic luminosity of the object, and the distance from the object. It shows

[4] As the size of star is very small in comparison to the distance between the star and an observer, the star may be treated as a point object.

that the radiant flux received from an object varies as the luminosity of luminosity of inverse square of its distance from the object. Thus, if the distance of an object is large from the observer, the radiant flux received from the object is small. For example, if we increase the distance between the Sun and observer, the Sun (keeping its luminosity unchanged) may appear even fainter than a star. The radiant flux from an object is a measure of the brightness of the object.

Example 2. The luminosity of Sun is $L = 3.826 \times 10^{26}$ J/s. Calculate the radiant flux received on the surface of Earth from the Sun, assuming that the absorption of energy in the atmosphere between the Sun and Earth is negligible. If the Sun had been at a distance of 10 pc from the Earth, what had been radiant flux received on the surface of the Earth. Given, average distance between the Sun and Earth is 1 AU = 1.5×10^{11} m, and 1 pc = 3.09×10^{16} m.

Solution: Radiant flux received from the Sun on the surface of the Earth is

$$F = \frac{L}{4\pi(1\text{AU})^2} = \frac{3.826 \times 10^{26}}{4 \times 3.14 \times (1.5 \times 10^{11})^2} = 1.35 \times 10^3 \text{ J/s m}^2$$

If the Sun had been at a distance of 10 pc = 3.09×10^{17} m. The radiant flux received at a distance of 10 pc is

$$F = \frac{3.826 \times 10^{26}}{4 \times 3.14 \times (3.09 \times 10^{17})^2} = 3.19 \times 10^{-10} \text{ j/s m}^2$$

Notice that with the increase of the distance, the flux is decreased.

3.2 Solar constant

Value of radiant flux from the Sun received on the surface of the Earth is known as the solar constant. Here, it is assumed that the atmosphere between the Sun and Earth does not absorb any part of the energy generated by the Sun. The amount of energy received from the Sun on the surface of the Earth per unit surface area per unit time (solar constant) is 1.35×10^3 J/s m^2.

Example 3. Calculate the solar constant. Given, 1 AU = 1.5×10^{11} m, and solar luminosity 3.826×10^{26} J/s.

Solution: The solar constant is s

$$F = \frac{L}{4\pi(1\text{AU})^2} = \frac{3.826 \times 10^{26}}{4 \times 3.14 \times (1.5 \times 10^{11})^2} = 1.35 \times 10^3 \text{ J/s m}^2$$

3.3 Comparison of two stars

Let us have two cosmic objects of luminosities L_1 and L_2, situated at the distances r_1 and r_2, respectively, from an observer. The radiant fluxes received by the observer from the objects are, respectively,

$$F_1 = \frac{L_1}{4\pi r_1^2} \quad \text{and} \quad F_2 = \frac{L_2}{4\pi r_2^2}$$

Therefore, the ratio of the radiant fluxes is

$$\frac{F_2}{F_1} = \frac{L_2 r_1^2}{L_1 r_2^2}$$

(i) When the luminosities of the two objects are equal, $L_1 = L_2$, the ratio of the radiant fluxes is

$$\frac{F_2}{F_1} = \frac{r_1^2}{r_2^2}$$

(ii) When the distances of the two objects are equal, $r_1 = r_2$, the ratio of the radiant fluxes is

$$\frac{F_2}{F_1} = \frac{L_2}{L_1}$$

4. Magnitude scale

When we observe stars (objects) in a clear night, the brightness (radiant flux) received from different stars is not the same. Some of them appear very bright and are visible without any efforts, whereas some are very dim and are barely visible to the naked eyes. (A large number of stars are not visible to the naked eyes, but become visible in the sky with the help of a telescope.) Other visible stars have brightness in between these two cases.

The stars which are visible to the naked eyes, depending on their apparent brightness, have been classified in terms of a parameter, called the apparent magnitude. The most bright stars in the sky were assigned the apparent magnitude $m = +1$, whereas the dimmest ones (which are barely visible to the naked eyes) were assigned the apparent magnitude $m = +6$. Notice that with the increase of m, the brightness of star decreases. Hence, all the stars, visible to the naked eyes, have magnitude in the range from $+1$ to $+6$.

The correspondence between the magnitude and brightness of stars was found to be logarithmic. That is, if m and F are the apparent magnitude and radiant flux, respectively, of a star, then we have

$$\log_{10} F = -k\, m + C$$

where k and C are constant. The negative sign on the right side indicates that with the increase of magnitude m, the brightness F decreases. Suppose, the radiant fluxes F_1 and F_2 received from two stars of apparent magnitudes m_1 and m_2, respectively. Then, we have

$$\log_{10} F_2 = -k\, m_2 + C \qquad \text{and} \qquad \log_{10} F_1 = -k\, m_1 + C$$

Therefore, we have

$$\log_{10} F_2 - \log_{10} F_1 = -k(m_2 - m_1) \qquad \text{or} \qquad \frac{F_2}{F_1} = 10^{k(m_1 - m_2)}.$$

According to the present definition, a difference of 5 magnitudes corresponds exactly to a ratio of 100 in the brightness of stars. That is, for $m_1 - m_2 = 5$, we have $(F_2/F_1) = 100$. On substituting these values, we get

$$100 = 10^{k5}$$

On taking logarithm base 10, we get

$$2 = k\,5 \qquad \text{or} \qquad k = 0.4$$

Thus, the relation between the radiant fluxes received from the stars and their apparent magnitudes is

$$\frac{F_2}{F_1} = 10^{0.4(m_1 - m_2)}.$$

Magnitude scale of stars extends on both sides. That is, we have stars having negative value of magnitude, and we have stars having magnitude more than 6. For example, the apparent magnitude of the star Sirius is −1.4, the full Moon is −12.6, and the Sun is −26.73. The Sun is obviously the brightest object, visible to us. On the other side, with the help of telescopes, we can see stars having magnitude up to + 30. Hubble telescope has located stars with magnitudes of +30 at visible wavelengths.

With the help of sensitive instruments (*e.g.*, photometers) it is possible to measure apparent magnitude of an object with an accuracy of ±0.01 magnitude, and difference in the magnitudes with an accuracy of ±0.002 magnitude.

Example 4. Calculate the ratio of the fluxes received from two stars whose magnitudes differ by 2.5.

Solution: The ratio of the fluxes F_1 and F_2 received from two stars of magnitudes m_1 and m_2, respectively, is

$$\frac{F_2}{F_1} = 10^{0.4(m_1 - m_2)}.$$

Consider that a dim star is of magnitude m_1 and a bright star of m_2. Then, $m_1 - m_2 = 2.5$. Hence, the ratio of their fluxes is

$$\frac{F_2}{F_1} = 10^{0.4 \times 2.5} = 10$$

Thus, the brighter star is 10 times bright as compared to the dimmer one.

Example 5. The ratio of fluxes from two objects is 145. The magnitude of brighter object is 1.2. Calculate the magnitude of other object.

Solution: For two objects with magnitudes m_1 and m_2, and fluxes F_1 and F_2, respectively, we have

$$\frac{F_2}{F_1} = 10^{0.4(m_1 - m_2)} \qquad\qquad 145 = 10^{0.4(m_1 - m_2)}$$

Thus, $m_1 - m_2 = 5.40$. We have $m_2 = 1.2$. Therefore, $m_1 = 6.6$.

Example 6. Magnitude of the most bright object in appearance (Sun) is -26.81, and with the help of a good telescope, we can see the stars as faint as having magnitude of 28.19. Calculate ratio of brightness of the two limiting stars.

Solution: Let $m_1 = 28.19$ and $m_2 = -26.81$. Hence, the ratio of the fluxes received form the star (F_1) and by the Sun (F_2) is expressed as

$$\frac{F_2}{F_1} = 10^{0.4(m_1 - m_2)} = 10^{0.4(28.19+26.81)} = 10^{22}$$

Thus, the Sun is 10^{22} times brighter than the star.

4.1 Absolute magnitude

The radiant flux (and thus the magnitude) of a star (object) depends on its distance form the observer. Hence, the comparison of apparent magnitudes of the stars cannot provide meaning information about their relative luminosities. We know that when two stars are at equal distances from an observer, the ratio of their radiant fluxes provides information about their relative luminosities. Thus, when we bring theoretically all the stars at a standard distance from an observer, then the radiant fluxes received from the stars provide real comparison of the stars.

This standard distance of an object from an observer is taken as 10 pc, following the decision of International Astronomical Union. When an object is brought theoretically at a distance of 10 pc, its apparent magnitude is defined as the **absolute magnitude**. Thus, the apparent magnitude of an object situated at a distance of 10 pc from an observer is known as the absolute magnitude of the star.

Suppose, a star at distance d_1 is moved to a distance d_2, the ratio of the fluxes received F_1 and F_2, respectively, is

$$\frac{F_1}{F_2} = \frac{L/\pi d_1^2}{L/\pi d_2^2} = \left(\frac{d_2}{d_1}\right)^2$$

Let the apparent magnitude of an object situated at a distance d pc from an observer be m. When the object is moved theoretically to a distance of 10 pc, its apparent magnitude, known as the absolute magnitude, is denoted by M. Since the object is the same, and thus, its luminosity does not change. Hence, the ratio of the radiant fluxes is expressed as

$$\frac{F_{10}}{F_d} = \left(\frac{d}{10}\right)^2 \quad \text{and} \quad \frac{F_{10}}{F_d} = 10^{0.4(m-M)}$$

From these two relations, we have

$$\left(\frac{d}{10}\right)^2 = 10^{0.4(m-M)} \quad \text{or} \quad m - M = 5 \log_{10} d - 5$$

Thus, for a star situated at a distance $d < 10$ pc, the absolute magnitude is larger than the apparent magnitude. But, for a star situated at a distance $d > 10$ pc, the absolute magnitude is smaller than the apparent magnitude.

Example 7. Suppose, two stars of apparent magnitudes m_1 and m_2 are at the distances d_1 and d_2, respectively. Calculate the difference between the absolute magnitudes of the stars.

Solution: Suppose, M_1 and M_2 are absolute magnitudes of two stars, we have

$$m_1 - M_1 = 5\ \log_{10} d_1 - 5 \qquad \text{and} \qquad m_2 - M_2 = 5\ \log_{10} d_2 - 5$$

On taking difference, we get

$$(m_1 - m_2) - (M_1 - M_2) = 5\ \log_{10} d_1 - 5\ \log_{10} d_2$$

or

$$(M_1 - M_2) = (m_1 - m_2) - 5\ \log_{10}\left(\frac{d_1}{d_2}\right)$$

Example 8. Apparent magnitude of the Sun is $m_\odot = -26.81$. Calculate the absolute magnitude of Sun. Given, 1 AU = 1.50×10^8 km; 1pc = 3.09×10^{16} m.

Solution: The distance of Sun from the Earth is $d = 1.50 \times 10^8$ km = 4.848×10^{-6} pc, and $m_\odot = -26.81$. Absolute magnitude $M_\odot$ of the Sun is

$$M_\odot = m_\odot - 5\ \log_{10} d + 5$$

On substituting the values, we get

$$M_\odot = -26.81 - 5\ \log_{10}(4.848 \times 10^{-6}) + 5 = 4.76$$

Thus, from a distance of 10 pc, the Sun looks like a faint star with the magnitude +4.76.

Example 9. The absolute magnitude of a star is +1. Calculate its distance so that the apparent magnitude of the star is −1.

Solution: We have $m_1 = +1$, $d_1 = 10$ pc and $m_2 = -1$. Thus, we have

$$\left(\frac{d_1}{d_2}\right)^2 = 10^{0.4(m_1 - m_2)} \qquad \text{or} \qquad \left(\frac{10}{d_2}\right)^2 = 10^{0.4(1+1)}$$

Therefore, we have

$$d_2 = 10 \times 10^{-0.4} = 3.981 \text{ pc}$$

4.2 Relation between luminosities and absolute magnitudes

The fluxes F_1 and F_2 of two stars of apparent magnitudes m_1 and m_2, respectively, are expressed as

$$\frac{F_2}{F_1} = 10^{0.4(m_1 - m_2)} \qquad \text{or} \qquad \frac{L_2/4\pi r_2^2}{L_1/4\pi r_1^2} = 10^{0.4(m_1 - m_2)}$$

When each of the r_1 and r_2 are 10 pc, the magnitudes m_1 and m_2 are absolute magnitudes M_1 and M_2. Thus, we have

$$\frac{L_2}{L_1} = 10^{0.4(M_1 - M_2)} \qquad \text{or} \qquad M_2 = M_1 - 2.5 \log_{10}(L_2/L_1)$$

When $L_2 > L_1$, we have $M_2 < M_1$ and when $L_2 < L_1$, we have $M_2 > M_1$.

Note: When one of the stars is taken as the Sun. Then, $L_1 = L_\odot$, and $M_1 = M_\odot = 4.76$, and thus, luminosity L and absolute magnitude M of a star are related as

$$M = 4.76 - 2.5 \log_{10}(L/L_\odot)$$

Example 10: For the Sun, the luminosity is $L_\odot = 3.826 \times 10^{26}$ J/s and absolute magnitude is $M_\odot = 4.76$. Calculate the absolute magnitude of a star having luminosity 5.26×10^{26} J/s.

Solution: The absolute magnitude of star is

$$M_2 = M_1 - 2.5 \log_{10}(L_2/L_1)$$

$$M_2 = 4.76 - 2.5 \log_{10}(5.26 \times 10^{26}/3.826 \times 10^{26}) = 4.41$$

5. Features of a telescope

There are three features of a telescope which improve the quality of our vision.

(i) Light-gathering

(ii) Resolution

(iii) Magnification

5.1 Light-gathering

In a telescope, the light falling on the primary (objective) lens is collected. The light-gathering is probably the most important feature of a telescope. This makes the faint images brighter. The amount of light gathered by a lens is proportional to the diameter (or aperture) of the lens. In a telescope, the light is gathered by the primary (objective) lens. With the increase of the size of aperture a, dim objects may be detected, depending on the amount of light gathered. Therefore, the telescopes having larger effective aperture are constructed. The image brightness B is proportional the square of the aperture a. That is,

$$B \propto a^2$$

Example 11: Two optical telescopes have apertures of 2 m and 8 cm. Compare the brightnesses of images formed by the telescopes under normal conditions.

Solution: Under normal viewing conditions, the ratio of image brightnesses is expressed as

$$\frac{B_{\text{large}}}{B_{\text{small}}} = \frac{a^2_{\text{large}}}{a^2_{\text{small}}} = \frac{2^2}{0.08^2} = 625$$

The image produced by the large telescope is thus 625 times brighter than that by the small one.

5.2 Resolution

Suppose, two point sources (objects) are very close together. Then, two diffraction patterns formed by the objects may overlap with each other and hence may be difficult to distinguish them as separate. An optical instrument (microscope, telescope, etc.) is said to have resolved the two point sources when the two diffraction patterns are sufficiently separated from each other. The ability of an optical instrument just to resolve the images of two nearby point sources is known as the resolving power.

Rayleigh criterion

According to Rayleigh, two close spectral lines of equal intensities can be just resolved by an optical instrument when the position of the maximum of one spectral line coincides with the minimum of the other and vice versa. When they come further closer, the lines can no longer be resolved. The individual intensity distribution curves (line-profiles) of two spectral lines having equal intensities and wavelengths λ and $\lambda + d\lambda$ are shown in Figure 2 for three different situations.

In the situation (a), the separation between the two lines is much larger than their individual line-widths. In the situation (b), the maximum of one line coincides with the minimum of the other and vice versa. That is, the separation between the maxima of the two lines is equal to the half-width of any one of the lines. In the situation (c), the two lines have come further closer. That is, the separation between the maxima of the two lines is less than the half-width of any one of the lines.

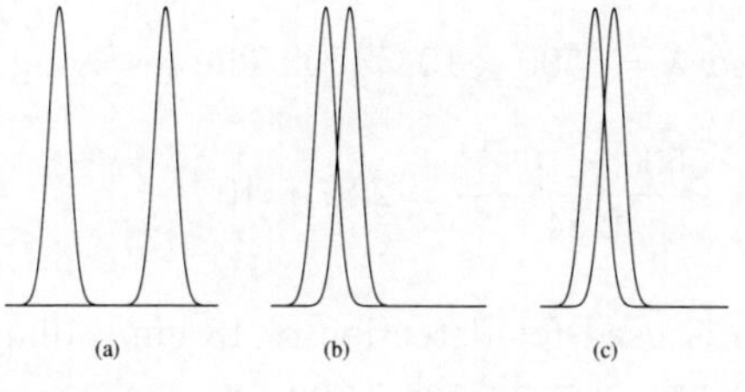

Figure 2: Individual intensities of two spectral lines having different wavelengths for three different situations.

In Figure 3, we have plotted individual intensities of two spectral lines along with the resultant intensity (sum of the intensities). In the situation (a), the resultant is the same as the intensities of the individual lines. In the situation (b), the resultant intensity curve shows a dip in the middle of maxima of the lines. Accordingly to

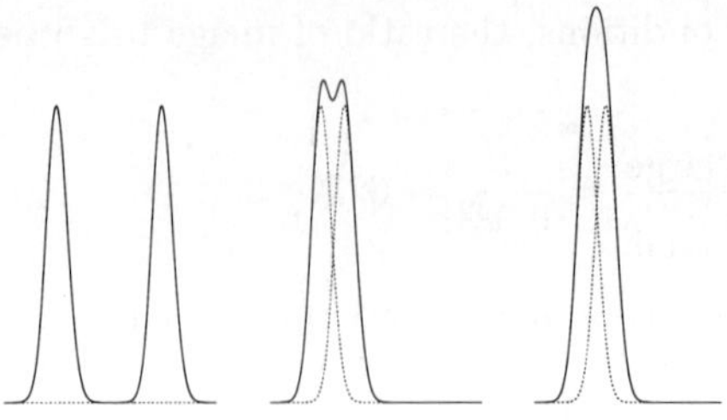

Figure 3: Individual intensities of two spectral lines having different wavelengths for three different situations are shown by the dotted lines. The sum of the intensities of the two lines are shown by the solid line.

Rayleigh, these spectral lines are distinguished from each other and said to be just resolved. In the situation (c), there is no more dip in the resultant intensity and the two lines cannot be resolved.

Resolving power of a telescope

The resolving power is measured as the smallest angle that can be resolved. For a telescope having diameter D, the resolving power θ (in radians) is expressed as

$$\theta = 1.22\,\frac{\lambda}{D}$$

where λ is the wavelength of radiation used for the detection.

As the wavelength λ of radiation cannot be changed, the only way to improve the resolving power of telescope is to increase its diameter. Another way to increase the resolution is to connect the telescopes together to make an interferometer. The principle of the Very Large Baseline Interferometer (VLBI) is discussed later on.

Example 12: Calculate the resolving power of a telescope having a diameter of 2.34 m when a radiation of wavelength 5500 Å is detected.

Solution: We have $D = 2.34$ m and $\lambda = 5500 \times 10^{-10}$ m. The resolving power is

$$\theta = 1.22\,\frac{\lambda}{D} = 1.22 \times \frac{5500 \times 10^{-10}}{2.34} = 2.87 \times 10^{-7}$$

Example 13: A 100 m radio dish is used for detection of 18 cm radiation of OH molecule. Calculate the resolving power of radio telescope.

Solution: We have $D = 100$ m and $\lambda = 18$ cm $= 0.18$ m. The resolving power is

$$\theta = 1.22\,\frac{\lambda}{D} = 1.22 \times \frac{0.18}{100} = 2.2 \times 10^{-3}$$

Example 14: What size of telescope aperture is required to resolve the bands which run around the Jupiter. Mean radii of the orbits of Earth and Jupiter revolving around the Sun are 150 $\times 10^6$ km and 780 $\times 10^6$ km, respectively; diameter of Jupiter is 140,000 km.

Solution: At the closest approach, the separation between the Earth and Jupiter is 630 $\times 10^6$ km. For convenience, let us consider that there are 14 strips, each of thickness 10,000 (=140,000/14) km. Hence, the telescope must be capable to resolve 10,000 km from a distance of 630 $\times 10^6$ km. As the angle θ is small and given in radian, we have

$$\theta = \frac{10,000}{630 \times 10^6} = 1.6 \times 10^{-5}$$

The average wavelength of visible spectrum is 5000 Å. Thus, the aperture of telescope is

$$D = \frac{1.22\lambda}{\theta} = \frac{1.22(5000 \times 10^{-10})}{1.6 \times 10^{-5}} = 0.038 \text{ m}$$

Hence, a telescope with an aperture of 4 cm is sufficient to make out surface features on the Jupiter.

Example 15: What size of telescope is required to resolve the size of a sun-like star situated at a distance of 4 $\times 10^{13}$ km. Diameter of star is 1.4 $\times$ 10^6 km.

Solution: The separation between the observer and object is 4 $\times$ 10^{13} km. The angular size of star in radian is

$$\theta = \frac{1.4 \times 10^6}{4 \times 10^{13}} = 3.5 \times 10^{-8} \text{ rad}$$

The average wavelength of visible spectrum is 5000 Å. Thus, the aperture of telescope is

$$D = \frac{1.22\lambda}{\theta} = \frac{1.22(5000 \times 10^{-10})}{3.5 \times 10^{-8}} = 17.4 \text{ m}$$

This value of aperture for an optical telescope is very large.

5.3 Magnification

An optical instrument (microscope, telescope, etc.) produces image of an object. The ratio of the size of image formed to the apparent size of object is known as the magnification. In case of the optical instrument, the magnification is larger than 1. The magnifying power of a telescope is described as

$$\text{magnifying power} = \frac{\text{focal length of objective}}{\text{focal length of eyepiece}}$$

Example 16: In an optical telescope, the focal lengths of objective and eyepiece are 24 cm and 12 mm, respectively. Calculate the magnifying power of telescope.

Solution: Magnifying power of telescope is

$$\text{magnifying power} = \frac{\text{focal length of objective}}{\text{focal length of eyepiece}} = \frac{24}{1.2} = 20$$

6. Atmospheric distortion

Many people believe that the astronomers want to build a telescope on high mountains or to put it in the space, above the Earth's atmosphere, so that it is "closer" to the objects to be observed and studied. This opinion is not correct, as the distance of a telescope from surface of the Earth is much smaller than the distance of object from the surface of Earth.

The reason why the telescopes are built on high mountains or put in the space is to get away from the distortion of starlight due to the atmosphere around the Earth. The atmospheric distortion is in the from of poor seeing, reddening, extinction and the addition of absorption lines to stellar spectra.

There is no way out to control the atmosphere. However, one can minimize the effect of the atmosphere by selecting a proper site and by putting the telescopes on a high mountain, so that the radiation has to travel through the minimum amount of the atmosphere. This is why the optical telescopes are installed on the selected places on the Earth.

6.1 Seeing

The air is constantly in fast turbulent motion and light from a celestial object is bent randomly in many ways. Images dance about (twinkle) and are blurred. The atmospheric blurring distorts the view of the astronomical object. This atmospheric effect is known as the **seeing**. The light reaches different parts of the telescope's objective from slightly different directions and at different times. Good seeing is when the air is stable (little turbulence) and the twinkling is small. Poor seeing happens when the air is turbulent so the images dance about and the details smaller than 2 to 3 arc seconds in size cannot be seen. The more atmosphere there is above a telescope, the greater is the turbulent motion and the poorer is the seeing. This is one reason why the telescopes for research are located on very high mountains.

6.2 Reddening and extinction

The air also absorbs and scatters electromagnetic radiation by an amount that varies with the wavelength. The redder (longer wavelength) light is scattered less by the atmosphere (molecules and dust) than the bluer (shorter wavelength) light. This effect is known as the reddening.[5] All wavelengths of light are scattered or absorbed by some amount. This effect is called the extinction. Some wavelength bands suffer more extinction than others.

[5]This effect explains why the Sun appears orange or red when it is close to the horizon. The other colours of sunlight are scattered out of your line of sight so that only the orange and red colours make it through the atmosphere to your eyes. This effect also explains why the sky is blue. Since blue light is scattered more, you see more blue light scattered back to your eyes when you look in a direction away from the Sun.

6.3 Diffraction limit

The minimum angular separation of two sources that can be distinguished by a telescope depends on the wavelength of light being observed and the diameter of the telescope. This angle is called the diffraction limit.

Example 17: Find the largest telescope aperture that is limited by the diffraction rather than by the scintillation in the good viewing conditions.

Solution: Take scintillation to limit the resolution of a telescope to 1″ (4.8×10^{-6} radians). The average wavelength of visible spectrum is 5000 Å. The telescope aperture that provides the same resolution can be found as

$$D = \frac{1.22\lambda}{\theta} = \frac{1.22(5000 \times 10^{-10})}{4.8 \times 10^{-6}} = 0.13 \text{ m}$$

The resolution of a telescope with an aperture greater than 13 cm is thus likely to be limited by the atmosphere rather than by the diffraction.

7. Optical telescopes

The main purpose of an optical telescope is to collect the electromagnetic radiations spectrum lying in the optical window (*i.e.*, the visible light). The light collecting surface in an optical telescope may be either a lens or a mirror. Hence, the optical telescopes may be classified into two categories:

(i) Lens telescopes or refractors

(ii) Mirror telescopes or reflectors

The amount of light, from a source, collected by a telescope is proportional to the area of aperture $\pi D^2/4$, where D is the diameter of the aperture of telescope. Now, the linear size of image of the object is proportional to the focal length f of the lens/mirror. Therefore, the image area must be proportional to f^2. Hence, the amount of light energy per second focused onto a unit area of the resolved image is proportional to D^2/f^2. The focal-ratio n of the telescope is defined as

$$\text{focal-ratio} = \frac{\text{focal length}}{\text{diameter}} \qquad \text{or} \qquad n = \frac{f}{D}$$

For a telescope having 10-m diameter primary mirror with focal length of 17.5 m, the focal ratio is 1.75. This telescope is referred to as 10 m $f/1.75$ telescope.

Example 18: Suppose, the eyepiece of a refracting telescope has focal length of 50 mm. What is the minimum focal length of the objective, so that for the eye to resolve the bands which run around the Jupiter. Mean radii of the orbits of Earth and Jupiter revolving around the Sun are 150×10^6 km and 780×10^6 km, respectively. Diameter of Jupiter is 140,000 km.

Solution: At the closest approach, the separation between the Earth and Jupiter is 630×10^6 km. For convenience, let us consider that there are 14 strips, each of thickness 10,000 (=140,000/14) km. Hence, the telescope must be capable to resolve 10,000 km from a distance of 630 $\times 10^6$ km. As the angle θ is small and given in radian, we have

$$\theta = \frac{10,000}{630 \times 10^6} = 1.6 \times 10^{-5} \text{ rad}$$

The average wavelength of visible spectrum is 5000 Å. The resolving power of telescope is

$$R = \frac{1.22\ \lambda}{f} = \frac{1.22(5000 \times 10^{-10})}{5 \times 10^{-3}} = 1.2 \times 10^{-4} \text{ rad}$$

For the eye to resolve the bands, the image should to be magnified by a factor of $(1.2 \times 10^{-4})/(1.6 \times 10^{-5}) = 7.5$. The focal length of the objective is

$$f_o = M f_e = 7.5 \times 0.05 = 0.38 \text{ m}$$

To resolve the bands, the focal length of the objective is 0.38 m. The total length of the telescope needs to be half a meter with an aperture of 4 cm and such telescope is easy to build.

Example 19: Suppose, the eyepiece of a refracting telescope has focal length of 50 mm. What is the minimum focal length of the objective, so that for the eye to resolve the size of a sun-like star situated at a distance of 4 $\times 10^{13}$ km. Diameter of star is 1.4×10^6 km.

Solution: The separation between the observer and object is 4×10^{13} km. The angular size of star in radian is

$$\theta = \frac{1.4 \times 10^6}{4 \times 10^{13}} = 3.5 \times 10^{-8} \text{ rad}$$

The average wavelength of visible spectrum is 5000 Å. The resolving power of telescope is

$$R = \frac{1.22\lambda}{D} = \frac{1.22(5000 \times 10^{-10})}{5 \times 10^{-3}} = 1.2 \times 10^{-4} \text{ rad}$$

The angular magnification required is $(1.2 \times 10^{-4})/(3.5 \times 10^{-8}) = 3400$. The focal length of the objective is

$$f_e = M f_o = 3400 \times 0.05 = 170 \text{ m}$$

The focal length of the objective must therefore be about 170 m for the eye to see the star.

8. Lens telescope or refractor

A lens telescope, shown in Figure 4, consists of two lenses, called the objective (O) and eyepiece (E). The objective having large diameter is situated towards the object, and the eyepiece is situated towards the observer. The size of the objective cannot be increased to very large size, as its weight increases tremendously and sags down. Thus, the diameter of a lens telescope is limited. It may be some cm (say, 10 cm).

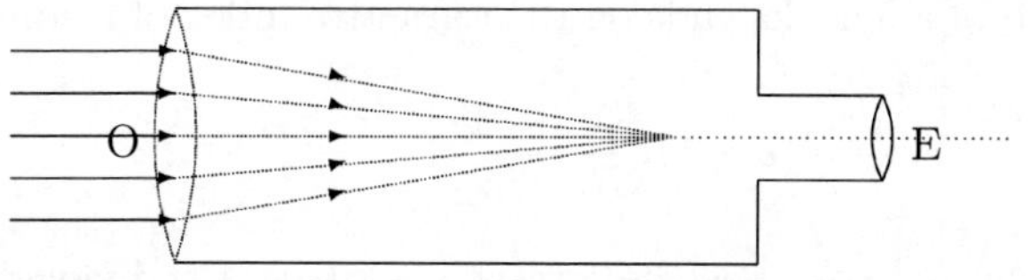

Figure 4: Lens telescope.

9. Aberrations

One of the basic problems of a lens is the imperfect quality of image formed by it. The deviations in size, shape, position of a colour (wavelength) in the actual image produced by a lens are known as the aberrations produced by the lens. The aberrations may be classified into two categories:

(i) chromatic aberrations, and

(ii) monochromatic aberrations

The **chromatic aberrations** are the distortions of the image due to dispersion of light in a lens used in a optical telescope, when white light is used. Thus, the chromatic aberration may be defined as a defect of colour image formed by the lens while using the white light. When a monochromatic light is used then the chromatic aberrations are automatically absent. However, there are some defects which are still present. Such defects are known as the **monochromatic aberrations**. The monochromatic aberrations are due to

(i) the large aperture of the lens,

(ii) the large angle subtended by the rays with the principal axis, and

(iii) the large size of the object.

Examples of monochromatic aberrations are (a) spherical aberration, (b) astigmatism, (c) coma, (d) curvature, and (e) distortion. In the present discussion we shall discuss about the chromatic aberration and spherical aberration.

9.1 Chromatic aberration

The rays of light of different wavelengths (colours) have different velocities in a refracting medium. The refractive index μ of medium for a colour is defined as $\mu = c/v$. Here, v and c are velocities of radiation in the medium and in the vacuum, respectively. Thus, the medium has different refractive index for each colour (wavelength). In the visible region of electromagnetic spectrum, the refractive index is the lowest for the red colour and the highest for the violet colour.[6] It shows that the velocity of the red colour is the largest and that of the violet colour the smallest.[7]

As the focal length of a lens depends on the refractive index of medium through the relation

$$\frac{1}{f} = (\mu - 1)\left(\frac{1}{R_1} - \frac{1}{R_2}\right) \tag{4.1}$$

the focal length for the red colour is the largest and that of the violet colour the smallest.[8] Thus, when the white light is used, a single lens will produce not only one image of an object but one image for each colour (wavelength) at different positions and of different sizes. The variation of distance of image from the lens with refractive index measures the axial or longitudinal chromatic aberration.

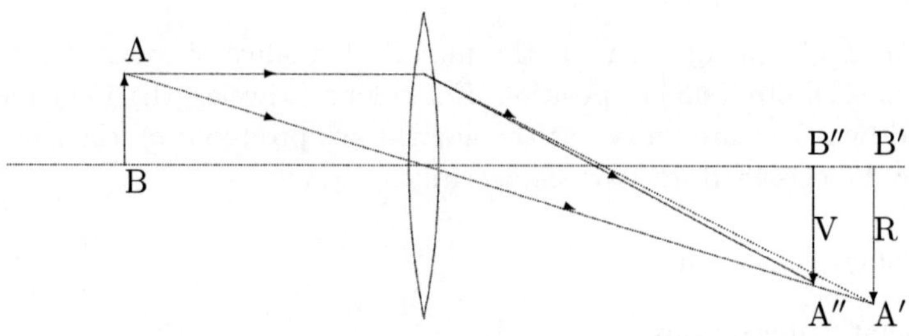

Figure 5: Chromatic aberration.

In Figure 5, AB is an object illuminated by white light. Its images corresponding to the red and violet colours are A′B′ and A″B″, respectively. The images corresponding to other colours are formed in between these two limiting images. The size of an image depends on the wavelength (colour). Hence, the magnification depends on the wavelength (colour). The difference B″B′ is the measure of longitudinal chromatic aberration and the difference (A′B′− A″B″) measures the lateral chromatic aberration.

Example 20: For a refracting medium, decide if the following statement is correct or not.

---

[6] The refractive index varies with wavelength (colour) as

$$\mu_v > \mu_i > \mu_b > \mu_g > \mu_y > \mu_o > \mu_r$$

[7] The velocity of radiation varies with wavelength (colour) as

$$v_v < v_i < v_b < v_g < v_y < v_o < v_r$$

[8] The focal length varies with wavelength (colour) as

$$f_v < f_i < f_b < f_g < f_y < f_o < f_r$$

(i) In the electromagnetic spectrum, the velocity of red-colour is larger than that of violet colour.

(ii) In the electromagnetic spectrum, the refractive index of red colour is larger than that of the violet colour.

(iii) Focal length of a lens for the red colour is larger than that for the violet colour.

Answer: (i) Yes, (ii) No, (iii) Yes

9.2 Chromatic aberration of a lens for an object placed at infinity

The chromatic aberration of a lens for an object placed at infinity can be calculated in the following manner.

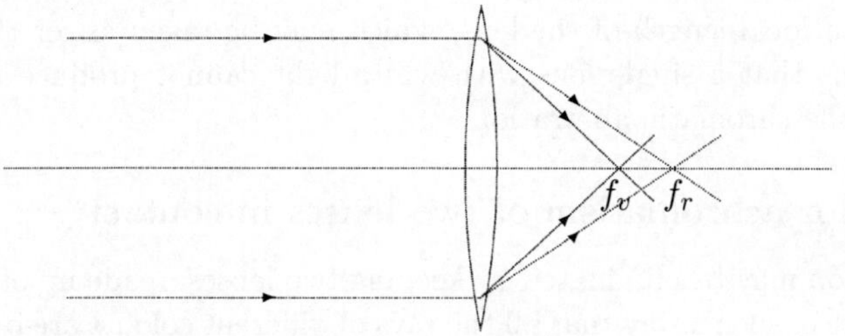

Figure 6: Longitudinal aberration due to a convex lens.

Let white light with plane wave front incident on a convex lens. After refraction through the lens, varies colours focus at different positions, as shown in Figures 6. Let f_r and f_v be the focal lengths, and μ_r and μ_v the refractive indices for red and violet colours, respectively. Now, for the red colour we have

$$\frac{1}{f_r} = (\mu_r - 1)\Big(\frac{1}{R_1} - \frac{1}{R_2}\Big) \tag{4.2}$$

and for violet colour we have

$$\frac{1}{f_v} = (\mu_v - 1)\Big(\frac{1}{R_1} - \frac{1}{R_2}\Big) \tag{4.3}$$

Here, R_1 and R_2 are radii of curvatures of two faces of the lens. For the average focal length f and refractive index μ of the lens, we have

$$\frac{1}{f} = (\mu - 1)\Big(\frac{1}{R_1} - \frac{1}{R_2}\Big) \tag{4.4}$$

From equations (4.2) and (4.4), we have

$$\frac{f}{f_r} = \frac{\mu_r - 1}{\mu - 1} \qquad \text{or} \qquad \frac{1}{f_r} = \frac{\mu_r - 1}{f(\mu - 1)} \tag{4.5}$$

and from equations (4.3) and (4.4), we have

$$\frac{f}{f_v} = \frac{\mu_v - 1}{\mu - 1} \quad \text{or} \quad \frac{1}{f_v} = \frac{\mu_v - 1}{f(\mu - 1)} \tag{4.6}$$

Subtracting equation (4.5) from (4.6), we get

$$\frac{1}{f_v} - \frac{1}{f_r} = \frac{(\mu_v - \mu_r)}{f(\mu - 1)} = \frac{\omega}{f} \tag{4.7}$$

where $\omega = (\mu_v - \mu_r)/(\mu - 1)$ is known as the dispersion power of material of the lens. Equation (4.7) can be rearranged as

$$\frac{f_r - f_v}{f_r f_v} = \frac{\omega}{f} \quad \text{or} \quad f_r - f_v = \frac{\omega}{f} f^2 = \omega f \tag{4.8}$$

where $f = \sqrt{f_r f_v}$ is the geometrical average focal length of the lens. Thus, the longitudinal chromatic aberration $(f_r - f_v)$ is equal to the product of the dispersive power and the mean focal length of the lens, which may be taken as for the yellow colour. It is obvious that a single lens with white light cannot produce an image which is free from the chromatic aberration.

9.3 Condition for achromatism of two lenses in contact

Chromatic aberration may be eliminated by keeping two lenses, made up of different materials, in contact in such a way that all the rays of different colours are brought to focus by the combination at a common point. In practice, the chromatic aberration cannot be removed completely. Usually, achromatism is achieved for two prominent colours.

A achromatic combination is made by placing in contact two lenses of different materials and of suitable focal lengths, so that the focal length of the combination is equal for both the extreme colours, red and violet. The refraction formula for a lens is

$$\frac{1}{f} = (\mu - 1)\Big(\frac{1}{R_1} - \frac{1}{R_2}\Big) \tag{4.9}$$

where μ is the refractive index of the material of lens, and R_1 and R_2 are radii of curvatures of two surfaces of the lens. On differentiating equation (4.9), we get

$$-\frac{\mathrm{d}f}{f^2} = \mathrm{d}\mu\left(\frac{1}{R_1} - \frac{1}{R_2}\right) \tag{4.10}$$

Diving equation (4.10) by (4.9), we get

$$-\frac{\mathrm{d}f}{f^2} = \frac{\mathrm{d}\mu}{f(\mu - 1)} \tag{4.11}$$

If μ_v and μ_r are the refractive indices for violet and red colours, respectively, then $\mathrm{d}\mu$ is the change in the refractive index such that $\mathrm{d}\mu = \mu_v - \mu_r$. Therefore, the dispersive power of the lens is

$$\omega = \frac{\mu_v - \mu_r}{\mu - 1} = \frac{\mathrm{d}\mu}{\mu - 1} \tag{4.12}$$

Using equation (4.12) in (4.11), we have

$$-\frac{\mathrm{d}f}{f^2} = \frac{\omega}{f} \tag{4.13}$$

When two lenses of focal lengths f_1 and f_2 are placed in contact, the focal length f of the combination is

$$\frac{1}{f} = \frac{1}{f_1} + \frac{1}{f_2}$$

On differentiating this equation, we get

$$-\frac{\mathrm{d}f}{f^2} = -\frac{\mathrm{d}f_1}{f_1^2} - \frac{\mathrm{d}f_2}{f_2^2}$$

In order that the rays of different colours are brought to focus at a common point, the change in focal length of the combination should be zero, *i.e.*, $\mathrm{d}f = 0$. Thus, we have

$$-\frac{\mathrm{d}f_1}{f_1^2} - \frac{\mathrm{d}f_2}{f_2^2} = 0$$

Using equation (4.13) here, we get

$$\frac{\omega_1}{f_1} + \frac{\omega_2}{f_2} = 0 \tag{4.14}$$

where ω_1 and ω_2 are dispersive powers for the materials of lenses, respectively, and are always positive. Equation (4.14) shows that f_1 and f_2 must have opposite signs, *i.e.*, the combination should consist of one concave lens and the other convex lens, generally made up of different materials. Such combination is often known as the achromatic doublet.

9.4 Achromatic doublet

An achromatic doublet consists of one convex lens and one concave lens, placed in contact with each other, as shown in Figure 7. One of them is made up of, say, crown glass and the other of, say, flint glass. The dispersive powers of the materials of these lenses are ω_1 and ω_2, respectively.

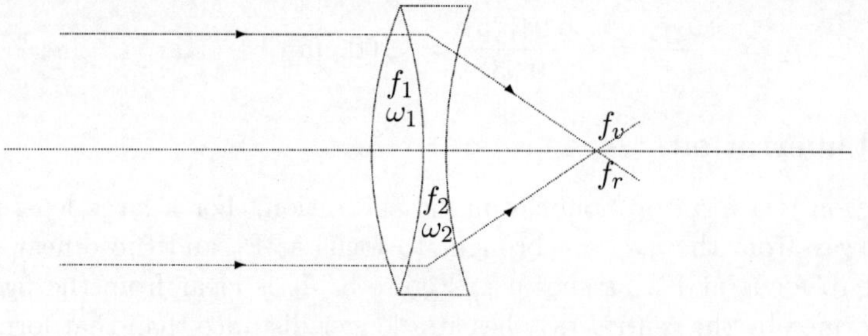

Figure 7: Achromatic doublet.

From the condition of achromatism, we have

$$-\frac{f_2}{f_1} = \frac{\omega_2}{\omega_1}$$

A negative sign here appears as the focal lengths of the lenses are of opposite sign. We know that focal length of a convex lens is positive whereas that of a concave lens is negative.

Example 21: Calculate the focal length of a lens having dispersive power 0.03 which should be kept in contact with a convex lens of focal length 80 cm and dispersive power 0.023 so that the combination is achromatic.

Solution: Given, $f_1 = 80$ cm, $\omega_1 = 0.023$, $\omega_2 = 0.03$. For achromatism, we have

$$-\frac{f_2}{f_1} = \frac{\omega_2}{\omega_1}$$

Using the values here, we get

$$-\frac{f_2}{80} = \frac{0.03}{0.023} \qquad \text{so that} \qquad f_2 = -104.34 \text{ cm}$$

Thus, the required focal length of a concave lens is -104.34 cm.

Example 22: A convex lens of focal length 80 cm and dispersive power 0.041 is kept in contact with a concave lens of focal length 60 cm so that the combination is achromatic. Calculate the dispersive power of concave lens.

Solution: We have $f_1 = -80$ cm and $\omega_1 = 0.041$, and $f_2 = 60$ cm. The dispersive power ω_2 of concave lens is

$$\omega_2 = -\frac{\omega_1 f_2}{f_1} = -\frac{0.041(-60)}{80} = 0.031$$

Example 23: Dispersive powers of convex and concave lenses in a achromatic combination are 0.03 and 0.04, respectively. The focal length of the convex lens is 75 cm. Calculate the focal length of the concave lens.

Solution: We have $\omega_1 = 0.03$, $\omega_2 = 0.04$ and $f_1 = 75$ cm. The focal length of the concave lens is

$$f_2 = -\frac{\omega_2 f_1}{\omega_1} = -\frac{0.04(75)}{0.03} = -100 \text{ cm}$$

9.5 Spherical aberration

Spherical aberration is a form of monochromatic aberration. For a large lens, the rays which are away from the axis are brought to focus at F_p and those near the axis are brought to focus at F_c, as shown in Figure 8. It is clear from the figure that the image formed by the central rays lies at a longer distance than that formed by the marginal rays. The image formed is, therefore, not sharp at any point on the axis. If a screen is placed perpendicular to the axis at F_p, the outer portions of the object are in focus, whereas at F_c, the inner portions of the object are in focus. The spherical aberration for a convex lens is positive. When the screen is placed in between F_p and F_c, the image appears to be a circular patch. This patch is known as the circle of least confusion and corresponds to the position of the best image.

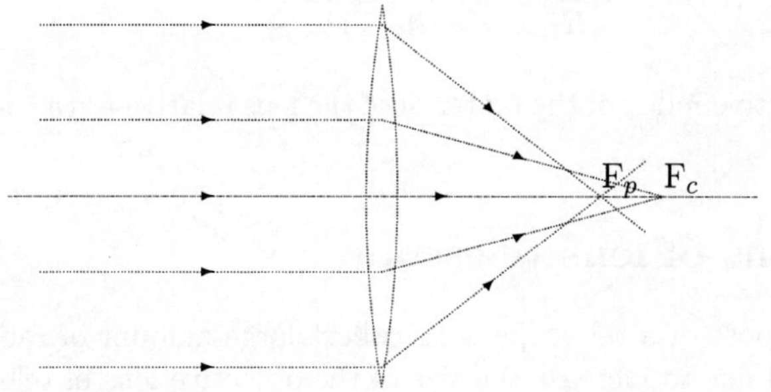

Figure 8: Spherical aberration due to convex lens.

The distance F_cF_p measures the axial or longitudinal spherical aberration. The radius of the circle of least confusion, lying between F_c and F_p, is a measure of the lateral spherical aberration.

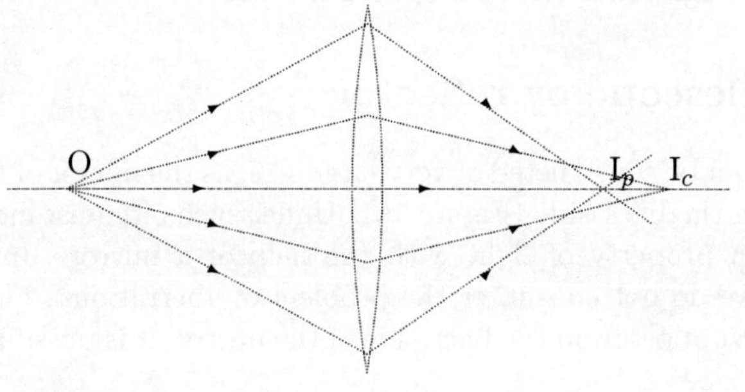

Figure 9: Formation of image of a point object O with the help of a convex lens using monochromatic light.

It is obvious that for an point-object O placed on the axis, the image formed by a central ray is I_c whereas formed by the marginal rays is I_p (Figure 9). Thus, the image extends from I_p to I_c. That is, the image is not sharp. This defect is known as the spherical aberration and is due to the fact that different zones (parts) of the lens have different focal lengths.

9.6 Removal of spherical aberration

For a single lens, spherical aberration can be minimized under the following conditions.

(i) The spherical aberration depends on the aperture of lens. It can therefore be reduced by reducing the aperture of the lens with the help of a co-axial aperture stop and therefore using only the central portion of lens.

(ii) By using a specially designed lens, called the crossed lens, the radii of curvature R_1 and R_2 of the two surfaces of the lens should satisfy the relation

$$\frac{R_2}{R_1} = -\frac{\mu(2\mu + 1)}{\mu(2\mu - 1) - 4}$$

where μ is the refractive index of the material of the lens relative to the outer medium (air).

10. Limitations of lens telescope

Since the main purpose of a telescope is to collect large amount of radiations from an object, we would like to increase the size of the objective lens of telescope. With the increase of the size not only the spherical aberration appears in the picture, but the weight of lens increases tremendously. For large weight, it becomes difficult to hold the lens with the help of a frame around it, and the lens breaks down.

Moreover, it is not possible to make the telescope free from the chromatic aberration for full range of visible light. In a lens telescope, the radiations are collected after refraction through the lens. We are aware of the fact that the radiations can also be collected through reflection by a spherical mirror.

11. Mirror telescope or reflector

A mirror telescope can be constructed of very large size, as the mirror of the telescope can be supported on the back-side (Figure 10). Under such circumstances, we make use of the reflection property of light with the help of a mirror. In case of the reflection of light, we do not encounter the problem of aberrations. Further, as we can put an external support from the back side of the mirror, it is possible to increase the size of the telescope.

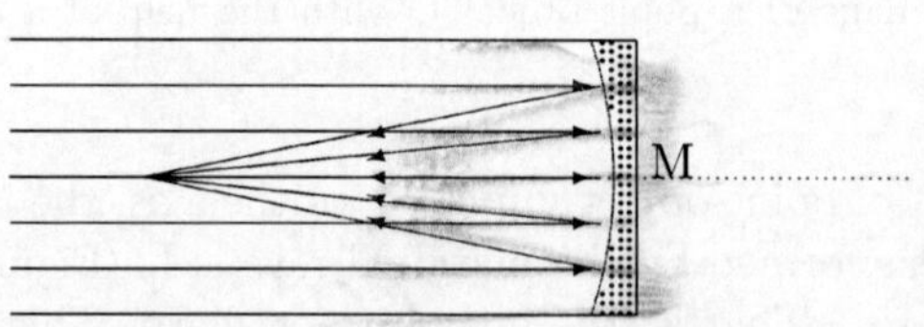

Figure 10: In a mirror telescope, all the rays coming parallel to the main axis are reflected by the mirror to a common point.

The most common telescope type used in astronomical research is the mirror telescopes or reflectors. As a light-collection surface, it employs a mirror coated with a thin layer of aluminum. The form of the mirror is generally parabolic, which reflects all the light rays entering the telescope parallel to the main axis into a common focal point. The image formed at this point is observed with the help of an eyepiece or is registered on a photographic plate.

One of the advantages of the reflectors is the absence of chromatic aberration, since all the wavelengths are reflected to the same point. Examples of large size reflectors are the Soviet 6-m BTA telescope in the Caucasus (1975) and the American 5-m Hale telescope on Mount Palomar (1947).

In a very large telescope, the observer can sit with his instruments in a special cage at the primary focus (Figure 12) without eclipsing too much of the incoming light. In a smaller telescope, this is not possible, and the image must be inspected from outside the telescope. Information about the sizes of some big optical telescopes is given in Table 4.

11.1 Prime focus

All the rays coming parallel to the prime axis, after reflection from the mirror, reach a common point, called the prime focus, as shown in Figure 11.

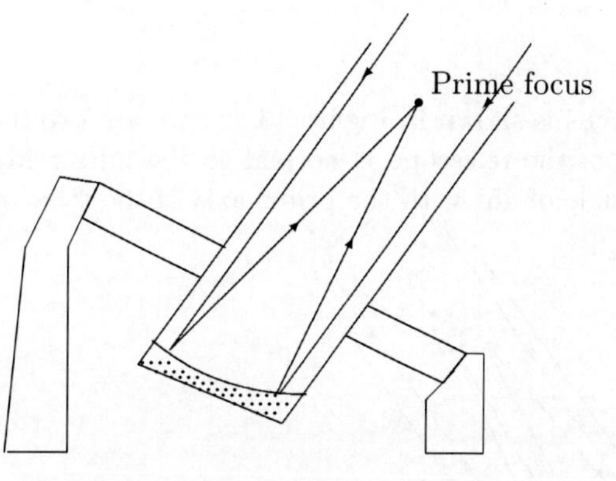

Figure 11: Prime focus of reflection telescope.

11.2 Newton focus

Isaac Newton, who invented the reflector, guided the light perpendicular out from the telescope with the help of a small flat mirror M_1. The mirror is inclined at an angle of 45° with the prime axis of the telescope. Therefore, the focus of the image in such a system is outside the tube of the telescope. The arrangement of Newton focus is shown in Figure 12.

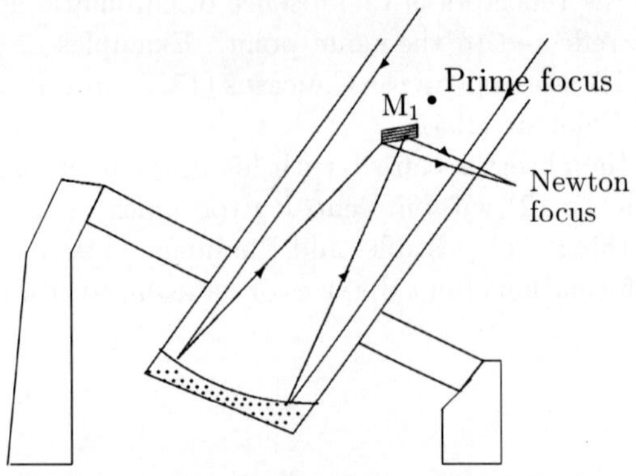

Figure 12: Newton focus of reflection telescope.

11.3 Coude focus

The arrangement of Coude focus is shown in Figure 13. There are two flat mirrors M_1 and M_2. The prime axis of the telescope is normal to the mirror M_1 and the mirror M_2 is inclined at an angle of 45° with the prime axis of the telescope.

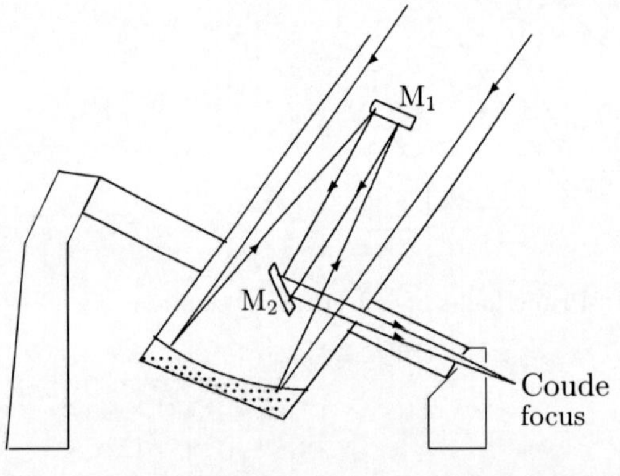

Figure 13: Coude focus of a reflection telescope.

11.4 Cassegrain focus

The arrangement of Cassegrain focus is shown in Figure 14. The prime axis of the telescope is normal to the mirror M_1. The final image of an object is formed behind the telescope.

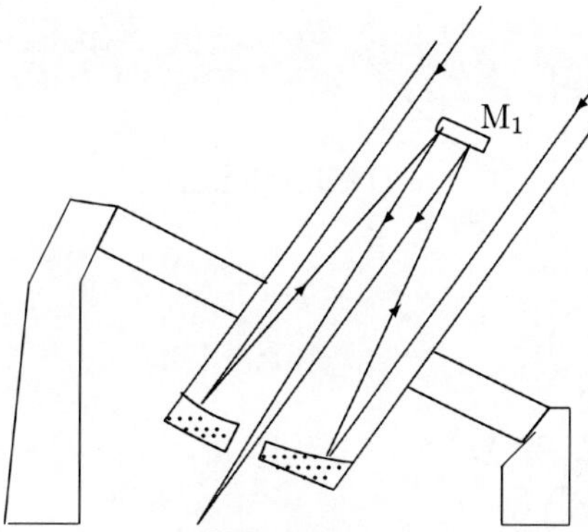

Figure 14: Cassegrain focus of a reflection telescope.

12. Mounting of telescopes

A telescope is to be mounted on a steady support to prevent shaking, and it must be smoothly rotated during the observations. There are two principal types of mounting:

(i) equatorial mounting, and

(ii) azimuthal mounting

12.1 Equatorial mounting

In the equatorial mounting (Figure 15), one of the axis is directed towards the celestial pole. It is known as the polar axis or the hour axis. The other one, known as the declination axis, is perpendicular to the polar axis. As the hour axis is parallel to the axis of the Earth, the apparent rotation of sky is compensated by turning the telescope around that axis at a constant rate.

12.2 Azimuthal mounting

In the azimuthal mounting (Figure 16), one of the axis is vertical and the other one is horizontal. This mounting is easier to construct than the equatorial mounting, and is more stable for very large telescopes. In order to follow the rotation of sky, the telescope must be turned around both the axes with changing velocities. The field of view also rotates, and is compensated when the telescope is used for the photography.

Most of the telescopes in the world were equatorially mounted until the development of computers which made possible the more complicated guidance needed for the azimuthal mounting.

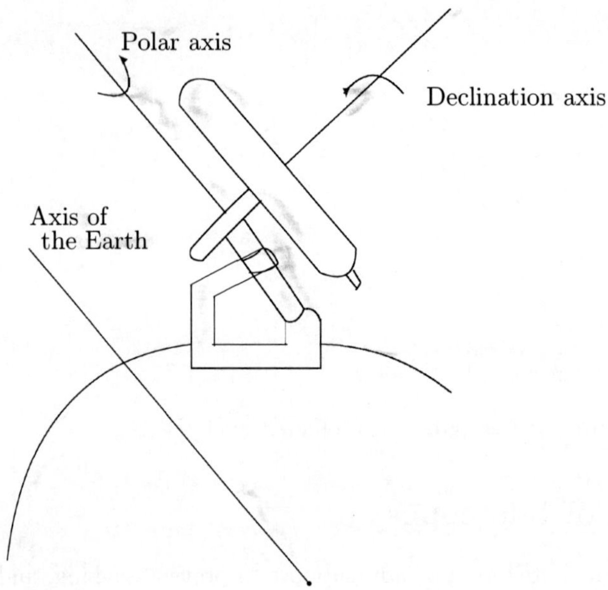

Figure 15: Equatorial mounting of an optical telescope.

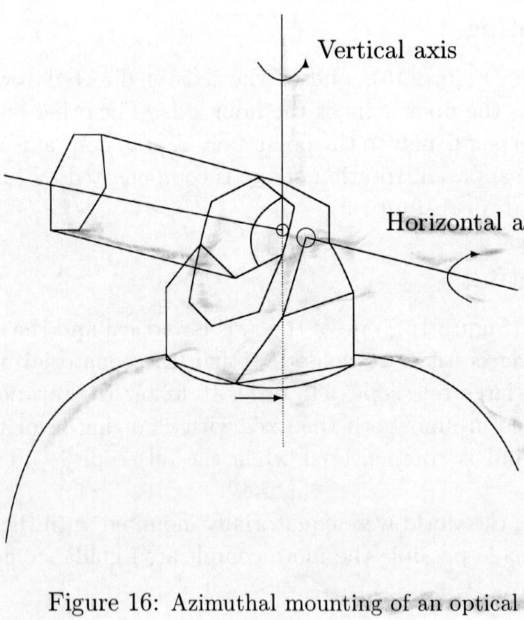

Figure 16: Azimuthal mounting of an optical telescope.

Table 4. Some big optical telescopes

Name	Size	Sponsor
Gran Telescopio Canarias	10.4 m	Maxico, USA
Keck-1	10 m	USA
Keck-2	10 m	USA
Southern African Large Telescope	9.2 m	South Africa, USA, UK, Germany, Poland, New Zealand
Hobby-Eberly Telescope	9.2 m	USA, Germany
Large Binocular Telescope	8.4 m × 2	USA, Italy, Germany
Subaru (INLT)	8.2 m	Japan
VLT UT1 (Antu)	8.2 m	ESO countries, Chile
VLT UT2 (Kueyen)	8.2 m	ESO countries, Chile
VLT UT3 (Melipal)	8.2 m	ESO countries, Chile
VLT UT4 (Tepun)	8.2 m	ESO countries, Chile
Gemini North (Gillett)	8.1 m	USA, UK, Canada, Chile, Australia, Argentina, Brazil
Gemini South	8.1 m	USA, UK, Canada, Chile, Australia, Argentina, Brazil
MMT	6.5 m	USA
Magellan 1 (Walter Baade)	6.5 m	USA
Magellan 2 (London Clay)	6.5 m	USA
BTA-6	6 m	USSR/Russia
Large Zenith Telescope (ZLT)	6 m	Canada, France, USA
Hale Telescope	5.08 m	USA
LAMOST	4.9 m - 3.6 m	PRC (China)
MMT	4.7 m	USA
Discovery Channel Telescope	4.3 m	USA
William Herschel Telescope	4.2 m	UK, Netherlands, Spain
SOAR	4.1 m	USA, Brazil

13. Radio telescopes

Schematic diagram of a radio telescope is shown in Figure 17. The paraboloidal dish (Figure 17) serves several important functions. First, it is a collector which reflects radiation from a beam equal to its own diameter to the receiver. As we always deal with very weak signals, a reflector with a large diameter focuses large amounts of energy.

13.1 Limitation of single-dish radio telescope

When a radio telescope has a single dish, the resolving power of the radio telescope is very poor. For example, for an optical radiation of 5000 Å observed by an optical telescope of diameter 2 m, the resolving power of optical telescope is

$$R.P. \approx \frac{\lambda}{D} = \frac{5000 \times 10^{-8}\ \text{cm}}{2 \times 10^{2}\ \text{cm}} = 2.5 \times 10^{-7}\ \text{rad}$$

Now, for a radio wave of 18 cm wavelength observed by a radio telescope of diameter 100 m, the resolving power of telescope is

$$R.P. \approx \frac{\lambda}{D} = \frac{18 \text{ cm}}{100 \times 10^2 \text{ cm}} = 1.8 \times 10^{-3} \text{ rad}$$

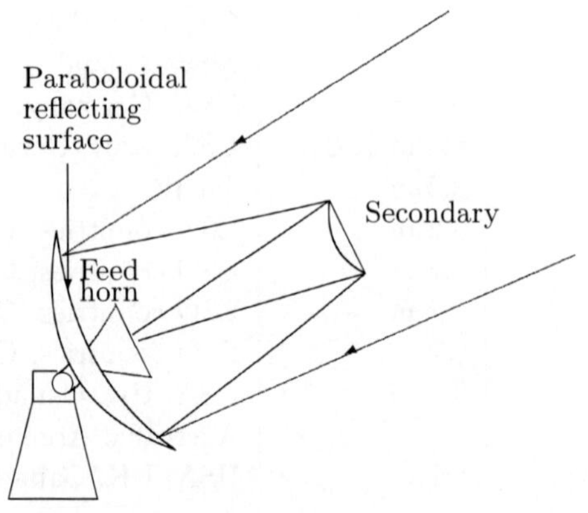

Figure 17: Schematic diagram of a radio telescope.

Obviously, the resolving power of a single dish radio telescope is very poor in comparison to that of an optical telescope. Thus, the resolving power of an optical telescope is 10^4 times better than that of a radio telescope. However, resolving power of a radio telescope may be enhanced tremendously by making the use of the Very Long Baseline Interferometry (VLBI) technique in which two radio telescopes situated at two different places, separated by a large distance, are used to behave like two outer segments of a large telescope. The two telescopes behave like two outer most segments of a single large telescope. The distance between the two telescopes is known as the baseline. The largest baseline on the surface of the Earth can be equal to the diameter of the Earth, which is about 12,700 km. Now, the resolving power of the telescope using the VLBI technique becomes

$$R.P. \approx \frac{\lambda}{D} = \frac{18 \text{ cm}}{12,700 \text{ km}} = 1.42 \times 10^{-8} \text{ rad}$$

which is better than that of an optical telescope. For further improvement in the resolving power of a radio telescope, one of the radio telescopes used in the VLBI technique may be installed in the sky on an geostationary satellite, orbiting around the Earth.

13.2 Principle of VLBI

The principle of VLBI helps for improving the resolving power of a radio telescope. The idea of VLBI can be understood in the following manner. When the angular size of an object is very small (few milli arc seconds), in order to study it in detail, radio telescope with high spatial resolution is required. The resolving power of a telescope varies as λ/D, where λ is wavelength of radiation observed and D the size of the disc (mirror) of the telescope. The only possibility to improve the resolving power of a telescope is to vary the size D of the disc. The VLBI experiments involving radio telescopes across the United States were performed for the first time in 1968. The principle of VLBI is shown in Figure 18.

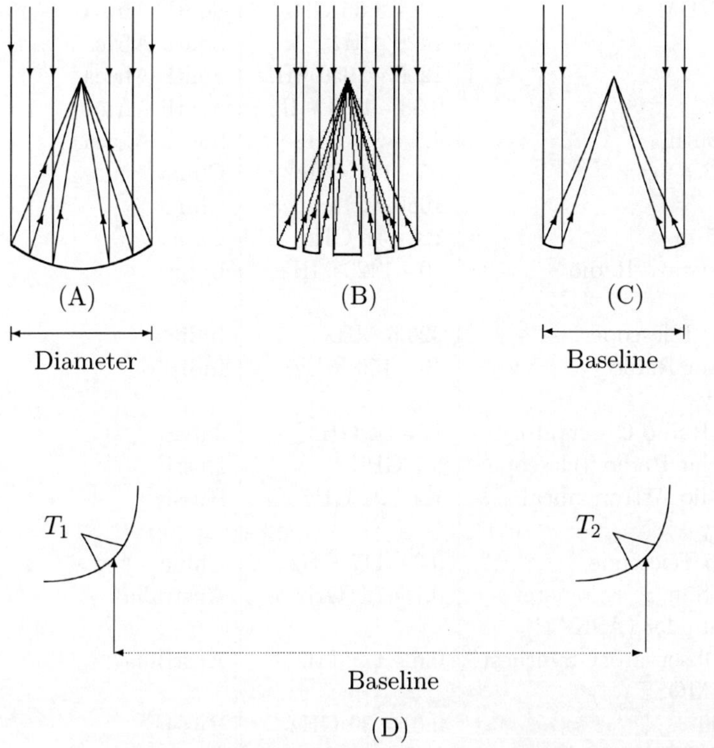

Figure 18: Principle of very long baseline interferometry.

A large single mirror (shown in A) can be assumed as a set of a number of smaller mirrors (shown in B). Since the resolving power of a telescope, for a given wavelength, depends only on its aperture, therefore, retaining only the outermost segments (shown in C) would give the same resolving power as of the full mirror (shown in A). However, the intensity of radiations decreases as the less number of photons are received. The separation between the outer most segments is known as the baseline. In practice, the two telescopes T_1 and T_2 situated at two distant stations (shown in D) behave as two outer most segments.

In the VLBI experiments, signals from a common source are recorded simultaneously at two stations separated by continental or intercontinental baselines and later on correlated. The maximum baseline which can be used on the Earth's surface is equal to the Earth's diameter (12,700 km). It corresponds to a resolution 10^{-4} arc seconds for a wavelength of few centimeters, which is better than the resolution that can be achieved with the optical telescopes.

Information about some radio telescopes is given in Table 5.

Table 5. Some radio telescopes

Name	Frequency range	Location
HartRAO 26 m	1.66 - 23 GHz	South Africa
HartRAO XDM	2.3 - 8.65 GHz	South Africa
Indlebe	1420 MHz	South Africa
KAT-7	1200 - 1950 MHz	South Africa
MeerKAT	0.58 - 14.5 GHz	South Africa
C-BASS South	4.5 - 5.5 GHz	South Africa
Delingha 13.7 m	85 - 115 GHz	China
Sheshan	1660 MHz	China
Nanshan 25 m	1.4 - 18 GHz	China
Giant Meterwave Radio Telescope	50 - 1420 MHz	India
Ooty Radio Telescope	326.5 MHz	India
Gauribidnaur Radio Observatory	40 - 150 MHz	India
Nobeyama Radio Observatory	17 - 34 GHz	Japan
Siberian Solar Radio Telescope	5.7 GHz	Russia
Badary Radio Astronomical Observatory	1.4 - 22 GHz	Russia
Qitai Radio Telescope	0.3 - 117 GHz	China
Australian Square Kilometer Array Pathfinder (ASKAP)	0.7 - 1.8 GHz	Australia
Molonglo Observatory Synthesis Telescope (MOST)	0.6 - 1.2 GHz	Australia
RATAN-600	0.61 - 30 GHz	Russia
Nancy Radio Heliographe	150 - 450 MHz	France
Allen Telescope Array	0.5 - 11.5 GHz	USA
Dominion Radio Astrophysical Observatory	0.408 - 1.42 GHz	Canada
OVRO 40 meter Telescope	15 GHz	USA
Very Large Array	0.1 - 0.2 GHz	USA

14. Space based telescopes

The telescopes on the Earth suffer from the limitations caused by light pollution, atmospheric turbulence and weather systems. Moreover, the Earth's atmosphere absorbs γ-rays, x-rays, UV-radiations emitted by the objects in the universe. In order to study γ-rays, x-rays, UV-radiations, highly sensitive telescopes are placed in the orbits around the Earth. These space based telescopes can get a clearer view of an object in the universe without any restriction of night and day or weather.

14.1 Hubble space telescope (HST)

The most famous space telescope is the Hubble space telescope (HST) which was launched aboard the space shuttle Discovery on 24 April 1990 and its orbit lies at 569 km above the Earth's surface. It completes one rotation around the Earth in 95.47 minutes (15.08 rev/day), traveling at a speed of approximately 7.59 km/s. Some orbital parameters of the HST are given in Table 6. During its journey, its mirrors capture light and other radiations and direct them onto its several instruments.

Table 6. Orbital parameters

Semi-major axis	6919 km
Inclination	28.47°
Period	95.47 min
Mean motion	15.08 rev/day
Velocity	7.59 km/s

The telescope has a Cassegrain reflector with a primary mirror of diameter 2.4 m. Although this mirror diameter is smaller than those of the earth-based telescopes (which are up to 10 meters in diameter) its location above the Earth's atmosphere gives it excellent clarity. The resolution of the images produced is at least 10 times better than those from any telescope on the Earth.

Besides a reflector, the HST carries other instrumentation also. One of them is the Wide Field Camera 3 (WFC3). This allows it to see the radiations from three different wavelengths; near ultra violet, visible light and near infra-red, however not at the same time. In order to avoid contamination of Earth's atmosphere, the radiations from the interstellar medium can be studied by boarding an instrument on a geo-stationary satellite orbiting the Earth.

14.2 X-ray telescope

X-ray sources in the universe are of great interest to the astronomers as they are associated with highly energetic processes involving very high temperatures *e.g.*, exploding starts, a vast cloud of hot gases in a galaxy cluster. Ordinary mirrors do not work for the wavelengths in the x-ray region of electromagnetic spectrum. This is because x-rays have so small wavelengths that they pass right through standard mirrors rather than being reflected by them.

For their study, the x-rays need to be bounced of specially constructed mirrors at a very low angle. This technique is known as the grazing incidence and thus the x-ray telescopes are also referred to as the grazing incidence telescopes. In these telescopes, the mirrors are precisely shaped and aligned with their surfaces nearly parallel to incoming x-rays. The x-rays strike the mirrors at the grazing angles and are reflected. The surfaces of the mirrors are coated with a thin film of gold or iridium, so that the x-rays are reflected and not absorbed.

14.3 Gamma ray telescope

Gamma rays have the shortest wavelengths and most energy of all radiations in the electromagnetic spectrum. The sources of celestial gamma rays are important to the astronomers as they are connected to the violent events, such as the supernova explosion and the destruction of atoms. They are also linked to less violent events, such as the decay of radioactive material in space.

The gamma rays cannot be captured and reflected by the mirrors in the manner as the visible light and x-rays. The low wavelengths associated with the gamma rays makes them pass right through such devices. The gamma rays are detected by using a specially designed piece of equipment called the scintillation detector. The gamma rays transfer their high energies to the particles in the material of the scintillator. It results in the generation of high energy charged particles which interact with the crystalline structure of the scintillator material producing low energy photons. These photons are collected by the detector and give the intensity of the received gamma rays.

15. Detectors

In astronomy, a detector is an instrument used for capturing radiations coming from the outer space. Our eyes can be considered as one of the detectors. However, our eyes can see the radiations having wavelengths from 4000 Å to 7500 Å. Radiations of other wavelengths cannot be seen by our eyes. Further, the image of an object recorded in the memory of an observer cannot be shown to others.

In order to see the dim objects, one takes the help of a telescope. In a telescope, image of an object is formed. Now, the prime object is to record the information received in the form of an image of the object.

By looking through a telescope, only a limited amount of information about the object, from which the radiations are coming, can be obtained. Detection with the help of the eyes has been the only way to make the observations up to the end of the nineteenth century.

The invention of photography brought a revolution in astronomy. The next important step in the forward direction in optical astronomy has been the development of photoelectric photometry in the 1940's and 1950's. A new revolution, comparable to that caused by the invention of photography, took place in the middle of the 1970's with the introduction of different semi-conductor detectors. The sensitivity of such detectors has grown tremendously large. The detectors used in the modern astronomy are the charge-coupled devices.

15.1 Charge-coupled device (CCD)

A charge-coupled device (CCD) is a device for the movement of electrical charge, usually, within the device, from a place to an area where the charge can be manipulated, *e.g.*, conversion into a digital value (Figure 19). This is achieved by *shifting* the signals between the stages within the device one at a time. A CCD moves the charge between capacitive bins in the device, with the shift allowing for the transfer of charge between the bins.

The CCD is a major piece of technology in digital imaging. In a CCD image sensor, pixels are represented by the p-doped MOS capacitors. These capacitors are biased above the threshold for inversion when the image acquisition begins, allowing the conversion of incoming photons into the electron charges at the semiconductor-oxide interface; the CCD is then used to read out these charges. Although the CCDs are not the only technology to allow for light detection, CCD image sensors are widely used in the medical and scientific applications, where high-quality image data are required. In the applications with less exacting quality demands, such as consumer and professional digital cameras, active pixel sensors (CMOS) are generally used; the large quality advantage CCDs enjoyed early on has narrowed over time.

Basics of operation

The charge packets (electrons) are collected in potential wells created by applying a positive voltage at the gate electrodes (G). Application of positive voltage to the gate electrode in the correct sequence transfers the charge packets. Visualization of the process can be found on the website:

https://en.wikipedia.org/wiki/Charge-coupled_device

In a CCD for capturing images, there is a photoactive region (an epitaxial layer of silicon), and a transmission region made out of a shift register (the CCD, properly speaking).

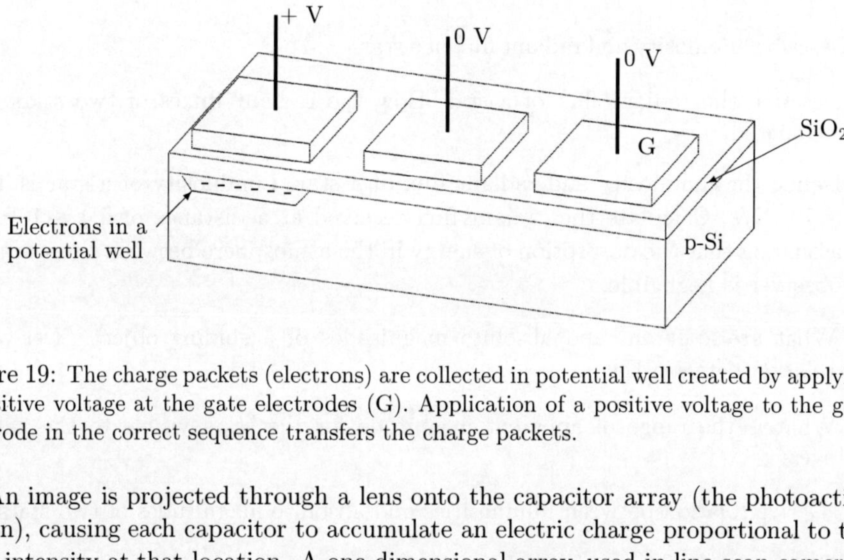

Figure 19: The charge packets (electrons) are collected in potential well created by applying a positive voltage at the gate electrodes (G). Application of a positive voltage to the gate electrode in the correct sequence transfers the charge packets.

An image is projected through a lens onto the capacitor array (the photoactive region), causing each capacitor to accumulate an electric charge proportional to the light intensity at that location. A one-dimensional array, used in line-scan cameras, captures a single slice of the image, whereas a two-dimensional array, used in the video and still cameras, captures a two-dimensional picture corresponding to the scene projected onto the focal plane of the sensor. Once the array has been exposed to the image, a control circuit causes each capacitor to transfer its contents to its neighbour (operating as a shift register). The last capacitor in the array dumps its charge into a charge amplifier, which converts the charge into a voltage. By repeating this process, the controlling circuit converts the entire contents of the array in the semiconductor to a sequence of voltages. In a digital device, these voltages are then sampled, digitized, and usually stored in the memory; in an analog device (such as an analog video camera), they are processed into a continuous analog signal (*e.g.*, by feeding the output of the charge amplifier into a low-pass filter), which is then processed and fed out to other circuits for transmission, recording, or other processing.

16. Review Questions

1. Describe the light year, astronomical unit and parsec. Out of them, which is the biggest.
2. Define the light year and calculate its value in meter.
3. Define the astronomical unit and calculate its value in meter.
4. Define the parsec and calculate its value in meter.
5. Describe the electromagnetic spectrum of radiation, and its absorption in the Earth's atmosphere.
6. Describe the optical window and radio window in the Earth's atmosphere. Which one is advantageous for investigation of distant cosmic objects.
7. Describe luminosity and radiant flux of a star.
8. Describe the radiant flux of a star. How the radiant fluxes of two stars are compared.
9. Define the luminosity and radiant flux of a star. Luminosity of a star is 4.26×10^{28} J/s. Calculate the radiant flux received at a distance of 5.4×10^{15} m, assuming that the obsorption of energy in the atmosphere between the star and observer is negligible.
10. What are apparent and absolute magnitudes of a shining object. Derive a relation between them.
11. What is the range of apparent magnitude for the stars visible to the naked eyes?
12. Derive relation between luminosities and absolute magnitudes of two stars.
13. Discuss about the magnitude scale for the stars and explain the absolute magnitude.
14. The absolute magnitude of a star is +1.5. Calculate its distance so that the apparent magnitude of the star is –1.5.
15. Describe the Rayleigh criterion for resolution of two objects.
16. Discuss about the atmospheric distortion of starlight.
17. Discuss in brief about the salient features of a telescope.

18. What is f-ratio of a telescope. The diameter of primary mirror of a telescope is 234 cm and its focal length is 4.2 m. Calculate the f-ratio of telescope.
19. Describe a lens telescope and its limitations.
20. Derive an expression for chromatic aberration of a lens for an object placed at infinite distance.
21. Describe a mirror telescope and discuss about its resolving power.
22. Describe about a radio telescope.
23. Discuss about the limitation of a single dish radio telescope. Describe the principle of Very Large Baseline Interferometry.
24. Write notes on the following:

    (i) Light year
    (ii) Astronomical unit
    (iii) Parsec
    (iv) Electromagnetic spectrum
    (v) Transparency of Earth's atmosphere
    (vi) Optical-window and radio-window
    (vii) Luminosity of a star
    (viii) Radiant flux of a star
    (ix) Solar constant
    (x) Magnitude scale
    (xi) Apparent magnitude and absolute magnitude
    (xii) Salient features of a telescope.
    (xiii) Rayleigh criterion for resolution
    (xiv) Atmospheric distortion
    (xv) Lens telescope
    (xvi) Mirror telescope
    (xvii) Prime focus
    (xviii) Newton focus
    (xix) Coude focus
    (xx) Cassegrain focus
    (xxi) Equatorial mounting
    (xxii) Azimuthal mounting
    (xxiii) Radio telescope
    (xxiv) VLBI
    (xxv) CCD
    (xxvi) Hubble space telescope
    (xxvii) X-ray telescope
    (xxviii) Gamma ray telescope

17. Multiple Choice Questions

1. Light-year is the unit of

   A. time B. reflection C. distance D. refraction

   Ans. C

2. 1 light-year is equivalent to (in km)

   A. 1.50×10^8 B. 9.47×10^{12} C. 3.09×10^{13} D. 6.02×10^{16}

   Ans. B

3. 1 astronomical unit is equivalent to (in km)

   A. 1.50×10^8 B. 9.47×10^{12} C. 3.09×10^{13} D. 6.02×10^{16}

   Ans. A

4. 1 parsec is equivalent to (in km)

   A. 1.50×10^8 B. 9.47×10^{12} C. 3.09×10^{13} D. 6.02×10^{16}

   Ans. C

5. Which radiations have the largest frequency.

   A. IR radiation B. UV radiation C. γ-rays D. x-rays

   Ans. C

6. Which radiations have the largest wavelength.

   A. IR radiation B. UV radiation C. γ-rays D. x-rays

   Ans. A

7. The light reaches the Earth from the Sun in

   A. 100 s B. 200 s C. 500 s D. 800 s

   Ans. C

8. Our eyes can see the radiations having wavelengths in the range

   A. 4000 – 7500 cm B. 4000 – 7500 mm

   C. 4000 – 7500 nm D. 4000 – 7500 Å

   Ans. D

9. Which radiations from the interstellar medium can be studied on the surface of the Earth.

   A. γ-rays B. x-rays C. UV D. visible light

   Ans. D

10. Which radiations from the interstellar medium can be studied on the surface of the Earth.

    A. γ-rays B. x-rays C. IR D. radio waves

    Ans. D

11. In which layer of the Earth's atmosphere, the γ-rays coming from the interstellar medium are absorbed

A. ionosphere B. troposphere C. exosphere D. mesosphere

Ans. A

12. In which layer of the Earth's atmosphere, the x-rays coming from the interstellar medium are absorbed

A. ionosphere B. troposphere C. exosphere D. mesosphere

Ans. A

13. In which layer of the Earth's atmosphere, the infra-red radiations from the interstellar medium are absorbed

A. ionosphere B. troposphere C. exosphere D. mesosphere

Ans. B

14. Two objects A and B of luminosities L and $2L$ are at the distances r and $2r$, respectively, from the observer. The ratio of their fluxes F_A/F_B is

A. 1/8 B. 1/4 C. 1 D. 2

Ans. D

15. The range of the magnitudes of stars which can be seen without the help of a telescope is from

A. from 1 to 6 B. from 0 to 6 C. from 1 to 10 D. from 0 to 9

Ans. A

16. The distance of a star is increased to 10 times of its initial distance. Then, its apparent magnitude

A. increases by 10 B. increases by 5 C. decreases by 10 D. decreases by 5

Ans. B

17. The distance of a star is increased by a factor of 2. Then, the flux of radiations received

A. increases by a factor 2 B. increases by a factor 4

C. decreases by a factor 2 D. decreases by a factor 4

Ans. D

18. For the absolute magnitude, the distance of objects from the observer is

A. 1 AU B. 10 AU C. 1 pc D. 10 pc

Ans. D

19. Which of the following features does not pertain to a telescope.

A. Light-gathering B. Resolution

C. Magnification D. Dispersion

Ans. D

20. A radiation of wavelength λ is detected by a telescope having diameter a. The resolving power depends on

    A. λ only B. λ and a both C. a only D. none of them

    Ans. B

21. A telescope has focal length 20 m and diameter 10 m. The focal-ratio of the telescope is

    A. 0.5 B. 30 m C. 2 D. 10 m

    Ans. C

22. The type of telescope having chromatic aberration is

    A. lens B. lens and mirror both
    C. mirror D. neither lens nor mirror

    Ans. A

23. The type of telescope having 2 m diameter is

    A. lens B. lens and mirror both
    C. mirror D. neither lens nor mirror

    Ans. C

5. Fundamentals

The universe has a large number of objects, generally called the cosmic objects. Examples of them are the stars, planets, satellites, comets, meteors, star clusters, galaxies, nebulae, etc. For the study of a cosmic object, one needs to understand basic things required for its investigation.

Based on the data gathered about the position of Mars observed on different days from the Earth, Kepler derived three laws of planetary motion, which may be used for the derivation of Newton's law of gravitation, for which an interesting story of fall of apple from a tree on the head/nose of Newton is famous. A plot of luminosity versus effective temperature of stars (H-R diagram) provides valuable information about the evolution of stars after their formation. The stars have hydrostatic equilibrium between the gravitational inward pull and the outward pressure produced in the core of star. After enjoying its life on the main sequence in the H-R diagram, a star evolves and the last stage, depending on its mass, of the star can be either a white dwarf, a neutron star or a black-hole. At that time, there are no nuclear reactions, and then degenerate matter balances the inward gravitational pull in white dwarfs and neutron stars. All these concepts and basic ideas used in understanding of physical phenomena are discussed in this chapter.

1. Parallax

It is also known as the trigonometric parallax. The parallax is a well known phenomenon from the experiments in the optics where the position of a real image is coincided with the position of a pin. When the two objects are at different distances, the closer one appears to move relative to the distant one (Figure 1). The same situation is applicable in case of the closer versus distant stars.

When observations are made from two opposite extremes in the Earth's orbit about the Sun, a nearby star appears displaced relative to the more distant stars available in the same part of the celestial sphere. The parallax p is defined as half the apparent angular displacement measured in radian in this way. The distance d to the star is then[1]

$$p = \frac{\text{astronomical unit}}{d} \qquad \text{or} \qquad d = \frac{1.5 \times 10^{11}\ \text{m}}{p}$$

[1]Since the angle is small, we may take p in radians instead of $\tan p$.

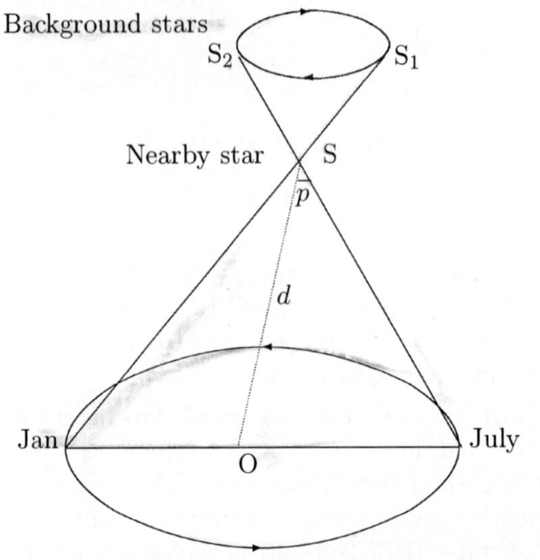

Figure 1: The parallax method for measuring distance of a nearby star.

Example 1: The angular separation in the apparent positions of a star relative to distant stars is 20° 15′ 18″. Calculate the distance of star from the Sun.

Solution: 1 AU $= 1.5 \times 10^{11}$ m. Half of the angular separation is $p = 10° \ 7' \ 39'' = 10.1275° = 0.17667$ rad. The distance of the star from the Sun is

$$d = \frac{1.5 \times 10^{11}}{0.17667} = 8.49 \times 10^{11} \text{ m}$$

Example 2: Distance of a star from the Sun is 7.2×10^{11} m. Calculate the parallax shown by the star.

Solution: We have parallax p

$$p = \frac{\text{astronomical unit}}{\text{distance}} = \frac{1.5 \times 10^{11} \text{ m}}{7.2 \times 10^{11} \text{ m}} = 0.2083 \text{ rad} = 11° \ 56' \ 33.6''$$

2. Two-body system

Let us consider a binary system having two components A and B of masses m_1 and m_2, respectively. In order to keep the system in equilibrium, the components revolve around the center-of-mass and hence the force of attraction between the components is balanced by the centrifugal forces. Let in an inertial reference frame, at time t, the components are situated at the positions $\vec{r}_1$ and $\vec{r}_2$, respectively, as shown in Figure 2.

Since the sizes of the components are much smaller than the distance under consideration, the masses may be treated as the point masses. The position of B relative to A is

$$\vec{r} = \vec{r}_2 - \vec{r}_1 \tag{5.1}$$

The position vector $\vec{R}$ of the center-of-mass C is

$$\vec{R} = \frac{m_1\,\vec{r}_1 + m_2\,\vec{r}_2}{m_1 + m_2} \tag{5.2}$$

Treatment of a binary system becomes convenient when we consider a coordinate system whose origin is at the center-of-mass of the binary system. In the center-of-mass system, we have $\vec{R} = 0$ and equation (5.2) gives

$$m_1\,\vec{r}_1 + m_2\,\vec{r}_2 = 0 \tag{5.3}$$

Now, $\vec{r}_1$ and $\vec{r}_2$ are the position vectors of the components of binary relative to the center-of-mass. Using the expression for $\vec{r}_1$ from equation (5.1) in (5.3), we get

$$m_1(\vec{r}_2 - \vec{r}) + m_2\,\vec{r}_2 = 0$$

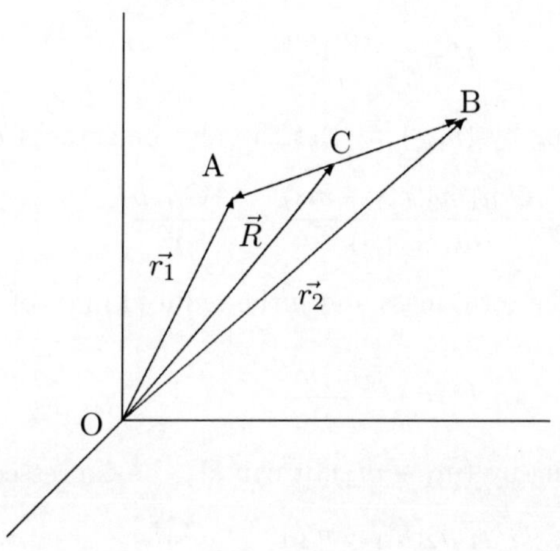

Figure 2: In an inertial reference frame, at time t, two components A and B of masses m_1 and m_2 of a binary system are situated at the positions $\vec{r}_1$ and $\vec{r}_2$, respectively. $\vec{R}$ is the position vector of the center-of-mass C.

On rearranging this relation, we get

$$\vec{r}_2 = \frac{m_1\,\vec{r}}{m_1 + m_2} = \frac{\mu}{m_2}\,\vec{r} \tag{5.4}$$

where $\mu = m_1 m_2/(m_1 + m_2)$ is the reduced mass of the binary system. Using the value of $\vec{r}_2$ in equation (5.1), we get

$$\vec{r} = \frac{m_1\,\vec{r}}{m_1 + m_2} - \vec{r}_1 \qquad \text{and} \qquad \vec{r}_1 = -\frac{\mu}{m_1}\,\vec{r}$$

It shows that with respect to the center-of-mass, the distances of bodies are inversely proportional to their masses.

2.1 Velocities of components

Velocity $\vec{v}_1$ of components A can be obtained by differentiating $\vec{r}_1$ with respect to time t, as

$$\vec{v}_1 = \frac{d\vec{r}_1}{dt} = -\frac{\mu}{m_1}\frac{d\vec{r}}{dt} = -\frac{\mu}{m_1}\,\vec{v} \tag{5.5}$$

where $\vec{v}$ is the velocity of B relative to that of A. Velocity $\vec{v}_2$ of components B can be obtained by differentiating $\vec{r}_2$ with respect to time t, as

$$\vec{v}_2 = \frac{d\vec{r}_2}{dt} = \frac{\mu}{m_2}\frac{d\vec{r}}{dt} = \frac{\mu}{m_2}\,\vec{v} \tag{5.6}$$

It shows that with respect to the center-of-mass, the velocities of bodies are inversely proportional to their masses.

2.2 Total energy

The gravitational force between the components is

$$F = -\frac{Gm_1m_2}{r^2}$$

By multiplying and dividing by $(m_1 + m_2)$, it can also be expressed as

$$F = -\frac{Gm_1m_2(m_1+m_2)}{(m_1+m_2)r^2} = -\frac{GM\mu}{r^2}$$

where $M = (m_1 + m_2)$ is the total mass and μ the reduced mass of the system. The potential energy is

$$U = -\frac{Gm_1m_2}{r}$$

By multiplying and dividing by $(m_1 + m_2)$, it can also be expressed as

$$U = -\frac{Gm_1m_2(m_1+m_2)}{(m_1+m_2)r} = -\frac{GM\mu}{r}$$

The kinetic energy of the binary system is

$$K = \frac{1}{2}\,m_1v_1^2 + \frac{1}{2}\,m_2v_2^2$$

Using the expressions for v_1 and v_2, we have

$$\begin{aligned} K &= \frac{1}{2}\,m_1\Big(-\frac{\mu}{m_1}\,v\Big)^2 + \frac{1}{2}\,m_2\Big(\frac{\mu}{m_2}\,v\Big)^2 \\ &= \frac{1}{2}\,\mu^2v^2\Big(\frac{1}{m_1} + \frac{1}{m_2}\Big) = \frac{1}{2}\,\mu v^2 \end{aligned}$$

Thus, the total energy E of the system is

$$E = K + U = \frac{1}{2}\,\mu v^2 - \frac{GM\mu}{r}$$

2.3 Angular momentum

Total orbital angular momentum $\vec{L}$ of the system is

$$\vec{L} = \vec{L}_1 + \vec{L}_2 = m_1\vec{r}_1 \times \vec{v}_1 + m_2\vec{r}_2 \times \vec{v}_2$$

Using the expressions for $\vec{r}_1$, $\vec{r}_2$, $\vec{v}_1$, $\vec{v}_2$, we get

$$\vec{L} = m_1\Big(-\frac{\mu}{m_1}\,\vec{r}\Big) \times \Big(-\frac{\mu}{m_1}\,\vec{v}\Big) + m_2\Big(\frac{\mu}{m_2}\,\vec{r}\Big) \times \Big(\frac{\mu}{m_2}\,\vec{v}\Big)$$

$$= \mu^2\vec{r} \times \vec{v}\Big(\frac{1}{m_1} + \frac{1}{m_2}\Big) = \mu\,\vec{r} \times \vec{v}$$

The expressions for the total energy and total angular momentum show that in the center-of-mass reference frame, a two-body problem may be treated as equivalent to one-body problem where a body of mass μ moves at a distance r around a stationary body of mass M.

Case I. System of Sun and planet: Suppose, the masses of Sun and a planet are m_s and m_p, respectively. Here, we have $M = m_s + m_p$ and $\mu = m_s m_p/(m_s + m_p)$. Since the mass of planet m_p is negligibly small as compared to the mass of Sun m_s, we can neglect the m_p as compared to m_s and therefore $M = m_s$ and $\mu = m_p$. Thus, we can take the Sun to be stationary and the planet revolving around the sun.

Case II. System of nucleus and electron: Suppose, the masses of neutron and electron are m_n and m_e, respectively. Here, we have $M = m_n + m_e$ and $\mu = m_n m_e/(m_n + m_e)$. Since the mass of electron m_e is negligibly small as compared to the mass of nucleus m_n, we can neglect the m_e as compared to m_n and therefore $M = m_n$ and $\mu = m_e$. Thus, we can take the nucleus to be stationary and the electron revolving around the nucleus.

Case III. Binary system of stars: Suppose, the masses of two stars A and B in the binary system are m_a and m_b, respectively. Here, we have $M = m_a + m_b$ and $\mu = m_a m_b/(m_a + m_b)$. Since the masses of stars are comparable, so the binary system is equivalent to the revolution of star of mass μ around another star of mass M. If $m_a = m_b = m$, we have $M = 2m$ and $\mu = m/2$.

2.4 Revolution period

The mass μ is revolving around the mass M at a distance r. The gravitational force between them is balanced by the centrifugal force and we have

$$\frac{GM\mu}{r^2} = \frac{\mu v^2}{r} \qquad \text{or} \qquad v = \sqrt{\frac{GM}{r}}$$

The time-period of revolution is

$$T = \frac{2\pi r}{v} = \frac{2\pi r^{3/2}}{\sqrt{GM}}$$

Example 3: In a binary system, two stars having masses 5 $\times 10^{26}$ kg and 2 $\times 10^{26}$ kg are at the positions $\vec{r_1} = 6\hat{\imath}+2\hat{\jmath}+5\hat{k}$ and $\vec{r_2} = 3\hat{\imath}-4\hat{\jmath}+2\hat{k}$, respectively. Calculate the position vector of the center-of-mass.

Solution: The position vector $\vec{R}$ of the center-of-mass is

$$\vec{R} = \frac{m_1\,\vec{r_1} + m_2\,\vec{r_2}}{m_1 + m_2}$$

On using the values, we get

$$\vec{R} = \frac{5\times 10^{26}(6\hat{\imath}+2\hat{\jmath}+5\hat{k}) + 2\times 10^{26}(3\hat{\imath}-4\hat{\jmath}+2\hat{k})}{5\times 10^{26} + 2\times 10^{26}} = \frac{36\hat{\imath}+2\hat{\jmath}+29\hat{k}}{7}$$

Example 4: Show that the reduced mass μ of two masses m_1 and m_2 is smaller than both the m_1 and m_2.

Solution: The reduced mass μ of masses m_1 and m_2 is

$$\frac{1}{\mu} = \frac{1}{m_1} + \frac{1}{m_2} \qquad \text{or} \qquad \mu = \frac{m_1 m_2}{m_1 + m_2}$$

Suppose, $m_1 < m_2$. We have

$$\mu = m_1\,\frac{m_2}{m_1 + m_2}$$

Since, $m_2/(m_1+m_2)$ is smaller than 1, we have $\mu < m_1$. Thus, μ is smaller than both the m_1 and m_2.

Example 5: Show that the reduced mass of Sun and a planet is nearly equal to the mass of the planet.

Solution: Suppose, the mass of Sun is M and that of planet m. The reduced mass of the combination is

$$\frac{1}{\mu} = \frac{1}{M} + \frac{1}{m} \qquad \text{or} \qquad \mu = \frac{Mm}{M+m}$$

Since $M >> m$, we can have

$$\mu \approx \frac{Mm}{M} = m$$

3. Kepler's laws of planetary motion

After analyzing the data about the position and velocity of the planet Mars on various dates, Kepler derived three fundamental laws, which have been found applicable to all the planets revolving around the Sun. These laws known as the Kepler's laws of planetary motion may be stated as the following. (The data used by Kepler were obtained by Tycho Brahe.)

1. A planet revolves around the Sun in an elliptical orbit. This is known as the **law of orbit**. Each of the planets has its elliptical orbit, which is distinct from those for the other planets.

2. The line joining one of the foci of the ellipse of orbit and the planet sweeps equal area in equal time. That is, the areal velocity of the planet is constant. This is known as the law of areal velocity.

As shown in Figure 3, a planet moves in equal times from P to P_1 and from P_2 to P_3, and the areas PSP_1 and P_2SP_3 are equal. Obviously, the speed of the planet during its motion form P to P_1 is fast as compared to that during the motion form P_2 to P_3.

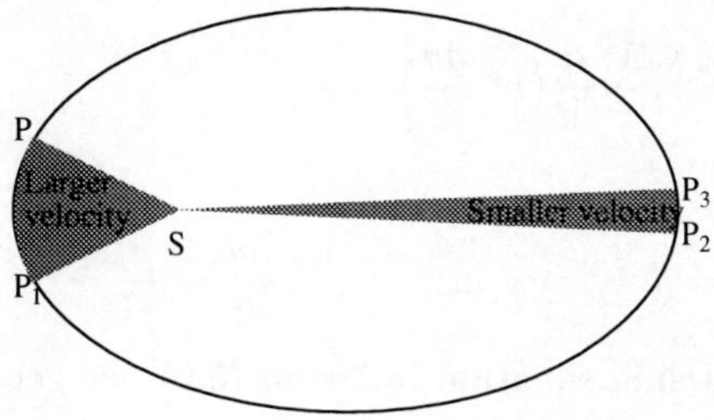

Figure 3: Elliptical motion of a planet around the Sun.

3. For a planet, the square of time-period for one revolution around the Sun is proportional to the cube of semi-major axis of the ellipse of path. This is known as the **law of time-period**.

It is interesting to note that though these three laws were first derived from the data recorded for the planet Mars, but they are found valid for the revolution of all planets around the Sun. Moreover, the laws are found valid for the revolution of satellites around its planet. These laws can be derived in the following manner.

As we discussed that Sun of mass M can be considered stationary and a planet of mass m is revolving around the Sun. Let at a given time, the position vector and velocity of the planet be $\vec{r}$ and $\vec{v}$, respectively. Then, the total energy E and total orbital angular momentum $\vec{L}$ of the planet are

$$E = \frac{1}{2}\, mv^2 - G\, \frac{mM}{r} \qquad \text{and} \qquad \vec{L} = m\, \vec{r} \times \vec{v}$$

3.1 Law of orbit

The force of attraction between the Sun and planet is

$$\vec{F} = -G\, \frac{Mm}{r^2}\, \hat{r}$$

where $\hat{r}$ is the unit vector along the radial distance. This is the case of central force. A planet around the Sun revolves under the central force. Thus, the planet has an acceleration

$$\vec{a} = -G\, \frac{M}{r^2}\, \hat{r} \tag{5.7}$$

When we use plane polar coordinates (r, θ) with the origin at the mass M, this acceleration can be resolved into the radial and transverse components as:

$$\vec{a} = \left[\frac{\mathrm{d}^2 r}{\mathrm{d}t^2} - r\left(\frac{\mathrm{d}\theta}{\mathrm{d}t}\right)^2\right]\hat{r} + \frac{1}{r}\frac{\mathrm{d}}{\mathrm{d}t}\left(r^2\frac{\mathrm{d}\theta}{\mathrm{d}t}\right)\hat{\theta} \tag{5.8}$$

Comparison of equations (5.7) and (5.8) gives

$$\frac{\mathrm{d}^2 r}{\mathrm{d}t^2} - r\left(\frac{\mathrm{d}\theta}{\mathrm{d}t}\right)^2 = -G\frac{M}{r^2} \tag{5.9}$$

and

$$\frac{1}{r}\frac{\mathrm{d}}{\mathrm{d}t}\left(r^2\frac{\mathrm{d}\theta}{\mathrm{d}t}\right) = 0$$

This equation gives

$$r^2\frac{\mathrm{d}\theta}{\mathrm{d}t} = h \qquad \text{or} \qquad \frac{\mathrm{d}\theta}{\mathrm{d}t} = \frac{h}{r^2} \tag{5.10}$$

where h is some constant. Using equation (5.10) in (5.9), we get

$$\frac{\mathrm{d}^2 r}{\mathrm{d}t^2} - \frac{h^2}{r^3} + G\frac{M}{r^2} = 0 \qquad \text{or} \qquad \frac{\mathrm{d}^2 r}{\mathrm{d}t^2} + \frac{\mathrm{d}}{\mathrm{d}r}\left(-G\frac{M}{r} + \frac{h^2}{2r^2}\right) = 0$$

Multiplying this equation by $\mathrm{d}r/\mathrm{d}t$ we get

$$\frac{\mathrm{d}^2 r}{\mathrm{d}t^2}\frac{\mathrm{d}r}{\mathrm{d}t} + \frac{\mathrm{d}}{\mathrm{d}r}\left(-G\frac{M}{r} + \frac{h^2}{2r^2}\right)\frac{\mathrm{d}r}{\mathrm{d}t} = 0$$

$$\frac{\mathrm{d}}{\mathrm{d}t}\left[\frac{1}{2}\left(\frac{\mathrm{d}r}{\mathrm{d}t}\right)^2\right] + \frac{\mathrm{d}}{\mathrm{d}t}\left(-G\frac{M}{r} + \frac{h^2}{2r^2}\right) = 0$$

$$\frac{\mathrm{d}}{\mathrm{d}t}\left[\frac{1}{2}\left(\frac{\mathrm{d}r}{\mathrm{d}t}\right)^2 - G\frac{M}{r} + \frac{h^2}{2r^2}\right] = 0$$

This equation therefore gives

$$\frac{1}{2}\left(\frac{\mathrm{d}r}{\mathrm{d}t}\right)^2 - G\frac{M}{r} + \frac{h^2}{2r^2} = \frac{E}{m} \tag{5.11}$$

where E/m is a constant. Here, E is the total (kinetic + potential) energy of the system.[2] Equation (5.11) can be rearranged as

$$\frac{\mathrm{d}r}{\mathrm{d}t} = \sqrt{2\left(\frac{E}{m} + \frac{GM}{r} - \frac{h^2}{2r^2}\right)} \tag{5.12}$$

[2]In the polar coordinates, radial and transverse components of velocity are, respectively,

$$v_r = \frac{\mathrm{d}r}{\mathrm{d}t} \qquad \text{and} \qquad v_\theta = r\frac{\mathrm{d}\theta}{\mathrm{d}t}$$

Thus, total (kinetic + potential) energy is

$$E = \frac{1}{2}m\left[\left(\frac{\mathrm{d}r}{\mathrm{d}t}\right)^2 + r^2\left(\frac{\mathrm{d}\theta}{\mathrm{d}t}\right)^2\right] + \left(-G\frac{Mm}{r}\right)$$

Using equation (5.10) we get

$$\frac{E}{m} = \frac{1}{2}\left(\frac{\mathrm{d}r}{\mathrm{d}t}\right)^2 + \frac{h^2}{2r^2} - G\frac{M}{r}$$

Dividing equation (5.12) by (5.10), we get

$$\frac{dr}{d\theta} = \frac{r^2}{h}\sqrt{2\Big(\frac{E}{m} + \frac{GM}{r} - \frac{h^2}{2r^2}\Big)}$$

or

$$d\theta = \frac{h}{r^2}\frac{dr}{\sqrt{2\Big(\frac{E}{m} + \frac{GM}{r} - \frac{h^2}{2r^2}\Big)}} \tag{5.13}$$

This equation can be solved by using the relation

$$r = \frac{1}{u} \qquad \text{so that} \qquad dr = -\frac{1}{u^2}\,du$$

and equation (5.13) gives

$$d\theta = -\frac{du}{\sqrt{\frac{2E}{mh^2} + \frac{2GMu}{h^2} - u^2}} = -\frac{du}{\sqrt{\Big(\frac{2E}{mh^2} + \frac{G^2M^2}{h^4}\Big) - \Big(u - \frac{GM}{h^2}\Big)^2}}$$

On integration, we get

$$\theta = \cos^{-1}\Bigg(\frac{u - \frac{GM}{h^2}}{\sqrt{\frac{2E}{mh^2} + \frac{G^2M^2}{h^4}}}\Bigg) + \theta_0$$

where θ_0 is the constant of integration. This equation can be rearranged as

$$u - \frac{GM}{h^2} = \Big[\sqrt{\frac{2E}{mh^2} + \frac{G^2M^2}{h^4}}\Big]\cos(\theta - \theta_0)$$

Therefore,

$$\frac{1}{r} = \frac{GM}{h^2} + \frac{GM}{h^2}\Big[\sqrt{1 + \frac{2h^2E}{mG^2M^2}}\Big]\cos(\theta - \theta_0)$$

$$= \frac{GM}{h^2}\Big\{1 + \Big[\sqrt{1 + \frac{2h^2E}{mG^2M^2}}\Big]\cos(\theta - \theta_0)\Big\}$$

Thus, we have

$$r = \frac{h^2/GM}{1 + \Big[\sqrt{1 + \frac{2h^2E}{mG^2M^2}}\Big]\cos(\theta - \theta_0)} = \frac{a(1 - e^2)}{1 + e\cos(\theta - \theta_0)}$$

This is equation of ellipse with semi-major axis a and eccentricity e, showing that the path of motion under a central force is elliptical. This is the Kepler's first law of motion of a planet revolving under the gravitational force. Here, θ_0 is value of θ when r is minimum, *i.e.*, at the periastron position and

$$e = \sqrt{1 + \frac{2h^2E}{mG^2M^2}} \tag{5.14}$$

and

$$a(1-e^2) = \frac{h^2}{GM} \tag{5.15}$$

Using equation (5.14) in (5.15) we get

$$a = -\frac{GMm}{2E} \tag{5.16}$$

It shows that the semi-major axis of the orbit is inversely proportional to the total energy of the system. As G, M, m, a all are positive quantities, the negative sign shows that the total energy of the planet is negative.

The total energy is the sum of kinetic energy and potential energy. As the kinetic energy can never be negative, it shows that the potential energy is negative and the magnitude of the potential energy is larger than that of the kinetic energy. The negative value of potential energy is consistent with the attractive force between the Sun and a planet.

3.2 Law of areal velocity

Suppose, the body of mass m moving in the elliptical path is at a position P at time t (Figure 4). After a time interval dt, the body reaches at Q. Thus, the area swept by the line, joining the focus O of the ellipse to the body, in time dt is

$$\text{Area} = \frac{1}{2}\ \text{OP} \times \text{PQ} = \frac{1}{2}\ r(r\ \text{d}\theta) = \frac{1}{2}\ r^2\ \text{d}\theta$$

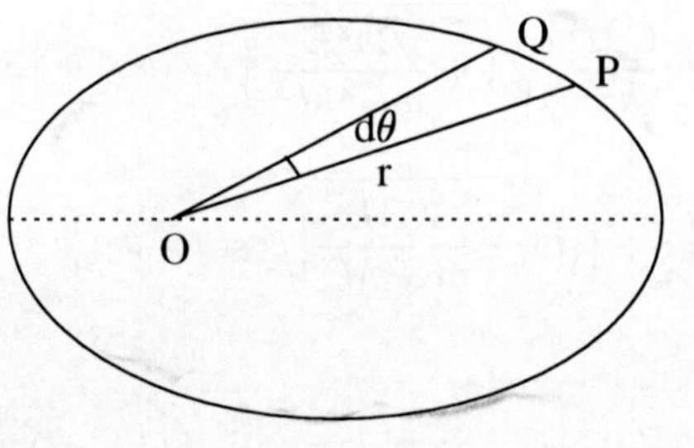

Figure 4: Body of mass m revolving in the elliptical path around a stationary body of mass M, is at a position P at time t. After a time interval dt the body reaches at the position Q.

Hence, the areal velocity is

$$\text{Areal velocity} = \frac{\text{area traversed}}{\text{time interval}} = \frac{1}{2}r^2\frac{\text{d}\theta}{\text{d}t} = \frac{h}{2} \tag{5.17}$$

where equation (5.10) is used. This shows that for a particle moving in an elliptical path, the areal velocity is constant. This is the Kepler's second law of motion.

3.3 Law of time-period

Area of an ellipse with semi-major axis a and semi-minor axis b is πab. If T is the time period of motion, we have

$$\frac{h}{2} = \frac{\pi ab}{T} = \frac{\pi a^2\sqrt{1-e^2}}{T} \tag{5.18}$$

Here, we have used the relation $b = a\sqrt{1-e^2}$. Using equations (5.16) and (5.14) in (5.18), we have

$$\frac{h}{2} = \frac{\pi a^2}{T}\sqrt{-\frac{2Eh^2}{mG^2M^2}} = \frac{\pi h}{T}\sqrt{\frac{a^3}{GM}}$$

Thus, we have

$$1 = \frac{2\pi}{T}\sqrt{\frac{a^3}{GM}} \qquad \text{or} \qquad \frac{a^3}{T^2} = \frac{GM}{4\pi^2} \tag{5.19}$$

This shows that square of the time-period is proportional to the cube of the semi-major axis a. This is the Kepler's third law of motion.

3.3.1 Mass of Sun

From the Kepler's third law of motion, we have

$$M = \frac{4\pi^2}{G}\frac{a^3}{T^2}$$

By knowing the time-period and average distance of a planet, the mass of the Sun can be estimated. For a circular orbit with radius a we have $T = 2\pi a/v$, where v is the velocity of revolution and thus, we get

$$M = \frac{4\pi^2}{G}\frac{a^3}{(2\pi a/v)^2} = \frac{av^2}{G}$$

It shows that with the increase of the radius of the orbit, the orbital velocity of the planet decreases in such a manner that av^2 remains constant.

Example 6: Earth completes one revolution around the Sun in 365.25 days at an average distance of 1.496×10^8 km. Calculate the mass of the Sun.

Solution: For the Earth revolving around the Sun, we have $a = 1.496 \times 10^8$ km and $T = 365.256$ days and thus

$$M = \frac{4 \times (3.14159)^2}{6.67 \times 10^{-11}}\frac{(1.496 \times 10^{11})^3}{(365.25 \times 24 \times 60 \times 60)^2} = 1.99 \times 10^{30} \text{ kg}$$

Example 7: A planet is revolving around the Sun in an elliptical orbit. If the minimum and maximum distances of the planet from Sun are r_{min} and r_{max}, respectively, show that the eccentricity e of the orbit is

$$e = \frac{r_{max} - r_{min}}{r_{max} + r_{min}}$$

Solution: Using the plane polar coordinates (r, θ), the equation of elliptical orbit is

$$r = \frac{a(1-e^2)}{1+e\cos\theta}$$

where e and a denote the eccentricity and semi-major axis, respectively, of the ellipse. The distance of planet from the Sun is the minimum when $\theta = 0$ and we have

$$r_{min} = \frac{a(1-e^2)}{1+e} = a(1-e)$$

The distance of planet from the Sun is the maximum when $\theta = \pi$ and we have

$$r_{max} = \frac{a(1-e^2)}{1-e} = a(1+e)$$

Thus, we have

$$r_{max} + r_{min} = 2a \qquad \text{and} \qquad r_{max} - r_{min} = 2ae$$

It gives

$$e = \frac{r_{max} - r_{min}}{r_{max} + r_{min}}$$

Example 8: A planet is revolving around the Sun in an elliptical orbit with eccentricity $e = \sqrt{1+\alpha E}$ and semi-major axis $a = -\beta/E$, where E is the total energy, and α and β are positive constants. Using Kepler's second law of planetary motion, derive the Kepler's third law of planetary motion.

Solution: For an ellipse with semi-major axis a, semi-minor axis b and eccentricity e, we have

$$b = a\sqrt{1-e^2}$$

and the area of the ellipse is πab. According to the Kepler's second law of planetary motion, the areal velocity is constant, say v_A. If T denotes the time-period of planetary motion, we have

$$v_A = \frac{\pi ab}{T} = \frac{\pi a^2\sqrt{1-e^2}}{T}$$

Using the value of e here, we get

$$v_A T = \pi a^2 \sqrt{1-(1+\alpha E)} \qquad \text{or} \qquad v_A^2 T^2 = -\pi^2 a^4 \alpha E$$

Using $E = -\beta/a$, we get

$$v_A^2 T^2 = \pi^2 a^3 \alpha\beta \qquad \text{or} \qquad T^2 \propto a^3$$

It gives the Kepler's third law of planetary motion.

Example 9: The mean distance of the planet Mars from the Sun is 1.533 times the mean distance between the Earth and Sun. Calculate the period of revolution of Mars around the Sun.

Solution: Following Kepler's third law of planetary motion, the time-periods T_M and T_E of Mars and Earth for revolution around the Sun and their average distances $a_M$and a_E of Mars and Earth, respectively, are related as the following.

$$\frac{T_M^2}{T_E^2} = \frac{a_M^3}{a_E^3} \qquad \text{or} \qquad T_M = T_E \left(\frac{a_M}{a_E}\right)^{3/2}$$

Given, $a_M/a_E = 1.533$. We know, $T_E = 365$ days. Using the values, we get

$$T_M = 365\ (1.533)^{3/2} = 692.8 \text{ days}$$

Thus, the time-period of revolution of Mars around the Sun is 692.8 days.

Note: Similar exercise can be considered for any pair of planets revolving around the Sun or for motion of a pair of satellites revolving around a common planet.

Example 10: The eccentricity of the Earth's orbit for revolution around the Sun is 0.0167. Calculate the ratio of maximum and minimum speeds of Earth in the orbit.

Solution: The angular momentum of Earth's motion around the Earth is constant. If at any instant the velocity is v and distance from the Sun is r, then we have

$$mvr = \text{constant} \qquad \text{or} \qquad vr = \text{constant}$$

This relation shows that the distance will be minimum when the velocity is the maximum, and vice versa. Thus, we have

$$v_{max} r_{min} = v_{min} r_{max} \qquad \text{or} \qquad \frac{v_{max}}{v_{min}} = \frac{r_{max}}{r_{min}}$$

For the motion in the elliptical path, the maximum distance r_{max} and minimum distance r_{min} are related to the eccentricity e as

$$\frac{r_{max}}{r_{min}} = \frac{1+e}{1-e}$$

Therefore, we have

$$\frac{v_{max}}{v_{min}} = \frac{1+e}{1-e} = \frac{1+0.0167}{1-0.0167} = 1.033$$

4. Newton's law of gravitation

About the Newton's law of gravitation, you might have heard a story about the falling of apple from a tree on the head of Newton, from which he got idea for the law of gravitation between two point-masses separated by a distance. The story may not be correct, but it is sounds interesting. On the scientific ground, it may not be ruled out that the law of gravitation would have been derived from the Kepler's laws for planetary motion, as discussed in the following.

According to Kepler, a planet (mass m) revolves around the Sun (mass M) in an elliptical orbit. When we use plane polar coordinates $(r,\ \theta)$ with the origin at the Sun, the areal velocity is

$$\text{Areal velocity} = \frac{1}{2}\, r^2\, \frac{d\theta}{dt}$$

According to Kepler's second law of planetary motion, the areal velocity is constant and therefore,

$$\frac{1}{2}\, r^2\, \frac{d\theta}{dt} = \frac{h}{2} \qquad \text{or} \qquad r^2\, \frac{d\theta}{dt} = h \tag{5.20}$$

where h is constant and factor 2 is introduced for convenience. For a motion along a curved path, the radial component a_r and transverse component a_θ of acceleration are expressed as

$$a_r = \frac{d^2r}{dt^2} - r\left(\frac{d\theta}{dt}\right)^2 \qquad \text{and} \qquad a_\theta = \frac{1}{r}\,\frac{d}{dt}\left(r^2\,\frac{d\theta}{dt}\right)$$

Use of equation (5.20) here gives $a_\theta = 0$, showing that the transverse component of acceleration is zero and there is acceleration only along the line joining the planet and Sun. Now, we want to calculate a_r. From equation (5.20), we have

$$\frac{d\theta}{dt} = hu^2 \tag{5.21}$$

where $r = 1/u$. On differentiation with respect to t, we have

$$\frac{dr}{dt} = -\frac{1}{u^2}\,\frac{du}{dt} = -\frac{1}{u^2}\,\frac{du}{d\theta}\,\frac{d\theta}{dt}$$

Using equation (5.21) here, we have

$$\frac{dr}{dt} = -h\,\frac{du}{d\theta}$$

Differentiation of this equation with respect to t gives,

$$\frac{d^2r}{dt^2} = -h\,\frac{d}{dt}\left(\frac{du}{d\theta}\right) = -h\,\frac{d}{d\theta}\left(\frac{du}{d\theta}\right)\frac{d\theta}{dt} = -h^2u^2\,\frac{d^2u}{d\theta^2} \tag{5.22}$$

Here, equation (5.21) is used. Using equations (5.21) and (5.22), the expression for radial component of acceleration a_r is

$$a_r = -h^2u^2\,\frac{d^2u}{d\theta^2} - \frac{1}{u}\,h^2u^4 = -h^2u^2\left[\frac{d^2u}{d\theta^2} + u\right] \tag{5.23}$$

The elliptical orbit of planet can be expressed as

$$\frac{l}{r} = 1 + e\cos\theta \qquad \text{or} \qquad lu = 1 + e\cos\theta$$

where e is eccentricity and $l = h^2/GM$. Differentiation of this equation with respect to θ twice gives,

$$l\,\frac{\mathrm{d}^2u}{\mathrm{d}\theta^2} = -e\cos\theta$$

Therefore, we have

$$l\,\frac{\mathrm{d}^2u}{\mathrm{d}\theta^2} + lu = 1 \qquad \text{or} \qquad \frac{\mathrm{d}^2u}{\mathrm{d}\theta^2} + u = \frac{GM}{h^2} \tag{5.24}$$

Using equation (5.24) in (5.23), we have

$$a_r = -GMu^2 = -\frac{GM}{r^2}$$

Thus, the force F on the planet is

$$F = ma_r = -G\frac{mM}{r^2}$$

This shows how the Newton's law of gravitation can be derived from the Kepler's law of planetary motion.

5. Virial theorem

This theorem helps us in understanding the dynamic behaviour of a system having a large number of bodies. It, however, does not tell about the precise behaviour of any individual body in the system. The theorem states that for a gravitationally bound system of bodies in equilibrium, total energy of the system is always one-half of the time averaged potential energy. If $\langle U\rangle$ and E denote the time-averaged potential energy and total energy, respectively, of a system of bodies, then according to the Virial theorem, $E = \langle U\rangle/2$. Since the system of bodies is necessarily bound, and therefore, the E and $\langle U\rangle$ both are negative.

In order to prove the Virial theorem, let us consider a system of large number of bodies of masses m_i, situated at the positions $\vec{r_i}$ and acted upon by the forces $\vec{F_i}$. The linear momentum $\vec{p_i}$ of a particle is expressed as $\vec{p_i} = m_i\ (\mathrm{d}\vec{r_i}/\mathrm{d}t)$, and the force $\vec{F_i}$ is expressed as $\vec{F_i} = (\mathrm{d}\vec{p_i}/\mathrm{d}t)$. Let us consider a quantity

$$Q = \sum_i \vec{p_i}.\vec{r_i} \qquad \text{so that} \qquad \frac{\mathrm{d}Q}{\mathrm{d}t} = \sum_i \left(\frac{\mathrm{d}\vec{p_i}}{\mathrm{d}t}\,.\,\vec{r_i} + \vec{p_i}\,.\,\frac{\mathrm{d}\vec{r_i}}{\mathrm{d}t}\right)$$

Now, we have

$$Q = \sum_i \vec{p_i}\,.\,\vec{r_i} = \sum_i m_i\,\frac{\mathrm{d}\vec{r_i}}{\mathrm{d}t}\,.\,\vec{r_i} = \sum_i \frac{m_i}{2}\,\frac{\mathrm{d}(\vec{r_i}\,.\,\vec{r_i})}{\mathrm{d}t}$$

$$= \frac{1}{2}\,\frac{\mathrm{d}(\sum_i m_i r_i^2)}{\mathrm{d}t} = \frac{1}{2}\,\frac{\mathrm{d}I}{\mathrm{d}t}$$

where, $I = \sum m_i r_i^2$ is the moment of inertia of the system about some axis. Thus, we have

$$\frac{\mathrm{d}Q}{\mathrm{d}t} = \frac{1}{2}\,\frac{\mathrm{d}^2I}{\mathrm{d}t^2}$$

Now, we have

$$\vec{p_i} \cdot \frac{\mathrm{d}\vec{r_i}}{\mathrm{d}t} = m_i\vec{v_i} \cdot \vec{v_i} = m_i v_i^2$$

Therefore, we get

$$\sum_i \vec{p_i} \cdot \frac{\mathrm{d}\vec{r_i}}{\mathrm{d}t} = \sum_i m_i v_i^2 = 2K$$

where, $K = \frac{1}{2}\sum m_i v_i^2$ is the kinetic energy of the system. Thus, we have

$$\frac{1}{2}\frac{\mathrm{d}^2 I}{\mathrm{d}t^2} = \sum_i \vec{F_i} \cdot \vec{r_i} + 2K \qquad \text{or} \qquad \sum_i \vec{F_i} \cdot \vec{r_i} = \frac{1}{2}\frac{\mathrm{d}^2 I}{\mathrm{d}t^2} - 2K \tag{5.25}$$

Let $\vec{F_{ij}}$ represent the force of interaction on the particle i due to the particle j in the system, then from the Newton's third law of motion we have $\vec{F_{ij}} = -\vec{F_{ji}}$, and the total force acting on the particle i due to all other particles in the system is

$$\vec{F_i} = {\sum_j}' \vec{F_{ij}}$$

where the prime over the summation sign denotes that the sum is taken over all i values of j except that $j = i$. This convention is followed later on also. Then,

$$\sum_i \vec{F_i} \cdot \vec{r_i} = \sum_i \Big(\sum_{j \neq i} \vec{F_{ij}}\Big) \cdot \vec{r_i} = \frac{1}{2} \sum_i \Big[{\sum_j}' \Big(\vec{F_{ij}} - \vec{F_{ji}}\Big)\Big] \cdot \vec{r_i}$$

After some manipulation, we can write[3]

$$\sum_i \vec{F_i} \cdot \vec{r_i} = \frac{1}{2} \sum_i {\sum_j}' \vec{F_{ij}} \cdot (\vec{r_i} - \vec{r_j}) \tag{5.26}$$

The force $\vec{F_{ij}}$ on the particle i due to the particle j is

$$\vec{F_{ij}} = G\,\frac{m_i m_j}{r_{ij}^2}\,\hat{r_{ij}} \tag{5.27}$$

---

[3]In order to understand it, let us consider a case of 3 bodies in a system

$$\sum_{i=1}^{3} \Big[\sum_{\substack{j=1 \\ j \neq i}}^{3} \Big(\vec{F_{ij}} - \vec{F_{ji}}\Big)\Big] \cdot \vec{r_i} = (\vec{F_{12}} - \vec{F_{21}}) \cdot \vec{r_1} + (\vec{F_{13}} - \vec{F_{31}}) \cdot \vec{r_1} + (\vec{F_{21}} - \vec{F_{12}}) \cdot \vec{r_2}$$

$$+(\vec{F_{23}} - \vec{F_{32}}) \cdot \vec{r_2} + (\vec{F_{31}} - \vec{F_{13}}) \cdot \vec{r_3} + (\vec{F_{32}} - \vec{F_{23}}) \cdot \vec{r_3} = \vec{F_{12}} \cdot (\vec{r_1} - \vec{r_2}) + \vec{F_{13}} \cdot (\vec{r_1} - \vec{r_3})$$

$$+\vec{F_{21}} \cdot (\vec{r_2} - \vec{r_1}) + \vec{F_{23}} \cdot (\vec{r_2} - \vec{r_3}) + \vec{F_{31}} \cdot (\vec{r_3} - \vec{r_1}) + \vec{F_{32}} \cdot (\vec{r_3} - \vec{r_2})$$

$$= \sum_{i=1}^{3} \sum_{\substack{j=1 \\ j \neq i}}^{3} \vec{F_{ij}} \cdot (\vec{r_i} - \vec{r_j})$$

where $\vec{r_i}$ and $\vec{r_j}$ are position vectors of the particles i and j, respectively, and $r_{ij} = |\vec{r_j} - \vec{r_i}|$ is the separation between the particles i and j, and $\hat{r_{ij}}$ is the unit vector in the direction from i to j:

$$\hat{r_{ij}} = \frac{\vec{r_j} - \vec{r_i}}{r_{ij}} \tag{5.28}$$

Using equations (5.27) and (5.28) in (5.26), we get

$$\sum_i \vec{F_i} \,.\, \vec{r_i} = \frac{1}{2} \sum_i {\sum_j}' G \frac{m_i m_j}{r_{ij}^2} \frac{(\vec{r_j} - \vec{r_i})}{r_{ij}} \,.\, (\vec{r_i} - \vec{r_j})$$

$$= -\frac{1}{2} \sum_i {\sum_j}' G \frac{m_i m_j}{r_{ij}^3} (\vec{r_i} - \vec{r_j})^2 = -\frac{1}{2} \sum_i {\sum_j}' G \frac{m_i m_j}{r_{ij}}$$

$$= -\frac{1}{2} \sum_i {\sum_j}' U_{ij} = U \tag{5.29}$$

where

$$U_{ij} = -G \frac{m_i m_j}{r_{ij}}$$

is potential between the bodies i and j, and U is the total potential energy of the system. Thus, from equations (5.25) and (5.29), we have

$$\frac{1}{2}\left\langle \frac{\mathrm{d}^2 I}{\mathrm{d}t^2} \right\rangle - 2\langle K \rangle = \langle U \rangle$$

Now, the average may be expressed as

$$\left\langle \frac{\mathrm{d}^2 I}{\mathrm{d}t^2} \right\rangle = \frac{1}{\tau} \int_0^\tau \frac{\mathrm{d}^2 I}{\mathrm{d}t^2} = \frac{1}{\tau} \left(\left.\frac{\mathrm{d}I}{\mathrm{d}t}\right|_\tau - \left.\frac{\mathrm{d}I}{\mathrm{d}t}\right|_0 \right)$$

If the system is periodic with a period τ, we have

$$\left.\frac{\mathrm{d}I}{\mathrm{d}t}\right|_\tau = \left.\frac{\mathrm{d}I}{\mathrm{d}t}\right|_0 \qquad \text{and thus,} \qquad \left\langle \frac{\mathrm{d}^2 I}{\mathrm{d}t^2} \right\rangle = 0$$

Even if the system is not strictly periodic, the average $\langle \mathrm{d}^2 I/\mathrm{d}t^2 \rangle$ is zero when it is evaluated over a sufficiently long period of time (*i.e.*, $\tau \to \infty$) assuming that $\mathrm{d}I/\mathrm{d}t$ is bound. This would describe, for example, a system that has attained steady state configuration (an equilibrium). In either case, we have $\langle \mathrm{d}^2 I/\mathrm{d}t^2 \rangle = 0$. Thus, we get

$$2\langle K \rangle + \langle U \rangle = 0 \qquad \text{so that} \qquad \langle K \rangle = -\frac{1}{2} \langle U \rangle$$

Now, we have

$$\langle E \rangle = \langle K \rangle + \langle U \rangle = -\frac{1}{2} \langle U \rangle + \langle U \rangle = \frac{1}{2} \langle U \rangle$$

This proves the Virial theorem, which applies to a wide variety of systems ranging from the particles in an ideal gas to a cluster of galaxies.

6. Cosmic abundances

Abundance of a chemical element (common or rare) is a measure how relatively presence of the element as compared to all other elements in a given environment. The abundance may be measured in terms of the mass-fraction (the same as weight fraction), mole-fraction (fraction of atoms by numerical count), or volume-fraction. A common way for representation of abundance is in terms of the volume-fraction in mixed gases, such as the planetary atmospheres, and is close to the molecular mole-fraction for ideal gas mixtures (*i.e.*, the gas mixtures at relatively low densities and pressures).

For example, in pure water (H_2O), the mass-fraction abundance of oxygen is about 89% because it is the fraction of water's mass which is oxygen. However, the mole-fraction abundance of oxygen atoms in water is 33%, because out of 3 atoms in water, there only 1 atom of oxygen.

In the Universe as a whole, and in the atmospheres of giant planets having gas, such as Jupiter, the mass-fraction abundances of hydrogen and helium are about 74% and 23 - 25%, respectively, whereas their (atomic) mole-fractions are closer to 92% and 8%, respectively. However, for example, in the outer atmosphere of Jupiter, hydrogen is in the molecular from whereas the helium is in the atomic form. Hence the molecular mole-fraction (fraction of total gas molecules), as well as the fraction of atmosphere by volume, of hydrogen is about 86%, and for helium is about 13%.

The mass-fraction of ten most common elements in the Milky Way Galaxy estimated spectroscopically are given in Table 1.

Table 1. Ten most common elements in Milky Way Galaxy

Z	Element	Mass fraction
1	Hydrogen	739,000
2	Helium	240,000
8	Oxygen	10,400
6	Carbon	4,600
10	Neon	1,340
26	Iron	1,090
7	Nitrogen	960
14	Silicon	650
12	Magnesium	580
16	Sulfur	440

Following standard cosmological model, the lightest elements were mostly produced shortly (*i.e.*, within a few hundred seconds) after the Big Bang through the nucleosynthesis. The heavier elements were mostly produced much later inside the stars.

Most of the baryonic mass in the universe is estimated to be in the form of hydrogen (74%) and helium (24%). Though the remaining heavy elements are of small fraction, but they can greatly influence the astronomical phenomena. Other than the hydrogen or helium, in astronomy, all the element are put under the category, named as the metal. This distinction is significant because hydrogen and helium are the only elements that were produced in significant quantities in the Big Bang, whereas the others were produced later on. Hence, the metalicity of a galaxy or other object is an indication of the past stellar activity. In the disk of the Milky Way galaxy, the mass of heavy elements is only about 2%.

Hydrogen and helium are most common elements, from the Big Bang. The next three elements (Li, Be, B) are rare because they were poorly synthesized in the Big Bang and also in the stars. The two general trends in the remaining elements produced in the stars are as the following.

(i) The abundances follow the Oddo-Harkins rule which states that an element with an even atomic number is more abundant than both elements with the adjacently smaller and larger atomic numbers. For example, the carbon (atomic number 6) is more abundant than both the boron (atomic number 5) and nitrogen (atomic number 7).

(ii) The abundance of element decreases in general with the increase of atomic number.

Iron is especially common because it represents the minimum energy nuclide that can be made by fusion of helium in supernovae.

7. Spectral classification

All the stars are not similar in their characteristics. Based on their spectral characteristics, the stars are classified into various categories. The basic idea behind the spectral classification is that for a given chemical composition, the pattern of absorption lines which are formed in the photosphere of a star depends on the temperature and pressure, in the photosphere. Indeed, a great variety of stellar spectra can be observed for different stars. This great variety arises from the variation of the effective temperature of star than they do from the variation of the chemical composition or surface pressure. Thus, to a first approximation, the spectral type of a star yields an estimate of its effective temperature T_e. That is, the spectral type is function of T_e.

$$\text{spectral type} = f(T_e)$$

Historically, the spectral types were assigned alphabetically in an order determined primarily by the decreasing strength of hydrogen absorption lines. The stars with the strongest hydrogen lines were called A type stars; The stars with the next strongest hydrogen lines were called B type stars; and so on. It was soon realized that this assignment did not give sequence to the stars according to their effective temperatures.

Through the work of Saha, astronomers learned that O stars have weak hydrogen lines because their photospheres are so hot that the hydrogen is almost ionized, whereas M stars, which are alphabetically close to O stars, have weak hydrogen lines for a completely different reason. The M stars have such cool photospheres that there is little atomic hydrogen in the first excited ($n = 2$) state from where an absorption of a stellar photon can produce a hydrogen line in the Balmer series, which is visible at the optical wavelengths.

At a relatively low temperature of an M star's photosphere, hydrogen atoms prefer to be in their ground electronic state. At the relatively high temperature of an O star's photosphere, they prefer to be ionized. When the spectral types are ordered in a sequence of decreasing effective temperature, they appear in the sequence:[4]

OBAFGKM

Table 2 shows the effective temperature for various classes. The Spectral classes are divided into subclasses denoted by the numbers 0, 1, . . . , 9; sometimes decimals are used, *e.g.*, B0.4.

The original classification scheme compiled in the Henry Draper (HD) Catalog,[5] included other spectral types, such as N, R, and S. It is now known that such stars have different-looking spectra, not because they are cooler than M stars, but because they have peculiar heavy-element abundances. The R and N stars are rich in carbon compounds. These carbon-stars have effective temperatures comparable to those of the K and M stars. The S stars have photospheres with enhanced abundances of s-process elements. The additional classes R, N, and S are often appended in parentheses to the main spectral types:

OBAFGKM(RNS)

To these classes finer subdivisions are introduced by appending the numbers, ranging from 0 to 9 to each spectral class. For example, G2 is the class of the Sun.

Table 2. Temperature versus class

Class	Effective temperature
O	$\geq$ 30,000 K
B	10,000 - 30,000 K
A	7,500 - 10,000 K
F	6,000 - 7,500 K
G	5,200 - 6,000 K
K	3,700 - 5,200 K
M	2,400 - 3,700 K

8. Black-body spectrum

A body which can absorb radiations of all wavelengths falling on it is known as the **black-body**. According to the Kirchhoff's law, a good absorber of radiations of particular wavelengths is also a good emitter of radiations of the same wavelengths. Thus, a black-body can emit radiations of all wavelengths. When the temperature of black-body is larger than that of the atmosphere around it, it emits the radiations,

[4]Some students memorize via the mnemonic: Oh Be A Fine Girl Kiss Me!

[5]The Henry Draper catalogue (HD) was published in 1938 - 1924. It contains 225,000 stars extending down to 9 magnitudes. Altogether more than 390,000 stars are published in the Harvard catalogue.

whereas it absorbs the radiations when the temperature of black-body is less than that of the atmosphere around it.

The distribution of energy versus wavelength emitted by a black-body for different values for the temperature of body was investigated experimentally in 1899 by Lummer and Pringsheim. A typical variation of intensity I_λ of radiation versus its wavelength λ is shown in Figure 5 for two different temperatures $T = 120$ K and $T = 100$ K of black-body. The black-body spectrum was found to have the following features.

(i) The intensity I_λ for a given wavelength λ of radiation increases with the increase of temperature T of the body.

(ii) For a given value of temperature T, the intensity I_λ becomes the maximum at some wavelength, denoted by λ_m.

(iii) With the increase of temperature T of the body, the value of λ_m is found to decrease in a manner such that $\lambda_m\, T =$ constant.

(iv) The area under the curve was found proportional to T^4.

In 1900, Max Planck made a theoretical attempt to explain the black-body spectrum. He modified the Wien's formula in such a way that the experimental curves could be fitted precisely. Then, he started looking for a sound basis for his modifications. Planck proposed a hypothesis where he considered that the atoms of the wall of a cavity behave like the oscillators, each of them oscillating with a characteristic frequency. Further, he assumed that these oscillators exchange electromagnetic energy with the cavity and for these atomic oscillators, he made the following assumptions:

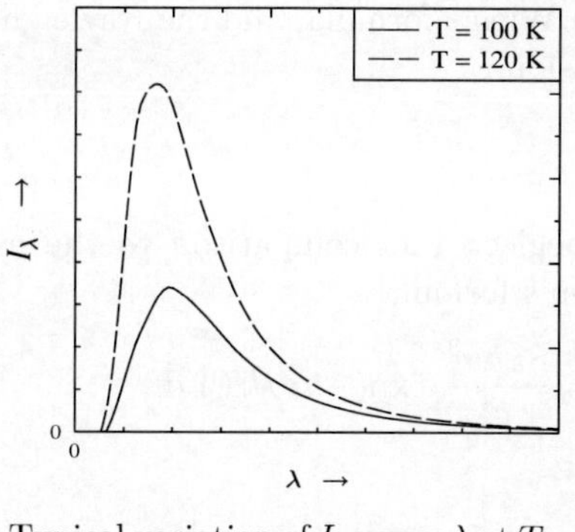

Figure 5: Typical variation of I_λ versus λ at $T = 100$ K and 120 K.

(i) An oscillator with a characteristic frequency ν, can have energies $E_n = nh\nu$, where h is the Planck constant and the integer n can have integer values 0, 1, 2, 3, ...

(ii) An oscillator can exchange energy with the cavity only in the discrete units. That is, $\Delta E_n = \Delta n\ h\nu$, where $\Delta n = 0,\ 1,\ 2,\ 3,\ \ldots$ Thus, the exchange of energy is in the form of the packets of energy, known as the quanta. These quanta were later on named as the photons.

In the Planck's hypothesis, the average energy $\langle E\rangle$ of an oscillator is[6]

$$\langle E\rangle = \frac{\sum_{n=0}^{\infty} E_n \exp(-E_n/kT)}{\sum_{n=0}^{\infty} \exp(-E_n/kT)}$$

$$= \frac{\sum_{n=0}^{\infty} nh\nu \exp(-nh\nu/kT)}{\sum_{n=0}^{\infty} \exp(-nh\nu/kT)} = \frac{h\nu}{\exp(h\nu/kT)-1}$$

In this hypothesis, the quantization condition invalidates the theorem of the equipartition of energy, which is based on the classical picture. The concept of quantum mechanics are introduced the first time.

We know that the number of oscillators per unit volume in the frequency range from ν to $\nu + \mathrm{d}\nu$ is $(8\pi\nu^2/c^3)\,\mathrm{d}\nu$. Thus, the spectral energy density I_ν is

$$I_\nu\,\mathrm{d}\nu = \frac{8\pi\nu^2}{c^3}\,\mathrm{d}\nu\,\langle E\rangle = \frac{8\pi h\nu^3}{c^3}\frac{1}{\exp(h\nu/kT)-1}\,\mathrm{d}\nu \tag{5.30}$$

Here, we have used the value of $\langle E\rangle$. This relation is known as the Planck's radiation formula, and it explains successfully the black-body spectrum for all the wavelengths. Thus, the energy density I_ν is

$$I_\nu = \frac{8\pi h\nu^3}{c^3}\frac{1}{\exp(h\nu/kT)-1}$$

This is Planck formula for the black-body spectrum.

Special cases

It can be easily shown that the Wien's formula, and the Rayleigh and Jeans formula are special cases of this expression.

(i) Wien's formula

When $h\nu >> kT$, one can neglect 1 in comparison to the exponential term in denominator and we have Wien's formula.

$$I_\nu\,\mathrm{d}\nu = \frac{8\pi h\nu^3}{c^3}\exp(-h\nu/kT)\,\mathrm{d}\nu$$

[6]Using $x = \exp(-h\nu/kT)$, we have

$$\langle E\rangle = \frac{h\nu\sum_{n=0}^{\infty} nx^n}{\sum_{n=0}^{\infty} x^n} = \frac{h\nu[x+2x^2+3x^3+4x^4+\ldots]}{1+x+x^2+x^3+x^4+\ldots}$$

$$= \frac{h\nu x[1+2x+3x^2+4x^3+\ldots]}{1+x+x^2+x^3+x^4+\ldots} = \frac{h\nu x(1-x)^{-2}}{(1-x)^{-1}} = \frac{h\nu x}{1-x}$$

Using the expression for x, we have

$$\langle E\rangle = \frac{h\nu\exp(-h\nu/kT)}{1-\exp(-h\nu/kT)} = \frac{h\nu}{\exp(h\nu/kT)-1}$$

(ii) Rayleigh and Jeans formula

When $h\nu << kT$, the exponential term can be expanded and we have

$$I_\nu \, d\nu = \frac{8\pi h\nu^3}{c^3}\left[1 + \frac{h\nu}{kT} + \frac{1}{2!}\left(\frac{h\nu}{kT}\right)^2 + \ldots - 1\right]^{-1} d\nu$$

$$= \frac{8\pi kT\nu^2}{c^3}\left[1 + \frac{1}{2!}\frac{h\nu}{kT} + \ldots\right]^{-1} d\nu$$

On neglecting the small terms, we get Wien's formula.

$$I_\nu \, d\nu = \frac{8\pi kT\nu^2}{c^3} \, d\nu$$

Planck's radiation formula in terms of wavelengths

Planck's radiation formula in terms of wavelengths can be obtained by using the relation $\nu = c/\lambda$ and $d\nu = -c \, d\lambda/\lambda^2$ in equation (5.30) as the following.

$$I_\lambda \, d\lambda = \frac{8\pi ch}{\lambda^5} \frac{1}{\exp(hc/\lambda kT) - 1} \, d\lambda$$

Since we are concerned with the intervals in frequency and wavelength, the negative sign is not considered. Thus, the energy density I_λ is

$$I_\lambda = \frac{8\pi ch}{\lambda^5} \frac{1}{\exp(hc/\lambda kT) - 1} \tag{5.31}$$

Special cases

It can be easily shown that the Wien's formula, and the Rayleigh and Jeans formula are special cases of this expression.

(i) Wien's formula

When $(hc/\lambda) >> kT$, one can neglect 1 in comparison to the exponential term in the denominator and we have the Wien's formula.

$$I_\lambda \, d\lambda = \frac{8\pi ch}{\lambda^5} \exp(-hc/\lambda kT) \, d\lambda$$

(ii) Rayleigh and Jeans formula

When $(hc/\lambda) << kT$, the exponential term can be expanded and we have

$$I_\lambda \, d\lambda = \frac{8\pi ch}{\lambda^5}\left[1 + \frac{hc}{\lambda kT} + \frac{1}{2!}\left(\frac{hc}{\lambda kT}\right)^2 + \ldots - 1\right]^{-1} d\lambda$$

$$= \frac{8\pi kT}{\lambda^4}\left[1 + \frac{1}{2!}\frac{hc}{\lambda kT} + \ldots\right]^{-1} d\lambda$$

On neglecting the small terms, we get the Rayleigh and Jeans formula.

$$I_\lambda \, d\lambda = \frac{8\pi kT}{\lambda^4} \, d\lambda$$

8.1 Wien's displacement law

For a given temperature T, if λ_m be the wavelength corresponding to the maximum value of I_λ, then we have

$$\left(\frac{\partial I_\lambda}{\partial \lambda}\right)_{\lambda=\lambda_m} = 0 \tag{5.32}$$

For convenience, we take logarithm of equation (5.31), and get

$$\ln I_\lambda = \ln\,(8\pi ch) - 5\,\ln\lambda - \ln[\exp(hc/\lambda kT) - 1]$$

Differentiation of this equation with respect to λ gives

$$\frac{1}{I_\lambda}\frac{\partial I_\lambda}{\partial \lambda} = -\frac{5}{\lambda} + \frac{hc}{\lambda^2 kT}\,\frac{\exp(hc/\lambda kT)}{\exp(hc/\lambda kT) - 1}$$

On applying the condition (5.32), we get

$$-5 + \frac{hc}{\lambda_m kT}\,\frac{\exp(hc/\lambda_m kT)}{\exp(hc/\lambda_m kT) - 1} = 0$$

Putting $x = hc/\lambda_m kT$, we have

$$\frac{x\,e^x}{e^x - 1} = 5 \qquad \text{or} \qquad \frac{x}{1 - e^{-x}} = 5$$

This equation can be solved numerically to get $x = 4.9651$. Thus, we have

$$\frac{hc}{\lambda_m kT} = 4.9651 \qquad \text{or} \qquad \lambda_m T = 2.898 \times 10^{-3}\ \text{m K}$$

where we have used $h = 6.62 \times 10^{-34}$ J s, $k = 1.38 \times 10^{-23}$ J/K and $c = 3 \times 10^8$ m/s. This relation is known as the Wien's displacement law.

Example 11. At what wavelength the Sun emits maximum number of photons. Surface temperature of the Sun is 6000 K.

Solution: According to the Wien's displacement law we have

$$\lambda_m T = 2.898 \times 10^{-3} \qquad \text{or} \qquad \lambda_m\, 6000 = 2.898 \times 10^{-3}$$

On solving, we have $\lambda_m = 4.83 \times 10^{-7}$ m = 4830 Å. Thus, the largest number of photons emitted from the Sun is of wavelength λ = 4830 Å.

Example 12. The spectrum of Moon gives two maxima at the wavelengths λ = 4800 Å and 14 micron. What conclusions can be drawn from this.

Solution: Two maxima in the spectrum show that the radiations are mixture from two sources at different temperatures. From the Wien's displacement law we have

$$\lambda_m T = 2.898 \times 10^{-3} \qquad \text{or} \qquad T = \frac{2.898 \times 10^{-3}}{\lambda_m}$$

Thus, we have

$$T_1 = \frac{2.898 \times 10^{-3}}{4800 \times 10^{-10}} = 6037.5 \text{ K} \qquad T_2 = \frac{2.898 \times 10^{-3}}{14 \times 10^{-6}} = 207 \text{ K}$$

$T_1 = 6037.5$ K is the temperature of Sun's surface and $T_2 = 207$ K is the temperature of Moon's surface. Thus, the spectrum contains the radiations emitted by the Moon itself and those emitted by the Sun and reflected by the Moon's surface.

Example 13. Calculate the wavelength of radiations emitted from the body of a person whose temperature is 36.89° C.

Solution: Temperature of the body of person = 36.89 + 273 = 309.89 K. The wavelength of radiations emitted from the body of person using Wien's displacement law is

$$\lambda_m = \frac{2.898 \times 10^{-3}}{T} = \frac{2.898 \times 10^{-3}}{309.89} = 9.3517 \times 10^{-6} \text{ m} = 93517 \text{ Å}$$

This wavelength lies in the far infrared region. Therefore, the radiations from the body of a person cannot be used to see the person. However, a person can be detected with the help of a detector that can detect the radiations of wavelength 93517 Å.

8.2 Stefan-Boltzmann law

The energy density I of all possible radiations for various wavelengths in a black-body enclosure at temperature T is

$$I = \int_0^\infty I_\lambda \, \mathrm{d}\lambda = \int_0^\infty \frac{8\pi ch}{\lambda^5} \frac{1}{\exp(hc/\lambda kT) - 1} \, \mathrm{d}\lambda$$

Putting $x = hc/\lambda kT$, we have

$$I = \frac{8\pi k^4 T^4}{h^3 c^3} \int_0^\infty \frac{x^3}{\mathrm{e}^x - 1} \, \mathrm{d}x$$

We know that

$$\int_0^\infty \frac{x^3}{\mathrm{e}^x - 1} \, \mathrm{d}x = \frac{\pi^4}{15}$$

Therefore, we have

$$I = \frac{8\pi k^4 T^4}{h^3 c^3} \frac{\pi^4}{15}$$

The total emissive power E is related to I at the same temperature T as $E = cI/4$ and therefore, we have

$$E = \frac{2\pi^5 k^4}{15 h^3 c^2} T^4 \qquad \text{or} \qquad E = \sigma T^4$$

where $\sigma = 2\pi^5 k^4/15h^3c^2 = 5.67 \times 10^{-8}$ J/m^2 s K^4 is the Stefan-Boltzmann constant and the relation is known as the Stefan-Boltzmann law.

Note: Planck's explanation of black-body radiation provided a crucial step in the development of quantum concept in Physics, that the energy of an electromagnetic radiation is not continuous, but discrete in the form of packets. This hypothesis of quantization of energy was put on a firm basis by the pioneering work of Einstein, Bohr and others later on. The concept of quantization of energy was however so radical that even Planck himself was reluctant to accept it.

Example 14. Total amount of energy radiated by a star per unit surface area per unit time is 16×10^4 times that radiated by the Sun. Estimate the surface temperature of the star when the surface temperature of Sun is 6000 K.

Solution: Total amount of energy E radiated per unit surface area per second by a body having surface temperature T is

$$E = \sigma T^4$$

Thus, for the star having surface temperature T_1 we have

$$E_1 = \sigma T_1^4$$

and for the Sun having surface temperature T_2 we have

$$E_2 = \sigma T_2^4$$

Therefore,

$$\frac{E_1}{E_2} = \frac{T_1^4}{T_2^4} \qquad \text{or} \qquad 16 \times 10^4 = \frac{T_1^4}{(6000)^4}$$

On solving, we have $T_1 = 1.2 \times 10^5$ K. Thus, the temperature of the surface of star is 1.2×10^5 K.

Example 15. Two bodies at the temperatures 227° C and 427° C are placed in a vessel maintained at the temperature 27° C. What is the ratio of losses of energy by the bodies.

Solution: The temperatures of the bodies are

$$T_1 = 227 + 273 = 500 \text{ K} \qquad \text{and} \qquad T_2 = 427 + 273 = 700 \text{ K}$$

The temperatures of environment inside the vessel

$$T_0 = 27 + 273 = 300 \text{ K}$$

The rates of energy losses are

$$E_1 = \sigma(T_1^4 - T_0^4) = \sigma(500^4 - 300^4)$$

and

$$E_2 = \sigma(T_2^4 - T_0^4) = \sigma(700^4 - 300^4)$$

Thus, the ratio of the energy losses is

$$\frac{E_1}{E_2} = \frac{(500^4 - 300^4)}{(700^4 - 300^4)} = 0.234$$

Example 16: An infrared telescope with an aperture area of 1.0 m^2 is placed in an orbit about the Earth, where the temperature of telescope is $-73°$ C. At what wavelength does the telescope emit most radiation, how much radiation does it emit and how much radiation does it collect from a sun-like star situated at 4 ly away. Given, $\sigma = 5.7 \times 10^{-8}$ J/m^2 s K^{4}i, luminosity of sun = 4×10^{26} J/s.

Solution: Consider a telescope to behave in a similar way as a black-body. The temperature of the telescope is 200 K and so the peak output wavelength is

$$\lambda_m = \frac{2.9 \times 10^{-3}}{T} = \frac{2.9 \times 10^{-3}}{200} = 1.5 \times 10^{-5} \text{ m}$$

The power radiated per unit area is

$$E = \sigma T^4 = (5.7 \times 10^{-8}) \times 200^4 \approx 100 \text{ W/m}^2$$

The telescope emits about 100 W/m^2 at a peak wavelength of 15 μm in the middle of the infrared part of electromagnetic spectrum. The star's power (4×10^{26} W) is radiated equally in all directions and so, at the telescope star separation, the power is spreaded over a sphere with a diameter of 8 ly ($\approx 8 \times 10^{16}$ m) and therefore,

$$\text{Power per unit area} = \frac{\text{Total power}}{\text{Total area}} = \frac{4 \times 10^{26}}{\pi(8 \times 10^{16})^2} = 2.0 \times 10^{-8} \text{ W/m}^2$$

The telescope collects 2.0×10^{-8} W/m^2 from the star as it has an aperture area of 1 m^2.

Note: The difficulty of observing in the infrared radiations is shown in this example. The telescope itself is radiating more power than can be collected from a nearby star.

9. Doppler red-shift

Following the Big Bang theory, the universe is expanding. Hence, each of the cosmic objects is receding away from us. Consequently, the spectrum from a cosmic object shows the Doppler red-shift. Suppose, λ_0 is the wavelength of a particular transition when measured in a laboratory on the Earth and λ the wavelength of the same transition when received from a cosmic object, receding away from us with velocity v. Following the Doppler effect, we have

$$\frac{\lambda}{\lambda_0} = \left(\frac{1 + v/c}{1 - v/c}\right)^{1/2}$$

The red-shift z is described as

$$z = \frac{\lambda - \lambda_0}{\lambda_0} \qquad \text{or} \qquad 1 + z = \frac{\lambda}{\lambda_0}$$

The wavelength of radiation from the galaxy is

$$\lambda = \lambda_0(1+z)$$

Thus, the relation between the red-shift z and receding velocity v is

$$1+z = \left(\frac{1+v/c}{1-v/c}\right)^{1/2} \tag{5.33}$$

9.1 For non-relativistic velocity

For a non-relativistic velocity, we have $v << c$, and from equation (5.33), we have

$$1+z = \left(1+\frac{v}{c}\right)^{1/2}\left(1-\frac{v}{c}\right)^{-1/2} = \left(1+\frac{v}{2c}\right)\left(1+\frac{v}{2c}\right)$$

Here, we have neglected smaller terms. Now, we have

$$1+z = 1+\frac{v}{c}+\frac{v^2}{4c^2} = 1+\frac{v}{c}$$

Here, also, we have neglected smaller term. Thus, we have

$$z = \frac{v}{c}$$

For a non-relativistic motion, the red-shift is proportional to the velocity of the object.

Example 17: A distant galaxy is receding away with a velocity $0.5c$, where c denotes the speed of light. Calculate the red-shift. What is the wavelength of 21-cm radiation of hydrogen atom, when received from the galaxy.

Solution: The red-shift z is

$$1+z = \left(\frac{1+v/c}{1-v/c}\right)^{1/2}$$

Therefore,

$$1+z = \left(\frac{1+0.5}{1-0.5}\right)^{1/2} \qquad \text{or} \qquad z = \sqrt{3}-1 = 0.732$$

The wavelength of 21-cm radiation received from the cosmic object is

$$\lambda = \lambda_0(1+z) = 21(1+0.732) = 36.37 \text{ cm}$$

Example 18: Observations of Sirius show it to be an A1 star with an apparent magnitude of -1.46 and a parallax of $0.37''$. Calculate the size of the Sirius.

Solution: A parallax of $0.37''$ implies that Sirius lies at a distance of (1/0.37 =) 2.7 parsec. The absolute magnitude of Sirius is

$$M = m - 5\log_{10}d + 5 = -1.46 - 5\log_{10}2.7 + 5 = 1.38$$

The absolute magnitude of Sirius is expressed using the standard relative scale but an absolute scale is required when the actual size of Sirius is to be calculated. The Sun's power output is 4 $\times 10^{26}$ W and the absolute magnitude is 4.9. The relative intensity of Sirius with respect to the Sun can be found as

$$I_{Si} = 10^{(M_{\odot} - M_{Si})/2.5} I_{\odot} = 10^{(4.9-1.38)/2.5}(4 \times 10^{26}) = 1.0 \times 10^{28} \text{ W}$$

As the Sirius is A1 star, its surface temperature must be about 11000 K. So, the power radiated per unit surface area is

$$P = \sigma T^4 = (5.7 \times 10^{-8}) \times 11000^4 = 8.3 \times 10^8 \text{ W/m}^2$$

Total surface area of Sirius is

$$\text{surface area} = \frac{1.0 \times 10^{28}}{8.3 \times 10^8} = 1.2 \times 10^{19} \text{ m}^2$$

For the radius r of Sirius, we have

$$4\pi r^2 = 1.2 \times 10^{19} \qquad r = \sqrt{\frac{1.2 \times 10^{19}}{4\pi}} = 9.8 \times 10^8 \text{ m}$$

10. Relation between B - V magnitude and surface temperature

For a star, we can calculate magnitudes corresponding to various colours, for example, corresponding to the blue (B), red (R), violet (V) colours. The temperature of a star is related to the difference $(B - V)$ in the magnitudes. Table 3 shows the temperature of a star corresponding to the value of $(B - V)$. The variation of $(B - V)$ versus temperature (K) is shown in Figure 6. Thus, for a known value of the difference $(B - V)$ for a star, the surface temperature of star can be obtained with the help of Figure 6 or through the numerical interpolation, using the data given in Table 3.

Table 3. (*B-V*) versus temperature

B-V	Temperature (K)
-0.31	34,000
-0.24	23,000
-0.20	18,500
-0.12	13,000
0.0	9500
0.15	8500
0.29	7300
0.42	6600
0.58	5900
0.69	5600
0.85	5100
1.16	4200
1.42	3700
1.61	3000

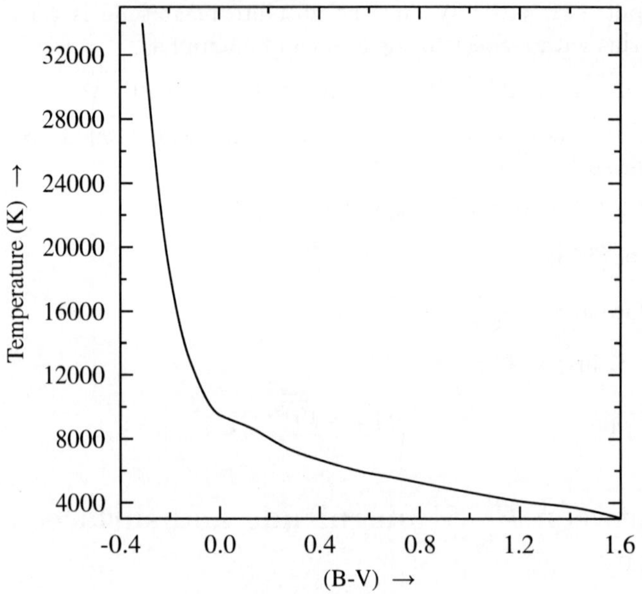

Figure 6: Variation of surface temperature (K) versus the magnitude difference $(B - V)$.

10.1 Colour-magnitude diagram

For the stars, the plot of B-V versus M_v is known as the colour-magnitude diagram.

11. Determination of radius of a star

Radius of a star cannot be measured directly, as even a big star is so far away from us that it appears as a single point from the Earth. However, there is an indirect method for determination of radius of a star. For a star having surface temperature T and radius R (assuming spherical in shape), the luminosity L is expressed as

$$L = 4\pi R^2 \sigma T^4 \tag{5.34}$$

where σ is the Stefan-Boltzmann constant. For our Sun, if temperature, radius and luminosity are denoted by $T_\odot$, $R_\odot$, $L_\odot$, respectively. Then, we have

$$L_\odot = 4\pi R_\odot^2 \sigma T_\odot^4 \tag{5.35}$$

The parameters for the Sun are known with a good accuracy. On dividing equation (5.34) by (5.35), we get

$$\frac{L}{L_\odot} = \left(\frac{R}{R_\odot}\right)^2 \left(\frac{T}{T_\odot}\right)^4 \qquad \text{or} \qquad \frac{R}{R_\odot} = \left(\frac{T_\odot}{T}\right)^2 \left(\frac{L}{L_\odot}\right)^{1/2}$$

Therefore, the radius of the star is

$$R = R_\odot \left(\frac{T_\odot}{T}\right)^2 \left(\frac{L}{L_\odot}\right)^{1/2} \tag{5.36}$$

From the spectrum of a star, the difference $(B - V)$ can be obtained, and corresponding temperature of the star can be determined. For finding the ratio $L/L_\odot$, we can use the absolute magnitudes of the stars. The $L/L_\odot$ can be found from the equation

$$\frac{L}{L_\odot} = 10^{0.4(M_\odot - M)},$$

where $M_\odot$ and M are the absolute magnitudes of the Sun and star, respectively. Using the values in equation (5.36), the radius of star can be determined.

Example 19: For the star Sirius, the visual magnitude is -1.44 and $(B - V) = 0.009$. The parallax of Sirius is 379.21 milli arc seconds. Calculate the radius of Sirius. Given, temperature of Sun = 5800 K, absolute magnitude of Sun = 4.83, the magnitude difference $(B - V) = .009$ corresponds to the temperature of 9500 K.

Solution: We have parallax $p = 0.37921''$. The distance d of Sirius is

$$d = \frac{1}{p} = \frac{1}{0.37921} = 2.637 \text{ pc}.$$

The absolute magnitude of Sirius is

$$M = m - 5 \log d + 5 = -1.44 - 5 \log(2.637) + 5 = 1.454$$

We know that the temperature of Sun is 5800 K. From the chart, the temperature of Sirius is about 9500 K. Our Sun has an absolute magnitude of 4.83. The difference in the magnitudes is 3.376. Putting the values, we get

$$\frac{R}{R_\odot} = \left(\frac{T_\odot}{T}\right)^2 \left(\frac{L}{L_\odot}\right)^{1/2} = \left(\frac{5800}{9500}\right)^2 \left(10^{0.4 \times 3.376}\right)^{1/2} = 1.764$$

Sirius has a radius approximately 1.764 times that of Sun.

12. Hertzsprung-Russell diagram

A diagram obtained by plotting the luminosity L of the stars versus their effective temperature T_e is known as the Hertzsprung-Russell (H-R) diagram. Let us plot $\log(L/L_\odot)$ on the ordinate and $\log(T_e)$ on the abscissa, where L and $L_\odot$ denote the luminosities of a star and the sun, respectively, and T_e is the effective surface temperature of the star (Figure 7). The luminosity on the ordinate is increasing from the down to the upward direction, whereas the effective surface temperature is decreasing from the left towards the right direction. The left side of diagram has hot stars, having extreme surface temperatures up to nearly 10^5 K. The right side has cool stars. At the top of the diagram, the stars are bright whereas at the bottom, the stars are faint.

Such diagram plays basic role in A&A, as a number of properties of stars can be inferred and discussed in terms of their locations in the diagram. In a diagram, most of the stars are found lying on a diagonal band, extending from the corner corresponding to the high L and high T_e (top-left) to the corner corresponding to the low L and low T_e (bottom-right). This band is commonly known as the main sequence. The stars on the main sequence are found to have the following common features:

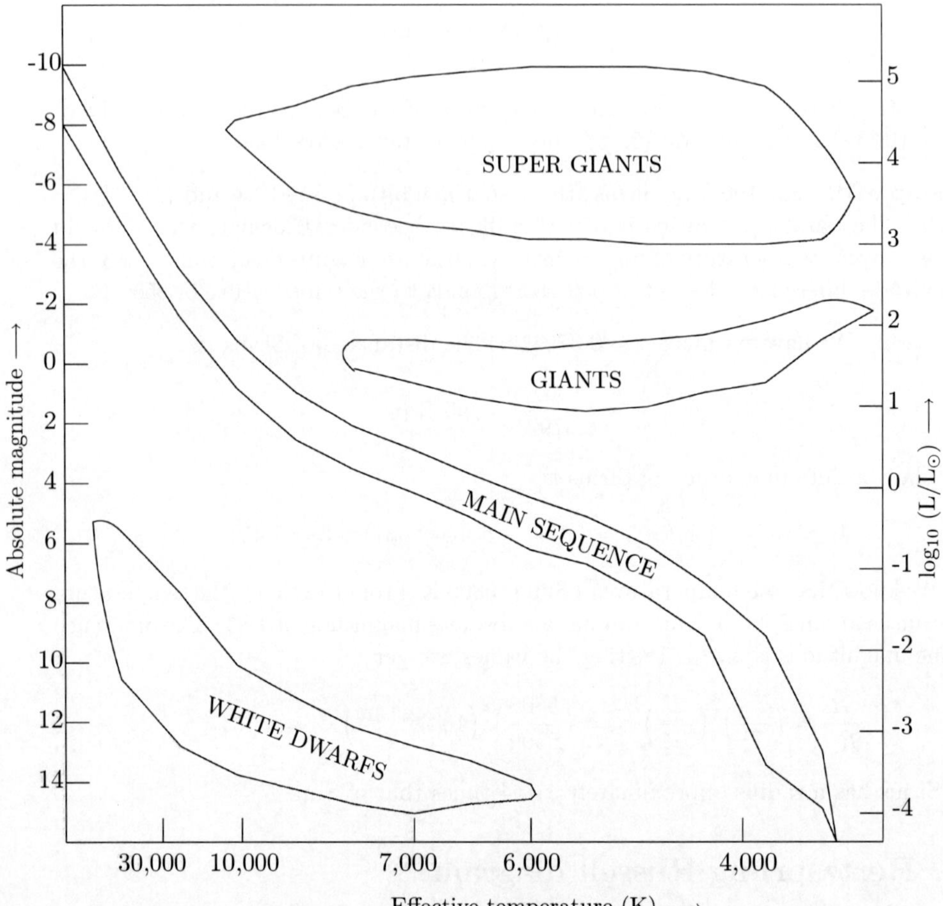

Figure 7: Hertzsprung-Russell diagram

(i) The stars are chemically homogeneous.

(ii) In the core of the stars, hydrogen is burning into helium. Fusion of four hydrogen nuclei into helium nucleus is the source of energy generated in the stars.

(iii) These stars having homogeneous distribution of hydrogen, helium, and other heavy elements can steadily shine for the longest period of their luminous lifetime.

In spite of these common features of the stars lying on the main sequence, there has to be some characteristics which make the stars to lie on different positions on the main sequence. One possibility could be the initial fractional amount of the elements heavier than hydrogen and helium in the stars. The gross properties of the main sequence stars are found to change relatively little on the basis of the initial fractional amount of heavier elements. This rules out the possibilities of initial fractional amount of heavier elements to be the reason for various positions of the stars in the main sequence.

The second possibility may be that the stars may not be of the same age and therefore the amount of hydrogen which has been burnt into helium is different for various stars. Estimates have shown that the ratio of hydrogen to helium in the star can shift its position a little. However, as long as the core hydrogen is not all used up, the star lies inside the band.

The major factor which is found responsible for the position of a star in the main sequence is its mass. Thus, it is the mass of a star which determines its position in the main sequence of the H-R diagram. The high-mass stars spend energy much faster than the low-mass stars. All the stars on the main sequence are found to have their luminosity L roughly proportional to the fourth power of the mass of star

$$L \propto M^4 \qquad \text{or} \qquad L = kM^4$$

where k is proportionality constant. However, this relation is not valid for very heavy and very light stars. For such stars, it varies as

$$L \propto M^{5.5}/R^{1/2} \quad \text{for low to medium mass stars}$$

$$L \propto M^3 \quad \text{for high mass stars}$$

$$L \propto M \quad \text{for very high mass stars}$$

Let us now discuss the consequence of this relation between the luminosity and mass of a star in the main sequence. Suppose, we have two stars, A and B, on the main sequence of masses 1 $M_\odot$ and 10 $M_\odot$, respectively. The ratio of their luminosities, denoted by L_A and L_B, respectively, is

$$\frac{L_B}{L_A} = \left(\frac{10M_\odot}{1M_\odot}\right)^4 = 10^4$$

The time t spent on the main sequence by the star is

$$t = \frac{M}{L} = \frac{M}{kM^4} = k'M^{-3} \qquad \text{or} \qquad t \propto M^{-3}$$

It shows that larger the mass of star, lesser the time spent on the main sequence. The ratio of time spent by the stars A and B on the main sequence is

$$\frac{t_B}{t_A} = \frac{(10M_\odot)^{-3}}{(1M_\odot)^{-3}} = 10^{-3}$$

Thus, the lifetime of the star B on the main sequence is one-thousandth of that of the star A. Assuming that the lifetime of the Sun on the main sequence be 10^{10} years, the star of 10 $M_\odot$ would stay on the main sequence for a period of 10^7 years.

Example 20: The luminosities of two stars A and B, having masses 1 $M_\odot$ and 10 $M_\odot$, respectively, vary as cube of its mass. Calculate the ratio of their luminosities and periods they spend on the main sequence of H-R diagram.

Solution: We have

$$L \propto M^3 \qquad \text{and} \qquad L = kM^3$$

where k is a constant. The ratio of luminosities of the stars is

$$\frac{L_B}{L_A} = \frac{k(10M_\odot)^3}{k(1M_\odot)^3} = 10^3$$

The time spent on the main sequence is

$$t = \frac{M}{L} = \frac{M}{kM^3} = \frac{M^{-2}}{k}$$

Thus, the ratio of times spent by the stars on the main sequence is

$$\frac{t_B}{t_A} = \frac{(10M_\odot/k)^{-2}}{(1M_\odot/k)-2} = 10^{-2}$$

Example 21: Obtain the law of variation of luminosity with mass of a star when two stars having different masses spend equal time on the main sequence.

Solution: Consider two stars A and B having different masses. Suppose, the mass of star B is β times that of star A and the law of variation of luminosity with the mass of star is

$$L \propto M^\alpha \qquad \text{and} \qquad L = kM^\alpha$$

where k and α are constants. The time spent on the main sequence is

$$t = \frac{M}{L} = \frac{M}{kM^\alpha} = \frac{M^{1-\alpha}}{k}$$

Thus, the ratio of times spent by the stars on the main sequence is

$$\frac{t_B}{t_A} = \frac{(M_B/k)^{1-\alpha}}{(M_A/k)^{1-\alpha}} = \left(\frac{M_B}{M_A}\right)^{1-\alpha}$$

Using $t_B/t_A = 1$ and $M_B/M_A = \beta$, we have

$$1 = \beta^{1-\alpha}$$

It gives $\alpha = 1$. Thus, the law is $L \propto M$.

Example 22: Luminosity L of a star varies as fourth power of its mass M. For two stars A and B, the mass of the first one is 10 times larger than that of the second. Calculate the ratio of their luminosities.

Solution: Given, $M_A = 10M_B$. The ratio of luminosities is

$$\frac{L_A}{L_B} = \frac{M_A^4}{M_B^4} = 10^4$$

13. Pauli exclusion principle

On the basis of their spin, the particles as well as atoms/ions are classified into two categories: (i) Fermi particles (Fermions) and (ii) Bose particles (Bosons). The particles having half-integer spin ($I = 1/2$, $3/2$, etc.) are known as the Fermi particles and they follow the Fermi-Dirac statistics. Some examples of Fermi particles, are the electron, proton, neutron, μ-meson, $_2He^3$ nucleus. The Pauli exclusion principle is applicable for the Fermi particles.

The particles having integer spin ($I = 0$, 1, 2, etc.) are known as the Bose particles and they follow the Bose-Einstein statistics. Some examples of Bose particles, are the photon, π-meson, $_2He^4$ nucleus. These particles do not obey the Pauli exclusion principle.

The Pauli exclusion principle states that no two fermions can occupy the same quantum state. In the quantum mechanics, a state is described by a set of four quantum numbers, n, l, m_l, m_s. The principal quantum numbers n can assume the value as a non-zero, positive integer, 1, 2, 3, etc. For a given value of n, the azimuthal quantum number l can assume the integer values from 0 to $(n-1)$ with the increment of 1. For example, for $n = 2$, the l can assume the values, 0 and 1. For a given value of n, the number of l-values is n.

For a given value of l, the magnetic quantum numbers m_l can assume the integer values from $-l$ to l with the increment of 1. For example, for $l = 2$, the m_l can assume the values, -2, -1, 0, 1, 2. For a given value of l, the number of m_l-values is $(2l + 1)$. The spin s of a Fermi particle can have half-integer value. That is, the spin s of a Fermi particle can assume the values 1/2, 3/2, 5/2, ... For a given s, the m_s can assume the values from $-s$ to s with the increment of 1.

Each of the electron, proton, neutron has spin $s = 1/2$. Hence, the m_s can assume two values, $-1/2$ and $1/2$. Some quantum states for the fermions with $s = 1/2$ are given in Table below. For a given principal quantum number n, the number of total possible states is $2n^2$.

Quantum states			
n	l	m_l	m_s
1	0	0	+1/2
1	0	0	-1/2
2	0	0	+1/2
2	0	0	-1/2
2	1	-1	+1/2
2	1	-1	-1/2
2	1	0	+1/2
2	1	0	-1/2
2	1	1	+1/2
2	1	1	-1/2

Each quantum state has corresponding energy. Some of the energies may be equal (degenerate state). Since the Pauli exclusion is obeyed by the fermions, even at 0 K temperature, all the fermions in a system cannot cascade down to the ground state. The highest level which is filled at 0 K temperature is known as the Fermi-level. With the increase of temperature, some particles from the states below the Fermi-level move to the states above the Fermi-level. Since all the fermions cannot be brought to the ground state, there is a pressure in the upward direction, called the **degeneracy pressure**. The electron degeneracy plays important role for the white dwarfs whereas the neutron degeneracy plays important role for the neutron stars.

14. Degenerate matter

Degenerate matter in physics is a collection of free, non-interacting Fermi-particles with a pressure and other physical characteristics determined by the quantum mechanical effects. The degenerate state of matter arises at an extraordinarily high density found in compact stars. It occurs for the matter particles such as electrons, neutrons, protons, and fermions in general and is referred to as the electron-degenerate matter, neutron-degenerate matter, etc. In a mixture of particles, such as ions and electrons in a white dwarf, the electrons may be degenerate, while the ions are not.

In quantum mechanical description, in a bound state, the energy levels are quantized. That is, the energies of the levels are discrete and are called the quantum states. The Pauli exclusion principle states that no two identical fermions can occupy the same quantum state. Consequently, even at a negligibly small thermal energy of the particles, all the lowest energy quantum states are filled. This state is referred to as the full degeneracy. The pressure (called the degeneracy pressure or Fermi pressure) remains finite even near absolute zero temperature. Addition of particles or reduce of volume, forces the particles into higher-energy quantum states. This requires a compression force, and is made manifest as a resisting pressure. The key feature is that this degeneracy pressure does not depend on the temperature and depends only on the density of fermions. It keeps dense stars in equilibrium irrespective of its thermal structure. The degeneracy pressure acts against the gravitational collapse.

A degenerate matter is also called a Fermi gas or a degenerate gas. A degenerate state with velocities of the fermions, close to the speed of light, is known as the relativistic degenerate matter.

Degenerate gases are composed of fermions at high densities. A fermion gas in which all the energy states below some energy level are filled is called a fully degenerate fermion gas. The electron gas in the interior of white dwarf stars constitute two examples of a degenerate electron gas. Most of the stars are supported against their own gravitation by normal thermal gas pressure. A white dwarf star is supported by the degeneracy pressure of electron gas in their interior, while for a neutron star the degenerate particles are neutrons.

14.1 Electron degeneracy

In an ordinary fermion gas in which thermal effects dominate, most of the available electron energy levels are unfilled and the electrons are free to move to these states.

As particle density is increased, electrons progressively fill the lower energy states and additional electrons are forced to occupy states of higher energy even at low temperatures. Degenerate gases strongly resist further compression because the electrons cannot move to already filled lower energy levels due to the Pauli exclusion principle. Since the electrons cannot give up energy by moving to lower energy states, no thermal energy can be extracted. The momentum of the fermions in the fermion gas nevertheless generates pressure, known as the degeneracy pressure.

Under high densities, the matter becomes a degenerate gas when the electrons are all stripped from their parent atoms. In the core of a star, once hydrogen burning in nuclear fusion reactions stops, it becomes a collection of positively charged ions, largely helium and carbon nuclei, floating in a sea of electrons, which have been stripped from the nuclei. Normal gas exerts higher pressure when it is heated and expands, but the pressure in a degenerate gas does not depend on the temperature. When gas becomes super-compressed, particles position right up against each other to produce degenerate gas that behaves more like a solid.

There is an upper limit to the mass of an electron-degenerate object, called the **Chandrasekhar limit**, beyond which electron degeneracy pressure cannot support the object against the gravitational collapse. The limit is approximately 1.44 $M_\odot$ for the objects with compositions similar to the Sun. Above this mass limit, a neutron star (supported by the neutron degeneracy pressure) or a black hole may be formed instead.

14.2 Proton degeneracy

Sufficiently dense matter containing protons experiences proton degeneracy pressure, in a manner similar to the electron degeneracy pressure in the electron-degenerate matter. Because the protons are much more massive than the electrons, the same momentum represents a much smaller velocity for the protons than that for the electrons. As a result, in a matter with approximately equal numbers of protons and electrons, the proton degeneracy pressure is much smaller than the electron degeneracy pressure, and proton degeneracy is usually modeled as a correction to the equations of state of electron-degenerate matter.

14.3 Neutron degeneracy

Neutron degeneracy is similar to the electron degeneracy and is demonstrated in the neutron stars, which are primarily supported by the pressure from a degenerate neutron gas. This happens when a stellar core above 1.44 $M_\odot$, the Chandrasekhar limit, collapses and is not halted by the degenerate electrons. As the star collapses, the Fermi energy of the electrons increases to the point where it is energetically favorable for them to combine with the protons to produce the neutrons. The result of this collapse is an extremely compact star composed of nuclear matter, which is predominantly a degenerate neutron gas with a small admixture of degenerate proton and electron gases.

There is an upper limit to the mass of a neutron-degenerate object, the Tolman-Oppenheimer-Volkoff limit, which is similar to the Chandrasekhar limit for the electron-degenerate objects. The precise limit is unknown, as it depends on the equations of state of the nuclear matter, for which a highly accurate model is not yet available. Above this limit, a neutron star may collapse into a black-hole, or into other, denser forms of degenerate matter.

15. Hydrostatic equilibrium

In the universe, most of the bodies are almost spherical in shape. For example, Sun, planets, satellites, stars, etc., are almost spherical in shape. Consider a spherical body of radius R and mass M. Let us further consider a spherical thin shell of radius r' and thickness $\mathrm{d}r'$. Volume of the shell is

$$\text{Volume of shell} = \frac{4\pi}{3}(r' + \mathrm{d}r')^3 - \frac{4\pi}{3}r'^3 = 4\pi r'^2 \mathrm{d}r' \tag{5.37}$$

where smaller terms are neglected. In general, the density of a body is not uniform. It can, however, be assumed to be radially symmetric, *i.e.*, varying along the radial distance. Suppose, the density of the shell is $\rho(r')$, the mass of the shell is

$$\mathrm{d}M(r') = \rho(r')\, 4\pi r'^2\, \mathrm{d}r' \tag{5.38}$$

Thus, the mass of a spherical volume of radius of r is

$$M(r) = \int_0^r \rho(r')\, 4\pi r'^2\, \mathrm{d}r' \tag{5.39}$$

Thus, from equation (5.38), we have

$$\frac{\mathrm{d}M(r)}{\mathrm{d}r} = 4\pi r^2 \rho(r) \tag{5.40}$$

For convenience, we assume that the density in the body is uniform and is denoted by $\overline{\rho}$, then from equation (5.39), we get

$$M(r) = \frac{4}{3}\pi r^3\, \overline{\rho} \tag{5.41}$$

Similarly, we get

$$M(R) = \frac{4}{3}\pi R^3\, \overline{\rho} \qquad \text{or} \qquad M = \frac{4}{3}\pi R^3\, \overline{\rho} \tag{5.42}$$

and thus,

$$\frac{M(r)}{M} = \frac{4\pi r^3 \overline{\rho}/3}{4\pi R^3 \overline{\rho}/3} \qquad \text{or} \qquad M(r) = M\left(\frac{r}{R}\right)^3 \tag{5.43}$$

We now consider a volume-element with base area $\mathrm{d}A$ and height $\mathrm{d}r$ at a distance r from the center of the sphere, as shown in Figure 8. Mass of the volume-element is

$$dm = \rho(r)\, \mathrm{d}A\, \mathrm{d}r$$

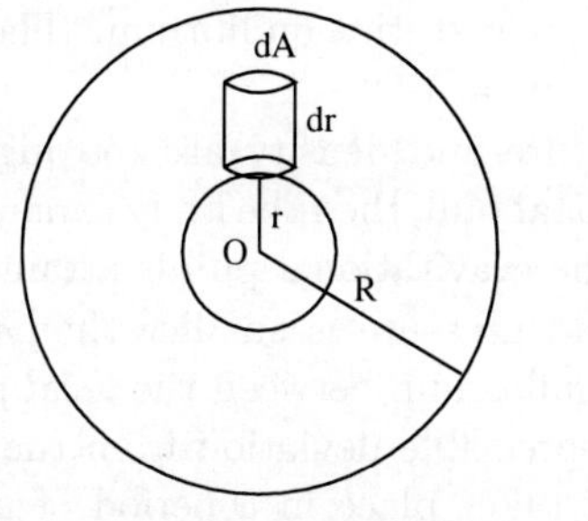

Figure 8: Spherical body with a volume element.

The mass $M(r)$, of sphere of radius r, produces a gravitational acceleration

$$g(r) = G\,\frac{M(r)}{r^2}$$

at its surface (*i.e.*, at a distance of r from its center) in the inward direction. Hence, the inward force on the volume-element is

$$\text{inward force} = \text{mass} \times \text{acceleration}$$

$$= \mathrm{d}m \times g(r) = \rho(r)\ \mathrm{d}A\ \mathrm{d}r\ g(r)$$

This inward force is balanced by some kind of force in the outward direction. This outward force is due to pressure developed inside the body, which decreases radially outward. This pressure depends on the density ρ, temperature T, and mass concentrations X_i of various elements present in the region. Hydrogen and helium mass concentrations are generally denoted by X and Y, respectively. Thus, we have

$$P = P(\rho, T, X, Y, X_i) \qquad i = 1, 2, \ldots, n$$

Here, X_i denote the concentration of elements other than hydrogen and helium.

In the absence of such balancing force (pressure), the body will continue to shrink and finally would reduce to a point. But, it is not the case because of the pressure gradient inside the body. Let P and $P - \mathrm{d}P$ be the pressures at the lower and upper faces of the cylinder. The pressure difference is $\mathrm{d}P$ over the distance $\mathrm{d}r$. The force acting on the volume-element in the outward direction is $\mathrm{d}P\ \mathrm{d}A$. Hence, in the equilibrium state, we have

$$-\mathrm{d}P\ \mathrm{d}A = \rho(r)\ \mathrm{d}A\ \mathrm{d}r\ g(r)$$

The negative sign appears as the directions of two forces are opposite to each other. Therefore,

$$\frac{\mathrm{d}P}{\mathrm{d}r} = -\rho(r)g(r) \quad \text{or} \quad \frac{\mathrm{d}P}{\mathrm{d}r} = -\rho(r)\,\frac{GM(r)}{r^2} \tag{5.44}$$

On using equation (5.43) in (5.44), we get

$$\frac{\mathrm{d}P}{\mathrm{d}r} = -\rho(r)\,\frac{GMr}{R^3} \tag{5.45}$$

This equation represents the hydrostatic equilibrium. The negative sign shows the decrease of pressure radially outward.

In a star, when the temperature and density are too high, the local pressure becomes larger than the gravitational pull, then the body expands till the equilibrium between the local pressure and the gravitational pull is attained.

On the other side, if the local pressure is smaller than the gravitational pull, then the body contracts till an equilibrium between the local pressure and the gravitational pull is attained. Any appreciable deviation from the equilibrium leads to a readjustment of equilibrium that takes place in a period of no more than an hour. Most of the stars have pressure equilibrium throughout unless they pulsate.

Example 23: Calculate the average density of Sun. Given, mass of Sun $M_\odot = 1.99 \times 10^{30}$ kg, radius of Sun $R_\odot = 6.96 \times 10^5$ km.

Solution: Average density of Sun is

$$\overline{\rho} = \frac{M_\odot}{4\pi R_\odot^3/3} = \frac{1.99 \times 10^{30}}{4\pi(6.96 \times 10^8)^3/3} = 1409.8 \text{ kg/m}^3$$

Example 24: Using the hydrostatic equilibrium, calculate the pressure at half the solar radius. Given, average density $\overline{\rho} = 1410$ kg/m^3.

Solution: Suppose, the density of Sun is constant, equal to the average density $\overline{\rho}$. Then the mass within the radius r is

$$M_r = \frac{4}{3}\pi\overline{\rho}r^3$$

and hydrostatic equilibrium can be written as

$$\frac{\mathrm{d}P}{\mathrm{d}r} = -\frac{GM_r\,\overline{\rho}}{r^2}$$

Using the expression for M_r, we have

$$\frac{\mathrm{d}P}{\mathrm{d}r} = -\frac{4\pi G\overline{\rho}^2 r}{3}$$

This equation can be integrated from half the solar radius to the surface, where the pressure varies from P to zero.

$$\int_P^0 \mathrm{d}P = -\frac{4\pi G\overline{\rho}^2}{3}\int_{R_\odot/2}^{R_\odot} r\,\mathrm{d}r \qquad \text{giving} \qquad P = \frac{1}{2}\pi G\overline{\rho}^2 R_\odot^2$$

Using the values, the pressure at half the solar radius is

$$P = \frac{1}{2}\pi(6.67 \times 10^{-11})(1410)^2(6.96 \times 10^8)^2 = 1.0 \times 10^{14} \text{ Pa}$$

This estimate is extremely rough, as the density increases strongly inwards.

15.1 Pressure at the center of a spherical body

In most of the stars, the pressure P is practically equal to the gas pressure P_g. But, in a hot and massive star, it is necessary to include the radiation pressure P_r, so that $P = P_g + P_r$.

If we assume that the density of the object is uniform $\overline{\rho}$, then for a star of radius R and mass M, we have

$$\frac{\mathrm{d}P}{\mathrm{d}r} = -\overline{\rho}\,\frac{GMr}{R^3}$$

For the pressure P_c at the center of star, we have

$$\int_{P_c}^{0} \mathrm{d}P = -\overline{\rho}\,\frac{GM}{R^3}\int_{0}^{R} r\,\mathrm{d}r \qquad \text{or} \qquad -P_c = -\overline{\rho}\,\frac{GM}{R^3}\,\frac{R^2}{2}$$

Thus, the pressure P_c at the center of star is

$$P_c = \frac{\overline{\rho}}{2}\,\frac{GM}{R}$$

This is a rough estimate for the pressure on the center of a spherical body of radius R, mass M and uniform density $\overline{\rho}$. Values of pressure at the center of some celestial bodies are calculated and are given in Table 4.

Table 4. Pressure at the center of body.

Body	Mass M (kg)	Radius R (m)	Density ρ (kg/m^3)	Pressure P_c (N/m^2)
Sun	1.99×10^{30}	6.96×10^{8}	1.40×10^{3}	1.34×10^{14}
Moon	7.35×10^{22}	1.74×10^{6}	3.34×10^{3}	4.71×10^{9}
Mercury	3.28×10^{23}	2.44×10^{6}	5.43×10^{3}	2.44×10^{10}
Venus	4.87×10^{24}	6.05×10^{6}	5.24×10^{3}	1.41×10^{11}
Earth	5.97×10^{24}	6.37×10^{6}	5.52×10^{3}	1.73×10^{11}
Mars	6.39×10^{23}	3.39×10^{6}	3.94×10^{3}	2.48×10^{10}
Jupiter	1.90×10^{27}	6.99×10^{7}	1.33×10^{3}	1.21×10^{12}
Saturn	5.68×10^{26}	5.82×10^{7}	7.00×10^{2}	2.28×10^{11}
Uranus	8.68×10^{25}	2.54×10^{7}	1.19×10^{3}	1.36×10^{11}
Neptune	1.02×10^{26}	2.46×10^{7}	1.66×10^{3}	2.30×10^{11}

Example 25: Estimate the pressure at the center of the Earth. Given, mass of Earth $M = 5.973\times10^{24}$ kg, radius of Earth $R = 6.378\times10^{6}$ m, average density of Earth $\rho = 5.5\ \bar{}\times10^{3}$ kg/m^3 and universal gravitational constant $G = 6.673\times10^{-11}$ m^3/s^2 kg.

Solution: The pressure at the center of Earth is

$$P_c = \frac{\overline{\rho}}{2}\,\frac{GM}{R} = \frac{5500}{2}\,\frac{(6.673\times10^{-11})(5.973\times10^{24})}{6.378\times10^{6}} = 1.7\times10^{11}\ \mathrm{N/m^2}$$

This value of pressure at the center of Earth is about a factor of two smaller than the actual value obtained from the detailed calculations.

Example 26: Estimate the pressure at the center of Sun. Given, mass of Sun $M_\odot$= 1.989 $\times 10^{30}$ kg, radius of Sun $R_\odot = 6.960 \times 10^8$ m, average density of Sun $\overline{\rho} = 1409$ kg/m^3 and universal gravitational constant $G = 6.673 \times 10^{-11}$ m^3/s^2 kg.

Solution: The pressure at the center of Sun is

$$P_c = \frac{\overline{\rho}}{2}\frac{GM}{R} = \frac{1409}{2}\frac{(6.673 \times 10^{-11})(1.989 \times 10^{30})}{6.960 \times 10^8} = 1.3 \times 10^{14}\ \text{N/m}^2$$

Exact model calculations for the Sun yield a central pressure 2.5 $\times 10^{16}$ N/m^2, which is about a factor of 100 larger than the value obtained from the calculations.

16. Thermal equilibrium

When the amount of energy emitted by a body is equal to the energy absorbed by the body, the body is said to be in the thermal equilibrium. In case of of non-thermal equilibrium the body warms up (absorbs energy) or cools down (emits energy) until the thermal equilibrium is attained.

16.1 Body hotter than atmosphere

Suppose, a hot body of temperature T_1 is placed in an atmosphere of temperature T_0, such that $T_1 > T_0$. The body gives energy to the atmosphere till both of them attain a common temperature. Since the content of atmosphere is quite large, there is no significant increase in temperature. That is, the temperature of the atmosphere remains T_0. If T is the temperature of body at time t, the rate of cooling is

$$-\frac{dT}{dt}$$

Now, the rate of cooling of the body is proportional to the temperature difference and therefore, we have

$$-\frac{dT}{dt} \propto (T - T_0) \qquad \text{or} \qquad -\frac{dT}{dt} = k(T - T_0)$$

where k, having dimension of s^{-1}, is proportionality constant. This equation can be rearranged as

$$\frac{dT}{(T - T_0)} = -k\ dt$$

On integration, we get

$$\ln(T - T_0) = -kt + c \tag{5.46}$$

The constant of integration c can be obtained by applying the condition that at $t = 0$, we have $T = T_1$ and therefore, we have

$$\ln(T_1 - T_0) = c \tag{5.47}$$

On substituting the value of c from equation (5.47) in (5.46), we have

$$\ln(T - T_0) = -kt + \ln(T_1 - T_0) \qquad \text{or} \qquad \ln\left(\frac{T - T_0}{T_1 - T_0}\right) = -kt$$

On taking anti-logarithm, we get

$$\frac{T - T_0}{T_1 - T_0} = \mathrm{e}^{-kt} \qquad \text{or} \qquad T = T_0 + (T_1 - T_0)\,\mathrm{e}^{-kt} \tag{5.48}$$

This relation gives the temperature T of body after a time t.

Note: Here, it does not matter if the temperatures are expressed in Celsius or Kelvin unit.

Example 27: A body having temperature of 127 C is placed in an atmosphere having temperature of 27 C. Considering $k = 0.005\ \mathrm{s}^{-1}$, calculate the temperature of body after 5 min.

Solution: We have $T_1 = 127 + 273 = 400$ K, $T_0 = 27 + 273 = 300$ K, $t = 5 \times 60 = 300$ s, and $k = 0.005\ \mathrm{s}^{-1}$. The temperature of body after 300 s is

$$T = T_0 + (T_1 - T_0)\mathrm{e}^{-kt} = 300 + (400 - 300)\ \mathrm{e}^{-0.005 \times 300} = 322.31\ \mathrm{K}$$

Temperature of the body is 322.31 – 273 = 49.31 C.

16.2 Body cooler than atmosphere

Suppose, a cool body of temperature T_1 is placed in an atmosphere of temperature T_0, such that $T_1 < T_0$. The body takes energy from the atmosphere till both of them attain a common temperature. Since the content of atmosphere is quite large, there is no significant decrease in temperature. That is, the temperature of the atmosphere remains T_0. If T be the temperature of body at time t, the rate of heating is

$$\frac{\mathrm{d}T}{\mathrm{d}t}$$

Now, the rate of heating of the body is proportional to the temperature difference and therefore, we have

$$\frac{\mathrm{d}T}{\mathrm{d}t} \propto (T_0 - T) \qquad \text{or} \qquad \frac{\mathrm{d}T}{\mathrm{d}t} = k(T_0 - T)$$

where k, having dimension of s^{-1}, is proportionality constant. This equation can be rearranged as

$$\frac{\mathrm{d}T}{(T_0 - T)} = k\ \mathrm{d}t$$

On integration, we get

$$-\ln(T_0 - T) = kt + c \tag{5.49}$$

The constant of integration c can be obtained by applying the condition that at $t = 0$, we have $T = T_1$ and therefore, we have

$$-\ln(T_0 - T_1) = c \tag{5.50}$$

On substituting the value of c from equation (5.50) in equation (5.49), we have

$$-\ln(T_0 - T) = kt - \ln(T_0 - T_1) \qquad \text{or} \qquad \ln\left(\frac{T - T_0}{T_1 - T_0}\right) = -kt$$

On taking anti-logarithm, we get

$$\frac{T - T_0}{T_1 - T_0} = \mathrm{e}^{-kt} \qquad \text{or} \qquad T = T_0 + (T_1 - T_0)\,\mathrm{e}^{-kt} \tag{5.51}$$

This relation gives the temperature T of body after a time t.

Note: Here, it does not matter if the temperatures are expressed in Celsius or Kelvin unit.

Example 28: A body having temperature of −23 C is placed in an atmosphere having temperature of 27 C. Considering $k = 0.005\ \mathrm{s}^{-1}$, calculate the temperature of body after 5 min.

Solution: We have $T_1 = -23 + 273 = 250$ K, $T_0 = 27 + 273 = 300$ K, $t = 5 \times 60 = 300$ s, and k 0.005 s^{-1}. The temperature of body after 300 s is

$$T = T_0 + (T_1 - T_0)\mathrm{e}^{-kt} = 300 + (250 - 300)\,\mathrm{e}^{-0.005\times 300} = 288.84\ \mathrm{K}$$

Temperature of the body is 288.84 − 273 = 15.84 C.

17. Gravitational energy

When the question about the source of energy being emitted by the Sun continuously was under discussion, scientists thought that the gravitational contraction of Sun itself could be an effective source of energy. In order to understand it, let us consider a point mass m situated at a distance r from a large mass M. The gravitational force of attraction acting on the mass m is

$$F = -G\,\frac{Mm}{r^2}.$$

Here, the negative sign indicates the attractive nature of the gravitational force. The potential energy of system U, which is equal to the amount of work required for moving the mass m from the distance r to an infinite distance, is expressed as

$$U = \int_r^{\infty} F\ \mathrm{d}r = \int_r^{\infty} -G\,\frac{Mm}{r^2}\ \mathrm{d}r = -G\,\frac{Mm}{r}$$

The work is performed when the motion is along the line joining the two bodies, and no work is performed when the motion is perpendicular to the line joining the two bodies. Further, in the conservative field, the amount of work done is independent of the path along which the mass is moved. Obviously, U is the amount of energy lost by the system in moving from an infinite distance to a distance r between the bodies. In other words, $-U$ is the amount of energy which is given to the system in order to separate the bodies at infinite distance from each other.

Let us consider a sphere of radius r, comprising of material of uniform density ρ. Mass of the sphere is

$$M(r) = \frac{4}{3}\,\pi r^3 \rho$$

Suppose, a thin layer of thickness $\mathrm{d}r$ is added to the sphere. Mass of the layer is $\mathrm{d}m$, expressed as

$$\mathrm{d}m = \frac{4}{3}\,\pi(r+\mathrm{d}r)^3\rho - \frac{4}{3}\,\pi r^3\rho = 4\pi r^2\rho\,\mathrm{d}r$$

Here, we have neglected smaller terms. Thus, the amount of energy released in adding the layer is

$$\mathrm{d}\phi = -G\,\frac{M(r)\mathrm{d}m}{r} = -G\,\frac{(4\pi r^3\rho/3)(4\pi r^2\mathrm{d}r\rho)}{r} = -G\,\frac{16}{3}\,\pi^2\rho^2 r^4\mathrm{d}r$$

Thus, the total amount of energy released in the formation of a sphere of radius R is

$$\phi = -\int_0^R G\,\frac{16}{3}\,\pi^2\rho^2 r^4\mathrm{d}r = -G\,\frac{16}{15}\,\pi^2\rho^2 R^5 = -\frac{3}{5}\,\frac{GM^2}{R}$$

where $M(=4\pi R^3\rho/3)$ is the mass of the sphere. Thus, the amount of energy released in the formation of sphere of radius R and mass M, known as gravitational energy E_G, is

$$E_G = \frac{3}{5}\,\frac{GM^2}{R}$$

The gravitational energy released during formation of various bodies is given in Table 5.

Table 5. Gravitational energy

Body	Mass (kg)	Radius (m)	Gravitational energy (J)
Sun	1.99×10^{30}	6.96×10^{8}	2.28×10^{41}
Moon	7.35×10^{22}	1.74×10^{6}	1.24×10^{29}
Mercury	3.28×10^{23}	2.44×10^{6}	1.77×10^{30}
Venus	4.87×10^{24}	6.05×10^{6}	1.57×10^{32}
Earth	5.97×10^{24}	6.37×10^{6}	2.24×10^{32}
Mars	6.39×10^{23}	3.39×10^{6}	4.82×10^{30}
Jupiter	1.90×10^{27}	6.99×10^{7}	2.07×10^{36}
Saturn	5.68×10^{26}	5.82×10^{7}	2.22×10^{35}
Uranus	8.68×10^{25}	2.54×10^{7}	1.19×10^{34}
Neptune	1.02×10^{26}	2.46×10^{7}	1.69×10^{34}

Example 29: Estimate the amount of energy released in the formation of Sun. Comment if it is sufficient for the existence of Sun as of today. Given, mass of Sun $M_\odot = 1.989 \times 10^{30}$ kg, radius of Sun $R_\odot = 6.960 \times 10^8$ m and universal gravitational constant $G = 6.673 \times 10^{-11}$ m^3/s^2 kg.

Solution: The amount of energy released in the formation of Sun is

$$E_G = \frac{3}{5}\frac{GM^2}{R} = \frac{3}{5}\frac{6.673 \times 10^{-11} \times (1.989 \times 10^{30})^2}{6.960 \times 10^8} = 2.28 \times 10^{41} \text{ J}$$

Thus, the gravitational energy of 2.28×10^{41} J is released in the formation of Sun. The Sun is emitting energy constantly at an average amount of $L_\odot = 3.85 \times 10^{26}$ J/s. Hence, the gravitational energy of the Sun would have been sufficient for a period

$$\text{Time} = \frac{2.28 \times 10^{41}}{3.85 \times 10^{26}} = 5.922 \times 10^{14} \text{ s} = 1.88 \times 10^7 \text{ years}$$

However, the age of Sun is of the order of 5×10^9 years, which is several hundreds times more than the period for which the gravitational energy has been sufficient for the illumination of the Sun. It was therefore concluded that the source of energy in the Sun is not its gravitational energy. Thus, the nuclear source of energy came in the picture.

Example 30: Estimate the amount of energy released in the formation of Earth. Given, mass of Earth $M = 5.973 \times 10^{24}$ kg, radius of Earth $R = 6.378 \times 10^6$ m and universal gravitational constant $G = 6.673 \times 10^{-11}$ m^3/s^2 kg.

Solution: The amount of energy released in the formation of Earth is

$$E_G = \frac{3}{5}\frac{GM^2}{R} = \frac{3}{5}\frac{6.673 \times 10^{-11} \times (5.973 \times 10^{24})^2}{6.378 \times 10^6} = 2.24 \times 10^{32} \text{ J}$$

Thus, the gravitational energy of 2.24×10^{32} J is released in the formation of Earth.

17.1 Estimation of kinetic temperature

Let us consider a celestial body consists of mono-atomic gas with atomic mass μ having radius R and total mass M in hydrostatic equilibrium. According to Avogadro hypothesis, μ kg of gas has $N = 6.023 \times 10^{26}$ particles. Thus, the number n of particles in the body of mass M is

$$n = \frac{N}{\mu} M$$

According to the thermodynamics, at temperature T, the thermal energy per degree of freedom is $kT/2$. Here, k is the Boltzmann constant. A particle of mono-atomic gas has three degrees of freedom. Thus, the energy of each particle is $3kT/2$. Hence, the total energy E_T of n particles is

$$E_T = \left(\frac{N}{\mu} M\right)\left(\frac{3}{2} kT\right) = \frac{3NMkT}{2\mu}$$

For a spherical body of mass M and radius R, the gravitational energy is

$$E_G = -\frac{3}{5}\frac{GM^2}{R}$$

Following the Virial theorem for a star in hydrostatic equilibrium, we have

$$2E_T + E_G = 0 \qquad \text{thus} \qquad 2\,\frac{3NMkT}{2\mu} + \left(-\frac{3}{5}\frac{GM^2}{R}\right) = 0$$

The temperature in the core of the body is

$$T = \frac{GM\mu}{5NkR}$$

Table 6. Temperature in the core of body

Body	Mass (kg)	Radius (m)	T/μ (K)
Sun	1.99×10^{30}	6.96×10^{8}	4.59×10^{6}
Moon	7.35×10^{22}	1.74×10^{6}	6.78×10^{1}
Mercury	3.28×10^{23}	2.44×10^{6}	2.16×10^{2}
Venus	4.87×10^{24}	6.05×10^{6}	1.29×10^{3}
Earth	5.97×10^{24}	6.37×10^{6}	1.50×10^{3}
Mars	6.39×10^{23}	3.39×10^{6}	3.03×10^{2}
Jupiter	1.90×10^{27}	6.99×10^{7}	4.36×10^{4}
Saturn	5.68×10^{26}	5.82×10^{7}	1.57×10^{4}
Uranus	8.68×10^{25}	2.54×10^{7}	5.49×10^{3}
Neptune	1.02×10^{26}	2.46×10^{7}	6.66×10^{3}

This is an expression for the temperature in the central core of a body. For calculating its value, we have $G = 6.67 \times 10^{-11}$ m^3/kg s^2, $k = 1.38 \times 10^{-23}$ J/K, $N = 6.023 \times 10^{26}$. Temperature in the core of some celestial bodies are calculated and are given in Table 6. For the Sun, we have found

$$T = 4.59 \times 10^6 \mu$$

Hence, the temperature of the central core of the Sun is of the order of $10^6 - 10^7$ K. Table 6 also shows that the core of Jupiter and Saturn are relatively hot and consequently, the material in them is in the fluid form.

18. Review Questions

1. Discuss the method of trigonometric parallax used for determination of distance of nearby celestial bodies.

2. The angular separation between two apparent positions, at the interval of six months, of a star relative to the distant stars is $30^\circ\,45'\,52''$. Calculate the distance of the star from the Sun.

3. Show that a system of two moving bodies of masses m_1 and m_2, separated by a distance r is equivalent to a system where one body of mass $M(= m_1 + m_2)$ is stationary and another body of mass $\mu = m_1 m_2/(m_1 + m_2)$ separated by a distance r is in motion.

4. Two components A and B of a binary system have masses m_1 and m_2, respectively. If the position vector and velocity of B relative to A are $\vec{r}$ and $\vec{v}$, respectively, show that total angular momentum of the system is

$$\vec{L} = \mu\, \vec{r} \times \vec{v}$$

where μ is the reduced mass of the system.

5. State the Kepler's laws of planetary motion and derive them.

6. For the motion of a planet around the Sun, show that the semi-major axis of the orbit is inversely proportional to the total energy of planet.

7. Show that the velocity of a planet revolving around the Sun is inversely proportional to the square root of its distance from the Sun.

8. The distance of Jupiter from the Sun is 5.2 AU. Calculate the time-period of revolution of Jupiter around the Sun.

9. The distance of Saturn from the Sun is 9.54 AU. Calculate the time-period of revolution of Saturn around the Sun.

10. State the Newton's law of gravitation, and derive it from the Kepler's laws of planetary motion.

11. State and prove the Virial theorem.

12. Describe the spectral classification of stars.

13. Describe the salient features of black-body spectrum. Write the Planck formula for black-body spectrum and show that Wien's formula, and Rayleigh and Jeans formula are special cases of it.

14. Using the Planck formula for the black-body spectrum

$$I_\lambda \, \mathrm{d}\lambda = \frac{8\pi ch}{\lambda^5} \frac{\mathrm{d}\lambda}{\exp(hc/\lambda kT) - 1}$$

discuss the Wien's displacement law and Stefan-Boltzmann law.

15. Describe the Doppler red-shift in the radiations coming from the cosmic objects.

16. Show that for non-relativistic motion, the red-shift is proportional to the velocity of object.

17. Describe the method for determination of radius of a star.

18. Describe the salient features of Hertzsprung-Russell diagram.

19. Show that a heavy star stays in the main sequence for a shorter period as compared to that for a lighter star when the luminosity L of star varies as $L \propto M^{\alpha}$, where M is the mass of star and $\alpha > 1$. Mass of star A is 2.5 times that of B. Calculate the ratio of their periods in the main sequence.

20. State the Pauli exclusion principle. Describe its importance in the context of white dwarfs and neutron stars.

21. Describe hydrostatic equilibrium inside a star.

22. Derive an expression for the pressure at the center of a spherical body of mass M, radius R and uniform density ρ.

23. Describe thermal equilibrium. A body having temperature T_1 is placed in an atmosphere having temperature T_0 such that $T_1 > T_0$. Derive expression for temperature of the body after a time t.

24. Describe thermal equilibrium. A body having temperature T_1 is placed in an atmosphere having temperature T_0 such that $T_1 < T_0$. Derive expression for temperature of the body after a time t.

25. A body having temperature of 177° C is placed in an atmosphere having temperature of 37° C. Considering $k = 0.004$ s^{-1}, calculate the temperature of the body after 6 min.

26. A body having temperature of −17° C is placed in an atmosphere having temperature of 37° C. Considering $k = 0.004$ s^{-1}, calculate the temperature of the body after 4 min.

27. Show that the amount of gravitational energy released in the formation of a celestial body of mass M and radius R is $3GM^2/5R$.

28. Derive an expression for the amount of energy released in the formation of sphere of radius R and mass M.

29. For the Sun, we have mass $M_{\odot} = 1.989 \times 10^{30}$ kg, radius $R_{\odot} = 6.960 \times 10^{8}$ m, luminosity $L_{\odot} = 3.85 \times 10^{26}$ J/s. Show that the amount of gravitational energy released during the formation of the Sun has not been sufficient for continuous shining of the Sun.

30. Derive an expression for the temperature inside a spherical body of radius R and mass M.

31. Write notes on the following:

    (i) Parallax
    (ii) Two body system
    (iii) Kepler's laws of planetary motion

(iv) Newton's law of gravitation
(v) Virial theorem
(vi) Cosmic abundances
(vii) Spectral classification
(viii) Black-body radiation
(ix) Wien's displacement law
(x) Stefan Boltzmann law
(xi) Doppler red shift
(xii) Relation between B - V magnitude and surface temperature
(xiii) Colour-magnitude diagram
(xiv) Determination of radius of a star
(xv) Hertzsprung-Russell diagram
(xvi) Pauli exclusion principle
(xvii) Degenerate matter
(xviii) Electron degeneracy
(xix) Proton degeneracy
(xx) Neutron degeneracy
(xxi) Hydrostatic equilibrium
(xxii) Thermal equilibrium
(xxiii) Gravitational energy

19. Multiple Choice Questions

1. The unit of the areal velocity is

A. cm/sec B. cm^2/sec C. cm/sec^2 D. cm^3/sec

Ans. B

2. Which does not have the unit of length.

A. Light year B. Astronomical unit C. Parsec D. Areal velocity

Ans. D

3. Which is not the Kepler's law of planetary motion.

A. A planet revolves around the Sun in an elliptical orbit.

B. The force between the Sun and planet is the gravitational force.

C. The line joining a planet and the Sun sweeps equal area in equal time.

D. The square of time-period of revolution of a planet around the Sun is proportional to the cube of the semi-major axis of the elliptical orbit.

Ans. B

4. In a binary of two stars A and B, having masses $2M_\odot$ and $5M_\odot$ have sizes E_A and E_B, respectively, of their ellipses of orbit. We have

A. $E_A > E_B$ B. $E_A = E_B$ C. $E_A < E_B$ D. None of them

Ans. A

5. In a binary of two stars A and B, having masses $5M_\odot$ and $1.4M_\odot$ have sizes E_A and E_B, respectively, of their ellipses of orbit. We have

A. $E_A > E_B$ B. $E_A = E_B$ C. $E_A < E_B$ D. None of them

Ans. C

6. The masses of the Sun and Earth are M and m, respectively. The reduced mass of the combination is nearly equal to

A. M B. m C. $M + m$ D. Mm

Ans. B

7. In a binary system, each of the two stars has the mass m. The reduced mass of the binary system is

A. $2m$ B. m C. $m/2$ D. $m/4$

Ans. C

8. The sequence of classification of stars is

A. OAFBGKM B. ABFGKMO C. OMABFGK D. OBAFGKM

Ans. D

9. Wien's displacement law is

A. $\lambda_m/T = \text{const}$ B. $\lambda_m T = \text{const}$ C. $\lambda_m + T = \text{const}$ D. $\lambda_m - T = \text{const}$

Ans. B

10. At temperature T, total energy emitted by a black-body per unit time, per unit surface area is proportional to

A. T B. T^2 C. T^3 D. T^4

Ans. D

11. For a red-shift of a spectral line, the

A. wavelength of radiation increases.

B. wavelength of radiation decreases.

C. source moves towards the observer.

D. relative velocity between the source and observer is zero.

Ans. A

12. For a blue-shift of a spectral line, the

    A. wavelength of radiation increases.

    B. wavelength of radiation decreases.

    C. source moves away from the observer.

    D. relative velocity between the source and observer is zero.

Ans. B

13. A planet moving with velocity v in an elliptical path is at a distance r from the Sun. When this planet is at a distance $3r/2$, its velocity is

    A. v B. $3v/2$ C. $2v/3$ D. $v/2$

Ans. C

14. A planet moving with velocity v in an elliptical path is at a distance r from the Sun. When this planet is moving with velocity $3v/2$, its distance is

    A. r B. $3r/2$ C. $2r/3$ D. $2r$

Ans. C

15. The speed of a planet revolving around the Sun in an elliptical increases when it comes closer. This fact is according to which of the Kepler's law of planetary motion.

    A. first only B. second only C. third only D. all three

Ans. B

16. The Earth at 1 AU distance from the Sun takes 1 year to revolve around the Sun. Some planet at a distance of 4 AU from the Sun would require how many years to complete one revolution around the Sun.

    A. 1 B. 4 C. 8 D. 16

Ans. C

17. The Earth at 1 AU distance from the Sun takes 1 year to revolve around the Sun. Some planet requires 8 years time to complete one revolution around the Sun. The distance of the planet in AU from the Sun is

    A. 1 B. 4 C. 8 D. 16

Ans. B

18. A graph between the luminosity and effective temperature of stars is known as

    A. phase diagram B. H-R diagram C. Lande diagram D. Planck diagram

Ans. B

19. For a group of stars, the luminosity is found to vary as the fourth power of the mass of star. When the ratio of masses of two stars is $\sqrt{2}$, the ratio of their luminosities is

    A. 2 B. 4 C. 8 D. 16

Ans. B

20. For a spherical body of radius R, mass M and average density ρ, the pressure at the centre of the body is

    A. $\rho GM/R$ B. $\rho GM/2R$ C. $3\rho GM/5R$ D. $5\rho GM/3R$

    Ans. B

21. For a spherical body of radius R and mass M, the gravitational energy is

    A. GM^2/R B. $3GM^2/5R$ C. $GM^2/2R$ D. $GM^2/4R$

    Ans. B

6. Stars

We believe that the universe has several hundred billion galaxies. However, it is difficult to estimate the number of galaxies in the universe, because it is difficult to find the boundaries of the universe. The name of our galaxy is the Milky Way. In a galaxy, more than 50% stars are in the form of groups of two or more stars. A majority of them is in the form of a binary system, *i.e.*, a group of two stars. Groups of three or more stars are very small in number. A well known group of three stars is the alpha-Centauri system whose third component is the Proxima Centauri, which is the next nearest star to us, after the Sun.

It is well established that the stars are made up of hot gases - mostly hydrogen and helium. Hydrogen is being converted continuously into helium in the shining stars. The stars are away from the Earth and the light from them passes through the atmosphere before reaching the earth. During this journey, they undergo the bending or refraction, due to the variation of density (refractive index) in the atmosphere. When there is wind, the magnitude of bending of light rays varies with time. This makes the stars to twinkle.

The stars which are relatively rich in the heavy elements - such as the Sun - are known as the population I stars. The stars which are relatively poor in the heavy elements - such as the globular clusters - are known as the population II stars. It is commonly believed that the population I stars represent a later generation of stars whereas the population II stars represent an earlier generation. That is, the population II stars are older than the population I stars. No stars with zero abundance of heavy elements are found. However, some astronomers theoretically consider their existence in the universe and they are known as the population III stars. In this chapter, we have discussed about various types of stars.

1. Formation of stars

The existing stars did not exist forever, is a common opinion. Therefore, about the birth of stars, the speculations may be made. Further, the stars which we consider to be young are found associated with the clouds of the interstellar dust and gas. It may therefore be argued that such clouds of dust and gas are contracting slowly, giving rise to an increasingly compact condensations, and some of them finally collapse down to a stellar size. This indicates the formation of stars out of the dust and gas. Such scenario may help for understanding the formation of stars. Some aspects about the formation of stars are as the following.

1.1 Cooling and collapsing of cloud

In a cloud out of which a star is formed, the initial gas is considered to consist of hydrogen atoms. In the cloud, the dust grains can radiate heat in a very efficient manner. When a hydrogen atom collides with a dust grain, the kinetic energy of hydrogen atom is transferred to the dust grain and the temperature of the grain increases slightly. This energy is radiated away fast by the dust grain in the form of infrared radiations, which can escape from the cloud easily. In this manner, the kinetic energy of gas is reduced. When the gas losses the kinetic energy, it falls towards the center of the cloud under the gravitational attraction. During the fall, the gas acquires some kinetic energy which is partly transferred to a dust grain, and the said cooling cycle repeats.

The atom transfers also some of its centrally directed momentum to the grain. This causes the grains to drift towards the center of the contracting cloud. As a result of many such interactions, the cloud as a whole contracts.

Radiation from the grain is not the only process responsible for the loss of energy of protostar. During the collapse, the temperature of the protostar cloud increases and due to the collisions, the molecules and atoms are excited. These excited molecules and atoms return back to their ground state, producing photons. This way the energy of the collapse is converted into radiations which can escape from the cloud into the outer space, giving rise to cooling. Though this explanation of cooling and collapse of a cloud looks attractive, but there are difficulties with this picture, as besides the lose of energy, there must be

(i) loss of angular momentum, and

(ii) conservation of magnetic field.

1.2 Problem of angular momentum

There are observational evidences about the rotation of interstellar molecular clouds where the star formation is in progress. Thus, the contracting cloud has large angular momentum from the early stage. In the process, the matter of interstellar cloud collapses into a star. For the formation of a star of density 10^9 atoms/m^3, there should be collapse of gas from a volume whose initial radius r should be $\sim 10^{16}$ m. Over such distances, for a cold galactic molecular cloud, the velocity v must be $\sim 3 \times 10^2$ m/s. Thus, the initial angular momentum of the cloud per unit mass is $l = mvr \sim 3 \times 10^{18}$ m^2/s.

But, the observations for the typical stars show that the angular momentum per unit mass varies from 10^{12} to 2×10^{14} m^2/s. It is many orders of magnitude lower than the initial value. Further, in case of the Sun, the angular momentum per unit mass is only $\sim 10^{11}$ m^2/s. Even if we consider the whole solar system, it is 10^{13} m^2/s. In the solar system, most of the angular momentum belongs to the motions of planets, and in particular, in the motion of Jupiter orbiting around the Sun.

It shows that for becoming a star or a planetary system, the initially contracting cloud of interstellar matter must lose almost all of its angular momentum and only less than 0.1% can remain. How the angular momentum is lost is not yet understood.

1.3 Problem of magnetic field

Another problem is about the conservation of magnetic field present in the initial cloud. The magnetic lines of force act as if they are frozen to the gaseous material, and the number of these lines of force threading through the cross-sectional area of the contracting cloud remain constant. Thus, the magnetic field B is proportional to r^{-2} in such a contraction.[1] Suppose, some weak magnetic field present in the initial cloud is predominantly oriented along some direction. Then, after the formation of a star or a planetary system, the magnetic field must have the same direction and strong intensity. Suppose, initially there is weak magnetic field B of the order of 10^{-7} Gauss (which is smaller than the observed value). In the contraction of a cloud from a radius of 10^{16} m to 10^9 m, the magnetic field becomes 10^{14} times stronger, *i.e.*, it becomes of the order of 10^7 Gauss.

The actual magnetic field found on the surface of stars are of the order of 1 Gauss; some peculiar stars have magnetic fields of the order 10^4 Gauss. It is very small as compared to the estimated value.[2] How this loss of magnetic field occurs is a major unsolved part of the problem of the star formation.

2. Evolution of stars

A star evolves with time after its formation. The lifetime of a star depends on its mass. A rough estimate shows that the lifetime may vary from a few million years for a most massive star to a few trillion years for the least massive star. Note that the age of the least massive star is considerably longer than even the age of the universe (13.6×10^9 years), showing no death of the least massive stars so far.

Stars are considered to be born out of the collapsing clouds having gas and dust, often called the nebulae or molecular clouds. Over a period of millions of years, these protostars settle down into a state of equilibrium, becoming what is known as a main-sequence star. Before reaching the main sequence, typical variation of paths, depending on the initial mass on the main sequence is shown in Figure 1, where luminosity versus effective temperature of star are plotted. In the HertzsprungRussell diagram, a star appears on the right side and after moving to left reaches the main sequence. For most of its life on the main sequence, a star is powered through nuclear fusion occurring in its core. In the initial phase, the energy is generated by the fusion of hydrogen atoms at the core of the main-sequence star. The fusion of hydrogen atoms generates helium atoms, and the stars like the Sun begin to fuse hydrogen along a spherical shell surrounding the core. During this process, the stars gradually grow in size, pass through the sub-giant stage until it reaches the red giant phase.

Stars having at least 0.5 $M_\odot$ can also begin to generate energy through the fusion of helium at their core, whereas a more massive star can fuse heavier elements along a series of concentric shells, producing a picture of onion shells, where hydrogen is

[1] The magnetic field is inversely proportional to the surface area of the spherical surface.

[2] Very compact white dwarf stars and the even denser neutron stars, respectively, have magnetic fields believed to be of the order of 10^5 Gauss and 10^{12} Gauss. These are the field strengths we expect when the Sun could shrink to the size of such compact stars.

burning in the outermost shell, and then helium, carbon, oxygen, neon, magnesium, silicon, and iron, in the inner shells in sequence. For a very heavy star (say, with mass 25 $M_{\odot}$), onion-like structure of its core, where the elements up to iron are formed, is shown in Figure 2. In the outermost shell, the hydrogen is burning and around that is the region containing hydrogen. Starting from the hydrogen burning shell, up to which shell in the inward direction goes, depends on the mass of the star.

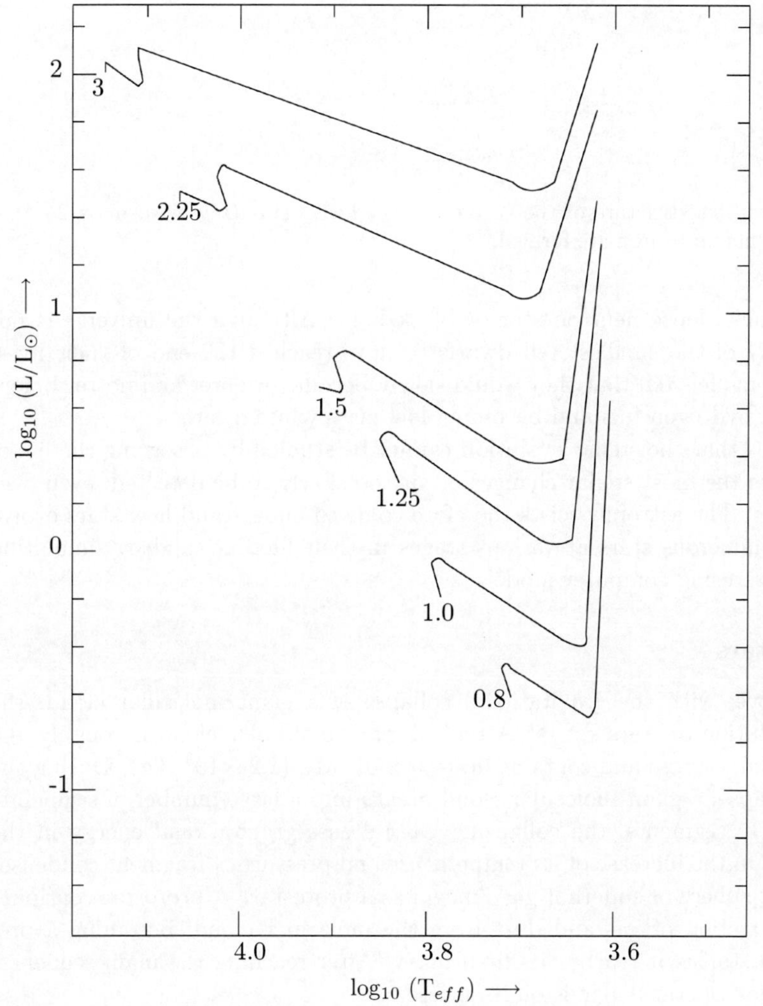

Figure 1: A diagram showing evolution of stars of various masses, written in the diagram in unit of solar mass $M_{\odot}$, till they arrive the main sequence.

After the exhaust of nuclear fuel, in a sun-like star, the core collapses into a dense white dwarf and the outer layers are expelled as a planetary nebula. A star having more than 10 $M_{\odot}$ can explode in a supernova as their inert iron cores collapse

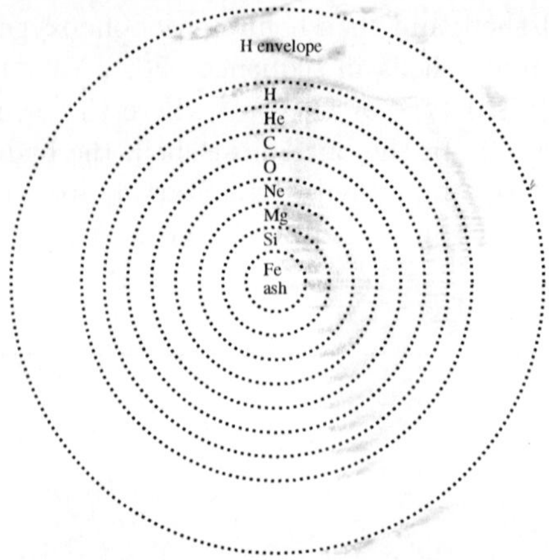

Figure 2: Onion like structure of the core of a very heavy star (say, with mass 25 $M_\odot$) where the elements up to iron are formed.

into an extremely dense neutron star or black hole. Although the universe is not so old that any of the smallest red dwarfs to have reached the end of their lives, but the stellar models tell that they would slowly become brighter and hotter before running out of hydrogen fuel and becoming low-mass white dwarfs.

It is obvious that the stellar evolution cannot be studied by observing the life of a single star, as the most stellar changes occur too slowly to be detected, even over many centuries. The astrophysicists therefore come to understand how stars evolve by observing numerous stars at various stages in their lifetime, and by simulating stellar structure using computer models.

2.1 Protostars

As stated above, with the gravitational collapse of a giant molecular cloud, the formation evolution of stars starts. A typical giant molecular cloud is roughly 100 ly (9.5×10^{14} km) across and contains up to 6×10^6 $M_\odot$ (1.2×10^{37} kg). During the process of collapse, a giant molecular cloud breaks into a large number of segments. In each of these segments, the collapsing gas releases gravitational energy in the form of heat. On the increase of its temperature and pressure, a fragment condenses into a rotating sphere of superhot gas, known as a protostar. A protostar continues to grow by accretion of gas and dust from the molecular cloud, becoming a pre-main-sequence star as it reaches its final mass. After reaching the main sequence, further evolution of star depends on its mass.

The protostars have dust as component, and therefore are easily visible at the infrared wavelengths.

2.2 Hayashi lines

The Hayashi track is a luminosity-temperature relationship obeyed by the infant stars of less than 3 $M_\odot$ in the pre-main-sequence phase of stellar evolution. In the

Hertzsprung-Russell (H-R) diagram, the track is a nearly vertical curve (Figure 1). The Hayashi line is defined as the locus in the H-R diagram of fully convective stars of given parameters (mass and chemical composition). Note that for each set of the parameters, there is a separate Hayashi line. After a protostar ends its phase of rapid contraction and becomes a T Tauri star, it is extremely luminous. The star continues to contract, but much more slowly. While slowly contracting, the star follows the Hayashi track downwards, becoming several times less luminous but staying at roughly the same surface temperature, until a radiative zone develops, at which point the star starts nuclear fusion, marking the beginning of main sequence.

The shape and position of the Hayashi track in the H-R diagram depends on the star's mass and chemical composition.

2.3 Brown dwarfs

A protostar having a mass less than 0.08 $M_\odot$ (1.6 $\times 10^{29}$ kg) never reaches a temperature high enough for the initiation of nuclear fusion of hydrogen into helium. These protostars are known as the brown dwarfs. As per definition given by the International Astronomical Union, a brown dwarf is a star having mass enough to fuse deuterium at some point in its life. Typical mass of a brown dwarf is 2.5 $\times 10^{28}$ kg or 0.0125 $M_\odot$. Objects having smaller than this mass are classified as sub-brown dwarfs (but when they orbit around another stellar object they are classified as planets). Both types, deuterium-burning and not, shine dimly and die away slowly, cooling gradually over hundreds of millions of years. These brown dwarfs never reach the main sequence in the H-R diagram.

2.4 Main sequence

For a more-massive protostar, the core temperature eventually reaches 10 million kelvin, sufficient for the initiation of proton-proton chain reaction which fuses hydrogen first into deuterium and then into helium. In a star having mass slightly larger than 1 $M_\odot$, the carbon-nitrogen-oxygen fusion reaction (CNO cycle) contributes a large portion of the energy generation. In the nuclear fusion, the energy released maintains a high gas pressure which balances the weight of the star's matter and prevents further gravitational collapse. The star thus evolves rapidly to a stable state, beginning the main-sequence phase of its evolution.

Depending on its mass, the newly born star arrives at a specific position on the main sequence in the H-R diagram. Small, relatively cold, low-mass red dwarfs fuse hydrogen slowly and remains on the main sequence for hundreds of billions of years or longer, whereas the massive, hot O-type stars leave the main sequence after just a few million years. A mid-sized yellow dwarf star, like the Sun, remains on the main sequence for about 10 billion years. The Sun is thought to be in the middle of its main sequence lifespan.

2.5 Mature stars

At the end of the life on the main sequence, the core exhausts its supply of hydrogen and the star begins to evolve off from the main sequence, where the outward pressure generated by the fusion of hydrogen is absent. The core of the star contracts under the force of gravity until either the electron degeneracy pressure becomes sufficient to oppose gravity or the core becomes hot enough (about 100 million kelvin) for helium fusion to begin. Which of these happens first depends upon the mass of the star as the following.

Low-mass stars

When there is no more production of energy through fusion, what happens to a low-mass star. The age of the universe (about 13.8 billion years) is less (by several orders of magnitude, in some cases) than it takes for stopping the fusion in such stars. Therefore, no further information about such stars is available. Astrophysical models suggest that the red dwarfs of 0.1 $M_{\odot}$ may stay on the main sequence for some six to twelve trillion years, gradually increasing in both temperature and luminosity, and take several hundred billion more to collapse, slowly, into a white dwarf. These stars never become red giants as they are fully convective and do not develop a degenerate helium core with a shell burning hydrogen. Instead, the hydrogen fusion proceeds until almost the whole star is helium.

Slightly more massive stars do expand into red giants, but their helium cores are not massive enough showing that their temperatures attain the value required for helium fusion. Thus, they never reach the tip of the red giant branch. At the end of the hydrogen burning, these stars move directly off the red giant branch like a post-asymptotic-giant-branch (AGB) star, but at lower luminosity, to become a white dwarf. A star with an initial mass above about 0.8 $M_{\odot}$ is able to attain the temperatures high enough to fuse helium, and these stars go on to further stages of evolution beyond the red giant branch.

Mid-sized stars

Stars having mass in the range of roughly 0.8 - 10 $M_{\odot}$ become red giants, which are large non-main-sequence stars of stellar classification K or M. Red giants due to their red color and large luminosity lie along the right edge of the H-R diagram.

The mid-sized stars are red giants during two different phases of their post-main-sequence evolution: red-giant-branch stars, with inert cores made of helium and hydrogen-burning shells, and asymptotic-giant-branch stars, with inert cores made of carbon and helium-burning shells inside the hydrogen-burning shells. In between these two phases, stars spend a period on the horizontal branch with a helium-fusing core. Many of these helium-fusing stars cluster towards the cool end of the horizontal branch as K-type giants and are known as the red clump giants.

Subgiant phase

After the exhaustion of hydrogen in its core, the star leaves the main sequence and begins to fuse hydrogen in a shell outside the core. As the shell produces more helium, the mass of the core increases. Depending on the mass of the helium core, this continues for several million to one or two billion years, with the star expanding and cooling at a similar or slightly lower luminosity to its main sequence state.

Finally, either the core becomes degenerate, in the stars around the mass of the sun, or the outer layers cool sufficiently to become opaque, in more massive stars. Either of these changes causes the hydrogen shell to increase in temperature and the luminosity of the star to increase, at which point the star expands onto the red giant branch.

2.6 Red-giant branch phase

The helium core continues to grow on the red-giant branch. It is no longer in the thermal equilibrium, so it increases in temperature which causes the rate of fusion in the hydrogen shell to increase. The luminosity of star increases towards the tip of the red-giant branch. Red-giant branch stars with a degenerate helium core all reach the tip with very similar core masses and very similar luminosities, although the more massive of the red-giants become hot enough to ignite helium fusion before that point.

2.7 Horizontal branch

In the helium cores of stars having mass from 0.8 to 2.0 $M_\odot$, which are largely supported by electron degeneracy pressure, the helium fusion ignites on a timescale of days in a helium flash. In the nondegenerate cores of more massive stars, the ignition of helium fusion occurs relatively slowly with no flash. The nuclear power released during the helium flash is very large, on the order of 10^8 $L_\odot$ for a few days and 10^{11} $L_\odot$ for a few seconds. However, the energy is consumed by the thermal expansion of the initially degenerate core and thus cannot be seen from outside the star. Due to the expansion of the core, the hydrogen fusion in the overlying layers slows and total energy generation decreases. The star contracts, although not all the way to the main sequence, and it migrates to the horizontal branch in the H-R diagram, gradually shrinking in radius and increasing its surface temperature.

2.8 Asymptotic-giant-branch phase

After consumption of helium at the core of star, hydrogen and helium fusion continues in shells around a hot core of carbon and oxygen. The star follows the asymptotic giant branch on the H-R diagram, paralleling the original red giant evolution, but with even faster energy generation (which lasts for a shorter time). Although helium is burning in a shell, the majority of energy is produced by the hydrogen burning in a shell further from the core of the star. From these hydrogen burning shells, the helium drops towards the center of the star and periodically the energy output from the helium shell increases dramatically. This is known as a thermal pulse and they occur towards the end of the asymptotic-giant-branch phase, sometimes even into the post-asymptotic-giant-branch phase.

There may be several to hundreds of thermal pulses, depending on the mass and composition of the star. On the ascent of the asymptotic-giant-branch, there is a phase where a deep convective zone forms and can bring carbon from the core to the surface of the star. This is known as the second dredge up. In some stars, there may be even a third dredge up. In this way a carbon star is formed. It is very cool and strongly reddened having strong carbon lines in their spectra. A process known as the hot bottom burning may convert carbon into oxygen and nitrogen before it is dredged to the surface.

Another well known class of asymptotic-giant-branch stars are the Mira variables, which pulsate with well-defined periods of tens to hundreds of days and large amplitudes up to about 10 magnitudes (in the visual, total luminosity changes by a much smaller amount). For a more massive star, the star becomes more luminous and the pulsation period is longer, leading to an enhanced mass loss, and the star becomes heavily obscured at visual wavelengths.

2.9 Post-AGB

The mid-range stars finally reach the tip of the asymptotic-giant-branch and run out of fuel for shell burning. They do not have sufficient mass so that full-scale carbon fusion could be started, and therefore, they contract again, going through a period of post-asymptotic-giant-branch superwind to produce a planetary nebula with an extremely hot central star. Then the central star cools to a white dwarf. The expelled gas is relatively rich in heavy elements generated within the star and depending on the type of star, may be particularly oxygen or carbon enriched. The gas forms an expanding shell called the circumstellar envelope, which cools as it moves away from the star, allowing the dust particles and molecules to form. With the high infrared energy input from the central star, ideal conditions are formed in these circumstellar envelopes for maser excitation.

In the massive stars, the core is already large enough at the onset of the hydrogen burning that the helium ignition occurs before electron degeneracy pressure has a chance to become prevalent. Therefore, when these stars expand and cool, they do not brighten as much as the lower-mass stars. They are however much brighter than the lower-mass stars to begin with, and are thus still brighter than the red giants formed from less-massive stars. These stars are unlikely to survive as red super giants; instead they would destroy themselves as type II supernovae.

Extremely massive stars (more than about 40 $M_\odot$), which are very luminous and thus have very rapid stellar winds, lose mass so rapidly due to radiation pressure that they tend to strip off their own envelopes before they can expand to become red supergiants, and thus retain extremely high surface temperatures from their main-sequence time onwards. The largest stars of the current generation are about 100 - 150 $M_\odot$ because the outer layers would be expelled by the extreme radiation.

The core grows hotter and denser as it gains material from fusion of hydrogen at the base of the envelope. In all massive stars, electron degeneracy pressure is insufficient to halt collapse by itself, so as each major element is consumed in the center, progressively heavier elements ignite, temporarily halting the collapse. If the core of the star is not too massive (less than approximately 1.4 $M_\odot$, taking into account mass loss that has occurred by this time), it may then form a white dwarf (possibly surrounded by a planetary nebula) as described above for the less massive stars, with the difference that the white dwarf is composed mainly of oxygen, neon, and magnesium.

Note: As an example, Figure 3 shows the post-main-sequence evolutionary track of Sun-like star.

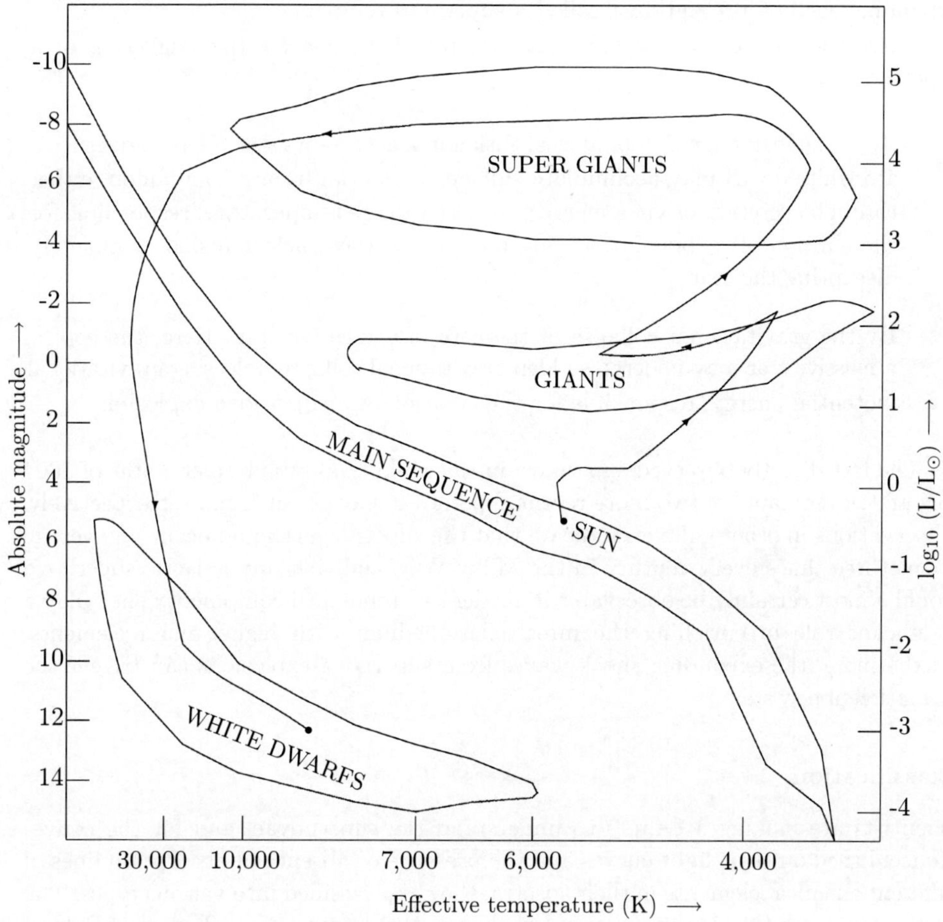

Figure 3: Post-main-sequence evolutionary track of sun-like star.

2.10 Supernova

Once the nucleosynthesis process arrives at iron-56, the continuation of this process consumes energy. If the mass of the core exceeds the Chandrasekhar limit, 1.4 $M_\odot$, the electron degeneracy pressure is insufficient to support its weight against the force of gravity, and the core undergoes a sudden, catastrophic collapse to form a neutron star or a black hole. Through a process that is not completely understood so far, some of the gravitational energy is released by this core collapse and is converted into a supernova.

A supernova (SN) is a stellar explosion that in a very short period outshines an entire galaxy, radiating energy equivalent to the total energy of the Sun or of an ordinary star, which later on fades out in a period of several weeks or months. The extremely luminous burst of radiation pushes outward much or all the material of the star at a velocity of up to 3×10^4 km/s (10% of the speed of light). It consequently drives a shock wave into the surrounding interstellar medium, which sweeps up an expanding shell of gas and dust called a supernova remnant.

There can be one of the following two possible ways for the triggering of a supernova.

1. By the sudden reignition of nuclear fusion in a degenerate star. Here, a degenerate white dwarf may accumulate sufficient material from a companion, either through accretion or via a merger, to raise its core temperature, responsible for the ignition of carbon fusion, and trigger runaway nuclear fusion, completely disrupting the star.

2. By the gravitational collapse of the core of a massive star. Here, the core of a massive star may undergo sudden gravitational collapse, releasing gravitational potential energy, responsible for the creation of a supernova explosion.

The last directly observed supernova in the Milky Way was Kepler's Star of 1604 (SN 1604); remnants of two more recent supernovae have been found retrospectively. Observations in other galaxies indicate that the supernovae should occur on average about three times every century in the Milky Way, and that any galactic supernova would almost certainly be observable in modern astronomical equipment. They play a significant role in enriching the interstellar medium with higher mass elements. Furthermore, the expanding shock waves from supernova explosions can trigger the formation of new stars.

Classification

Scientists are making attempts to understand the supernovae, and for the convenience, based on their light curves and the presence or absence of absorption lines of different chemical elements in their spectra, they are classified into various types. The first criterion for the classification of supernovae is the presence or absence of Balmer lines generated by hydrogen atom. When the supernova has Balmer lines of hydrogen atom, it is classified as Type II; otherwise it is Type I. In each of these two types there are subdivisions according to the presence of lines from other elements or the shape of the light curve.

The type I supernovae are subdivided on the basis of their spectra. When a strong ionized silicon absorption line is present, it is Type Ia, otherwise it may be Ib or Ic. When strong neutral helium lines are present, it is Ib, otherwise it is Ic. The light curve is not important for the classification of type I supernovae, as the all are similar, although type Ia are generally brighter at the peak luminosity.

The type II supernovae can also be subdivided on the basis of their spectra. Most of the Type II supernovae have very broad emission lines, showing the expansion velocities of many thousands of km/s. However, some have relatively narrow lines in their spectra and they are classified as Type IIn supernovae. Here, the 'n' indicates the 'narrow'.

A few supernovae, such as SN 1987K and SN 1993J, appear to change from Type Ib to Type II, as, at the early times, they show lines of hydrogen, but after a period of some weeks to months, they have dominated helium lines. Such supernovae are classified as Type IIb, indicating a combination of Ib and II.

Type II supernovae with normal spectra, dominated by broad hydrogen lines that remain for the life of the decline, are classified on the basis of their light curves. The most common type shows a distinctive "plateau" in the light curve shortly after peak brightness where the visual luminosity stays relatively constant for a period of several months before the decline starts. These are called the type II-P referring to the plateau. Some supernovae do not have distinct plateau, and they are classified as Type II-L, where the "L" signifies the "linear" although the light curve is not actually a straight line.

Supernovae that do not fit into the normal classification are designated as peculiar, or 'pec'. However, some scientists attempted to classify the supernovae into the Types III, IV and V, taking a few examples.

3. Compact stars

In view of the war between the gravity and thermodynamics in a star, regarding the end state of the star, astronomers believe that there are four possibilities.

(a) Nothing is left.

(b) A degenerate dwarf is left.

(c) A neutron star is left.

(d) A black hole is left.

The term compact star (sometimes compact object) is used to refer collectively to white dwarfs, neutron stars, and black holes. They are all small in volume for their mass, giving them a very high density. The white dwarfs and neutron stars may also be called the degenerate stars.

3.1 White dwarf

A white dwarf is a very dense stellar remnant which is composed mostly of electron-degenerate matter. A typical white dwarf has a mass comparable to that of the Sun, whereas its volume is comparable to that of the Earth. The nearest known white dwarf is the Sirius B, situated at a distance of 8.6 ly. A white dwarf is thought to be the final evolutionary state of a star whose mass is not high enough to become a neutron star. In our Milky Way, the masses of over 97% stars is such that they finally evolve into the white dwarfs.

During the process of fusion of hydrogen into helium, a low or medium mass star spends life on the main-sequence of HR diagram. When the process of fusion of hydrogen into helium is almost over, the star expands to form a red-giant where the fusion of helium to carbon and oxygen takes place in its core through the processes.

$$3\ {}_2\mathrm{He}^4 \rightarrow {}_6\mathrm{C}^{12} \qquad \text{and} \qquad 4\ {}_2\mathrm{He}^4 \rightarrow {}_8\mathrm{O}^{16}$$

After blowing out its outer layer to form a planetary nebula, the star leaves behind the core, called the white dwarf. Consequently, the white dwarfs generally consist of carbon and oxygen.

As the material in a white dwarf no longer undergoes fusion and therefore the star has no source of energy and has no support of heat, generated by fusion, against the gravitational collapse. It is supported only by electron degeneracy pressure, causing it to be extremely dense.

In a white dwarf, the electron degeneracy creates a barrier to the gravitational collapse. When the electrons are squeezed too close together, the Pauli exclusion principle requires them to have different energy levels. If we want to add another electron to a given volume, it raises an electron's energy level to make room and this requirement for energy to compress the material manifests as a pressure. The electron degeneracy pressure in a material is

$$P = \frac{3^{2/3}h^2\rho_e^{5/3}}{20\pi^{2/3}m_e} = \frac{0.048h^2\rho_e^{5/3}}{m_e}$$

where h stands for the Planck constant, m_e for the electron mass, ρ_e for the electron density. This expression is for non-relativistic case. For the relativistic case, a modified formula is required. This electron degeneracy pressure is always present and is in addition to the normal pressure $P = NkT/V$.

The electron degeneracy pressure works like a spring against the gravitational collapse. When the collapsing mass is large, the size of the spring is small. Consequently, with the increase of mass, the size of the white dwarf decreases. Chandrasekhar investigated the problem of electron degeneracy pressure and obtained a limiting mass of a white dwarf to be about 1.4 $M_\odot$. Beyond this, the white dwarf cannot be supported by the electron degeneracy pressure.

The relation between the mass M and radius R of a white dwarf is obtained as

$$R = 0.114\frac{h^2}{Gm_e m_p^{5/3}}\left(\frac{Z}{A}\right)^{5/3}M^{-1/3}$$

where m_e and m_p are masses of electron and proton, respectively, Z and A are the average atomic number and atomic weight of the ions in the white dwarf. The point to be noted here is that the volume of a white dwarf is inversely proportional to its mass. If we compare two white dwarfs of different masses, the more massive white dwarf has the smaller radius. This result is very different from that in case of an ordinary body, where the mass is proportional to its volume.

At the time of its formation, a white dwarf is very hot. As there is no more source of energy inside it, it gradually radiates away its energy and cools down with time. Consequently, its radiation which initially has a high colour temperature, decreases and reddens with time. After a long time, a white dwarf cools to a temperature at which it no longer emits significant heat or light, and it becomes a cold black dwarf. However, the estimated length of time it takes to convert from a white dwarf to a black dwarf is longer than the current age of the universe. As no white dwarf can be older than the age of the universe, it is thought that no black dwarfs yet exist. The older white dwarfs still radiate at temperature of few thousand kelvin.

Chandrasekhar limit

Chandrasekhar investigated theoretically that when the mass was big enough, the radius of white dwarf formally waits to shrink to zero. The limiting radius is reached to zero at a mass now known as the Chandrasekhar limit, expressed as

$$M_{ch} = 0.20\Big(\frac{Z}{Am_p}\Big)^2\Big(\frac{hc}{G}\Big)^{3/2}$$

where Z and A are the average atomic number and atomic weight of the ions in the white dwarf, and other symbols have their usual meanings.

White dwarfs are an end state of stellar evolution where we can expect all the hydrogen to have been converted to some heavier chemical elements. Since the ratio Z/A for the most heavier elements (until we approach and pass the iron group) is 0.5, the numerical value of Chandrasekhar limit can be calculated to be

$$M_{ch} = 1.4M_\odot$$

In other words, the maximum possible mass for a white dwarf is only 40% more than the mass of the Sun. All the white dwarfs whose mass have been observationally measured have masses less than this maximum value.

Example 1: We are given mass of proton $m_p = 1.67 \times 10^{-27}$ kg, universal gravitational constant $G = 6.673 \times 10^{-11}$ m^3/s^{2}kg, Planck constant $h = 6.62 \times 10^{-34}$ Js, speed of light $c = 3 \times 10^8$ m/s. Calculate the Chandrasekhar limit.

Solution: The Chandrasekhar limit is

$$\begin{aligned}M_{ch} &= 0.20\Big(\frac{Z}{Am_p}\Big)^2\Big(\frac{hc}{G}\Big)^{3/2}\\ &= 0.20\Big(\frac{0.5}{1.67\times 10^{-27}}\Big)^2\Big(\frac{6.62\times 10^{-34}\times 3\times 10^{8}}{6.673\times 10^{-11}}\Big)^{3/2}\\ &= 2.91\times 10^{30}\ \text{kg} = \frac{2.91\times 10^{30}}{1.99\ \times 10^{30}} = 1.4\ \text{M}_\odot\end{aligned}$$

3.2 Neutron stars

A neutron star is a type of stellar remnant that can result from a gravitational collapse of a massive star (with an initial mass of around 10 $M_\odot$ or above) and its almost entire composition is neutrons. Neutron stars are the densest (from 3.7×10^{17} to 5.9×10^{17} kg/m^3) and tiniest (with a radius of only about 12 - 13 km) stars known to exist in the universe. A typical neutron star has a mass between 1.4 and 3.2 solar mass with a surface temperature of about 6 $\times10^5$ K. That is, the mass of a neutron star is more than the Chandrasekhar limit (1.4 $M_\odot$) for the white dwarfs. A neutron star having diameter of about 10 km may have a mass of several times the solar mass. Obviously, the neutron stars are the end points of stars whose inert core's mass after nuclear burning is larger than 1.4 $M_\odot$ but the mass is not enough to overcome the neutron degeneracy pressure to become black holes. The neutron stars have mass between 1.4 $M_\odot$ and 3 $M_\odot$. The maximum observed mass of neutron stars is about 2 $M_\odot$. The gravitational field at the star's surface is about 2×10^{11} times stronger than that on Earth.

In case of a neutron star, the mass is larger than that of a white dwarf and the electron degeneracy pressure is insufficient to control the gravitational collapse. In the neutron stars, we have neurons, which also are fermions. There neutron degeneracy pressure comes in the picture. The neutron degeneracy pressure in a material is

$$P = \frac{3h^2 \rho_n^{5/3}}{4\pi^{2/3} m_n} = \frac{0.35 h^2 \rho_n^{5/3}}{m_n}$$

where m_n stands for the neutron mass and ρ_n for the neutron density. The neutron degeneracy pressure also works like a spring with large spring constant against the gravitational collapse. When the collapsing mass is large, the size of the spring is small. Consequently, with the increase of mass, the size of the neutron star decreases. The relation between the number of neutrons N and radius R of a neutron star is obtained as

$$R = \left(\frac{81\pi^{14}}{16}\right)^{1/3} \frac{h^2}{4Gm_n^3} N^{-1/3}$$

The point to be noted here is that the volume of a neutron star is inversely proportional to its mass (number of neutrons in the star). Neutron stars are very hot and are supported against further collapse by the quantum degeneracy pressure due to the neutrons, following the Pauli exclusion principle.

The discovery of pulsars in 1967 suggested the existence of neutron stars. They are born in supernova explosion and have radius of about 12 - 13 km. They spin about 642 times per second. Neutron stars have overall densities of 3.7 $\times 10^{17}$ to 5.9 $\times 10^{17}$ kg/m^3 which is comparable to the density of an atomic nucleus of $\approx$ 3 $\times 10^{17}$ kg/m^3. In case of neutron stars also it has been found that mass of a star is inversely proportional to its volume. That is, a neutron star having large mass is small in size.

The density of a neutron star is not uniform. Its density varies from below 1 $\times 10^9$ kg/m^3 in the crust - increasing with depth - to above 6 $\times 10^{17}$ kg/m^3 or 8 $\times 10^{17}$ kg/m^3 deeper inside.

Some neutron stars rotate very rapidly and emit beams of electromagnetic radiation as pulsars. Non-rotating and non-accreting neutron stars are virtually undetectable.

Pulsars

Radio astronomers have discovered first the celestial objects, called the pulsars, which correspond to the neutron stars. In 1967, scientists discovered pulsing radio sources which have been named the pulsars. Further investigations revealed that the pulsars are magnetized. These pulsars are spinning neutron stars. A new born neutron star can rotate several times a second; sometimes, the neutron stars absorb the orbiting matter from a companion star, increasing the rotation to several hundred times per second, reshaping the neutron star into an oblate spheroid.

3.3 Black holes

A black hole is defined as a region of space-time having so high gravitational field which does not allow escaping of any thing including light. The theory of general relativity says that a sufficiently compact mass deforms the space-time to form a black hole. Around a black hole, there is a mathematically defined surface, called an event horizon that marks the point of no return. That is, any thing which has gone inside the event horizon never comes back outside. The hole is given name "black" as it absorbs all the light that incident the horizon. It is similar to a black body in thermodynamics which absorbs all kinds of electromagnetic radiations falling on it. Quantum field theory in curved space-time predicts that event horizons emit radiation like a black body with a finite temperature. This temperature is inversely proportional to the mass of the black hole, making it difficult to observe this radiation for black holes of stellar mass or greater.

At the end of the life cycle of a very massive star, a black hole of stellar mass is expected to form. After the formation of a black hole, it continues to grow by absorbing mass from its surroundings. By absorbing other stars and merging with other black holes, supper massive black holes of millions of solar masses may form. There is general consensus that the super-massive black holes exist in the centers of most galaxies.

Despite its invisible interior, the presence of a black hole can be inferred through its interaction with other matter and with electromagnetic radiation, such as light. Matter falling into a black hole can form an accretion disk heated by the friction, forming some of the bright objects in the universe. If there are other stars orbiting a black hole, their orbit can be used to determine its mass and location. In this way, the astronomers have identified numerous stellar black hole candidates in binary systems, and established that the core of Milky Way contains a super-massive black hole of about 4.3 million solar masses.

In case of a neutron star, the escape speed from its surface is typically a good fraction of the speed of light. Now, in case of a black hole, the radius R of an object of mass M to have the escape speed from the surface equal to the speed of light has been obtained. This radius, known as the Schwarzschild radius, is

$$R_{sch} = \frac{2GM}{c^2}$$

The numerical value of the Schwarzschild radius for a 1 $M_\odot$ is $R_{sch} = 3$ km.

Example 2: For a star having 1 solar mass, calculate the Schwarzschild radius. Given, solar mass $M_\odot = 1.99 \times 10^{30}$ kg, gravitational constant $G = 6.673 \times 10^{-11}$ m^3/s^{2}kg and speed of light $c = 3 \times 10^8$ m/s.

Solution: The Schwarzschild radius is

$$R_{sch} = \frac{2GM}{c^2} = \frac{2(6.673 \times 10^{-11})(1.99 \times 10^{30})}{(3 \times 10^8)^2} = 2.9 \times 10^3 \text{ m}$$

Example 3: Estimate the smallest possible Schwarzschild radius for a black hole of 3 $M_\odot$ that forms as the radius of a supernovae. The mass of Sun is 2×10^{30} kg.

Solution: The Schwarzschild radius of a black hole is proportional to the mass of the collapsing core. The smallest possible Schwarzschild radius is therefore expressed by the smallest possible mass that is capable of overcoming neutron degeneracy pressure, thought to be about 3 solar mass. Thus, we have

$$R_{sch} = \frac{2GM}{c^2} = \frac{2(6.67 \times 10^{-11})(6 \times 10^{30})}{(3 \times 10^8)^2} \approx 9000 \text{ m}$$

A Schwarzschild radius of 9 km is the smallest possible.

Example 4: In the model for the core of an active galaxy an accretion disk forms around the super giant black hole. This is composed of former stellar material ripped apart by the tidal forces of the black hole and has a radius of about 10^{14} m. Calculate the upper limit for the power limited by the system when about one solar mass moves from the accretion disk the black hole (of 10^8 solar masses) per year.

Solution: The Schwarzschild radius of the black hole is

$$R_{sch} = \frac{2GM}{c^2} = \frac{2(6.67 \times 10^{-11})(10^8 \times 2 \times 10^{30})}{(3 \times 10^8)^2} = 3 \times 10^{11} \text{ m}$$

Energy released after the matter has entered within the Schwarzschild radius is not observable and so, in the calculations the release of gravitational potential energy, this represents the 'end point'. The rate at which matter enters inside the Schwarzschild radius is 2×10^{30} kg/year or $(2 \times 10^{30}/(365 \times 24 \times 60 \times 60) =)\ 6 \times 10^{22}$ kg/s. The gravitational energy released is given by the gravitational energy of matter while in the accretion disk (radius r_a) minus that at the Schwarzschild radius is

$$\text{energy released } E = \frac{GMm}{R_{sch}} - \frac{GMm}{r_a} = GMm\Big(\frac{1}{R_{sch}} - \frac{1}{r_a}\Big)$$

$$= (6.67 \times 10^{-11})(10^8 \times 2 \times 10^{30})(6 \times 10^{22})\Big(\frac{1}{3 \times 10^{11}} - \frac{1}{10^{14}}\Big) = 3 \times 10^{39} \text{ J}$$

This is equal to the amount of energy released per second, in other words, the active galaxy has a potential output of more than 10^{39} W.

Example 5: Compare the densities of a white dwarf, a neutron star and a main sequence star where all three bodies are of about 1.4 solar mass.

Solution: A white dwarf of 1.4 solar mass corresponds to the most massive white dwarf that can withstand electron degeneracy pressure and therefore has the smallest possible radius for a white dwarf. This is similar to the Earth's size (radius $\sim 10^7$ m). The density is

$$\text{density} = \frac{\text{mass}}{\text{volume}} = \frac{3 \times 10^{30}}{4\pi(10^7)^3/3} \approx 10^9 \text{ kg/m}^3$$

The largest neutron star (those of minimum mass 1.4 solar mass) have a radius of 20 km. The density is

$$\text{density} = \frac{\text{mass}}{\text{volume}} = \frac{3 \times 10^{30}}{4\pi(2 \times 10^4)^3/3} \approx 10^{17} \text{ kg/m}^3$$

A main sequence star of 1.4 solar mass has a similar density to the Sun (which shows that slightly more massive main sequence stars than the Sun have slightly larger radii). The Sun density is

$$\text{density} = \frac{\text{mass}}{\text{volume}} = \frac{2 \times 10^{30}}{4\pi(7 \times 10^8)^3/3} \approx 10^3 \text{ kg/m}^3$$

The densities of three bodies of the same mass are thus radically different.

4. Jeans criterion

In physics, the Jeans instability causes the collapse of an interstellar gas cloud and subsequent formation of star. It occurs when the internal gas pressure is not strong enough to prevent gravitational collapse of a region filled with the matter. For the stability, the cloud must be in the hydrostatic equilibrium, which in case of a spherical cloud is described through the relation

$$\frac{\mathrm{d}P}{\mathrm{d}r} = -\frac{G\rho(r)M(r)}{r^2}$$

where $M(r)$ is the mass enclosed in the sphere of radius r, P the pressure, $\rho(r)$ the spherically symmetric density of gas at a distance r from the center of star, G is the gravitational constant and r the radius. The equilibrium is stable if small perturbations are damped and is unstable if the perturbations are amplified. In general, the cloud is unstable if it is either very massive at a given temperature or very cool at a given mass for gravity to overcome the gas pressure.

4.1 Jeans mass

Jeans considered the process of gravitational collapse within a gaseous cloud, and he was able to show that, under the appropriate conditions, a cloud, or a part of cloud, would become unstable and begin to collapse when it does not have sufficient gaseous pressure support to balance the force of gravity. The cloud is stable for sufficiently small mass (at a given temperature and radius), but once this critical mass is exceeded, it begins a process of runaway contraction until some other force can oppose the collapse. He derived a formula for calculation of this critical mass as a function of its density and temperature. The greater the mass of the cloud, the smaller its size, and the colder its temperature, the less stable it is against the gravitational collapse.

The approximate value of the Jeans mass may be derived through a simple physical argument. One begins with a spherical gaseous region of radius R, mass M, and with a gaseous sound speed c_s. Imagine that we compress the region slightly. It takes a time,

$$t_{\text{sound}} = \frac{R}{c_s} \simeq (5 \times 10^5 \text{yr}) \left(\frac{R}{0.1 \text{ pc}}\right) \left(\frac{c_s}{0.2 \text{ km/s}}\right)^{-1}$$

for sound waves to cross the region, and attempt to push back and re-establish the system in pressure balance. At the same time, gravity attempts to contract the system even further, and does so on a free-fall time,

$$t_{\rm ff} = \frac{1}{\sqrt{G\rho}} \simeq (2 \text{ Myr}) \left(\frac{n}{10^9 \text{ m}^{-3}}\right)^{-1/2}$$

where G is the universal gravitational constant, ρ is the gas density within the region, and $n = \rho/\mu$ is the gas number density for mean mass per particle $\mu = 3.9 \times 10^{-27}$ kg, appropriate for the molecular hydrogen with 20% helium by number. Now, two situations may arise: (i) When the sound-crossing time is less than the free-fall time, pressure forces temporarily overcome the gravity, and the system returns to a stable equilibrium. (ii) When the free-fall time is less than the sound-crossing time, the gravity overcomes the pressure forces, and the region undergoes the gravitational collapse. The condition for the gravitational collapse is therefore:

$$t_{\rm ff} < t_{\rm sound}.$$

The resultant Jeans length λ_J is approximately:

$$\lambda_J = \frac{c_s}{\sqrt{G\rho}} \simeq (0.4 \text{ pc}) \left(\frac{c_s}{0.2 \text{ km/s}}\right) \left(\frac{n}{10^9 \text{ m}^{-3}}\right)^{-1/2}$$

This length scale is known as the Jeans length. All the scales larger than the Jeans length are unstable to the gravitational collapse, whereas the smaller scales are stable. The Jeans mass M_J is just the mass contained in a sphere of radius R_J (R_J is half the Jeans length):

$$M_J = \left(\frac{4\pi}{3}\right) \rho R_J^3 = \left(\frac{\pi}{6}\right) \frac{c_s^3}{G^{3/2}\rho^{1/2}} \simeq (2 \text{ M}_\odot) \left(\frac{c_s}{0.2 \text{ km/s}}\right)^3 \left(\frac{n}{10^9 \text{ m}^{-3}}\right)^{-1/2}$$

Later on, it was pointed out by other astrophysicists that in fact, the original analysis used by Jeans was flawed, for the following reason. In his formal analysis, Jeans assumed that the collapsing region of the cloud was surrounded by an infinite, static medium. In fact, because all the scales greater than the Jeans length are also unstable to collapse, any initially static medium surrounding a collapsing region in fact also is collapsing. Therefore, the growth rate of the gravitational instability relative to the density of the collapsing background is slower than that predicted by Jeans' original analysis. This flaw has come to be known as the Jeans swindle.

4.2 Jeans length

Jeans length is the critical radius of a cloud (typically a cloud of interstellar dust) where thermal energy, which causes the cloud to expand, is counteracted by gravity, which causes the cloud to collapse. The formula for the Jeans length is:

$$\lambda_J = \sqrt{\frac{15kT}{4\pi G\mu\rho}},$$

where k is the Boltzmann constant, T is the temperature of the cloud, r is the radius of the cloud, μ is the mass per particle in the cloud, G is the gravitational constant and ρ is the cloud's mass density (*i.e.*, the cloud's mass divided by the cloud's volume).

Perhaps the easiest way to conceptualize the Jeans length is in terms of a close approximation, where the factors 15 and 4π are discarded and ρ is rephrased as M/r^3. Then, the Jeans length is expressed as

$$\lambda_J \approx \sqrt{\frac{kTr^3}{GM\mu}}.$$

It follows immediately that when $kT = GM\mu/r$, we have $\lambda_J = r$ *i.e.*, the cloud's radius is the Jeans length when the thermal energy per particle equals gravitational work per particle. At this critical length the cloud neither expands nor contracts. It is only when the thermal energy is not equal to the gravitational work that the cloud either expands and cools or contracts and warms, a process that continues until equilibrium is attained.

5. Transport of energy inside a star

Prime source of energy in the stars is the thermonuclear reactions which produce energy in their deep interior as there are pressures and temperatures high enough to sustain thermonuclear reactions. However, most of the luminous energy of a star is radiated from a thin region at its surface, called the photosphere. Hence, an important issue in this context is the mode of transport of energy from the deep interior to the photosphere. There are 4 important processes discussed by the scientists for the transport of such energy:

(i) Radiation of photons

(ii) Conduction

(iii) Convection

(iv) Radiation of neutrinos

5.1 Radiation of photons

The most common method of energy transport in the normal stars is through the photons of various wavelengths, ranging from γ-rays to radio waves, depending on the kinetic temperature in the region. For the propagation of radiations, no medium is essentially required. However, a medium is a kind of obstacle and it reduces the speed of radiation. The stellar material in the deep interior is very opaque, and therefore a radiation travels only a small distance before it is absorbed. It is then re-emitted in a random direction, absorbed after a small distance, remitted, and so on until it reaches the photosphere. For example, in the case of the Sun, the average distance traveled by a photon between two consecutive absorptions is about a centimeter, and it takes perhaps hundreds of thousands of years for the energy released in the deep interior to make its way to the surface.

5.2 Conduction

Conduction is the way in which metals transport heat from one place to another. For the conduction process, a medium is required which should be a good conductor of heat. Conduction is not important in most normal stars because the atmosphere inside a star is approximately an ideal gas, which is a thermal insulator.

5.3 Convection

If the rate of production of energy in a star is much larger than the rate of transport through radiation, the stellar matter may boil. In the boiled material, the transport of energy is through the convection. The convection is a familiar process from the boiling water. The convection is a very efficient method of energy transport because it involves the vertical motion of large packets of gas.

Remark: In most normal stars the energy transport is by radiation unless the rate at which energy is being produced in the interior exceeds a critical value, in which case the transport becomes convective. In many stars both may operate: some regions of the interior may transport heat by convection and some by radiative transport.

5.4 Radiation of neutrinos

In a massive star, late in its life the amount of energy that must be transported is sometimes larger than either radiation of photons or convection. In these cases, significant amounts of energy may be transported from the center to space by the radiation of neutrinos. This is the dominant method of cooling of stars in advanced burning stages, and also plays a central role in the events like supernovae associated with the death of massive stars.

6. Simple stellar model

Here, we have developed a simple stellar model where the radial variation of density is expressed as

$$\rho(r) = \rho_c\left(1 - \frac{r}{R}\right)$$

where ρ_c is the density at the center of the star and R is the radius of star. For this spherically symmetric density profile, let us calculate various parameters of the star.

6.1 Mass

For spherically symmetric density $\rho(r)$, the mass of a star within a radius r is expressed by the equation of mass continuity.

$$M(r) = \int_0^r \rho(r) 4\pi r^2 \, dr$$

Using the expression for $\rho(r)$, we have

$$M(r) = \int_0^r \rho_c\Big(1 - \frac{r}{R}\Big)4\pi r^2 \, \mathrm{d}r = \frac{4\pi r^3}{3}\,\rho_c\Big(1 - \frac{3r}{4R}\Big)$$

The total mass of the star is

$$M = M(R) = \frac{\pi R^3}{3}\,\rho_c$$

Thus, the central density in terms of total mass is

$$\rho_c = \frac{3M}{\pi R^3}$$

6.2 Pressure

For the hydrostatic equilibrium, the pressure gradient is expressed as

$$\frac{\mathrm{d}P}{\mathrm{d}r} = -\frac{GM(r)\rho(r)}{r^2}$$

Using expressions for $M(r)$ and $\rho(r)$, we get

$$\frac{\mathrm{d}P}{\mathrm{d}r} = \pi G\rho_c^2\Big(-\frac{4r}{3} + \frac{r^2}{R} + \frac{4r^2}{3R} - \frac{r^3}{R^2}\Big)$$

On integration, we get

$$P(r) - P_c = \int_0^r \pi G\rho_c^2\Big(-\frac{4r}{3} + \frac{r^2}{R} + \frac{4r^2}{3R} - \frac{r^3}{R^2}\Big)\,\mathrm{d}r$$

$$P(r) = P_c - \pi G\rho_c^2\Big(\frac{2r^2}{3} - \frac{7r^3}{9R} + \frac{r^4}{4R^2}\Big)$$

The central pressure, can be obtained by considering the fact that the pressure at the surface must be zero. That is

$$P(R) = P_c - \pi G\rho_c^2\Big(\frac{2R^2}{3} - \frac{7R^3}{9R} + \frac{R^4}{4R^2}\Big) = 0$$

This relation gives

$$P_c = \pi G\rho_c^2\Big(\frac{5R^2}{36}\Big)$$

6.3 Temperature

The temperature $T(r)$ at the distance r from the center of the star, can be calculated in terms of the central density ρ_c and mean molecular weight by assuming that the star obeys the ideal gas relation. The ideal gas law says in terms of the mean molecular weight μ as

$$P(r) = \frac{\rho(r)kT(r)}{\mu m_H} \qquad \text{or} \qquad T(r) = \frac{P(r)\mu m_H}{\rho(r)k}$$

where $P(r)$ is the pressure at a distance r from the center of star, μ the mean molecular weight, m_H the mass of hydrogen atom and k the Boltzmann constant. On using the expressions for $\rho(r)$ and $P(r)$, we get

$$T = \frac{\mu m_H \left[P_c - \pi G\rho_c^2\left(\frac{2r^2}{3} - \frac{7r^3}{9R} + \frac{r^4}{4R^2}\right)\right]}{k\rho_c(1 - r/R)}$$

So the temperature at the center $(r = 0)$ is

$$T_c = T(0) = \frac{\mu m_H P_c}{k\rho_c}$$

Example 6: A gas contains, by number, 80% molecular hydrogen and 20% helium. Calculate the mean mass per particle.

Solution: In amu, mass of hydrogen molecule is 2 and that of helium is 4. Thus,

$$\text{Average atomic mass} = \frac{80 \times 2 + 20 \times 2}{100} = 2.4$$

$$\text{Mean mass per particle} = 2.4 \times (1.66 \times 10^{-27} = 3.984 \times 10^{-27} \text{ kg}$$

6.4 Polytropic stellar model

In astrophysics, a polytrope refers to a solution of the Lane-Emden equation in which the pressure P depends upon the density ρ in the form

$$P = K\rho^{(n+1)/n}$$

where K is a constant and the constant n is known as the polytropic index. This relation need not be interpreted as an equation of state. This is simply a relation that expresses an assumption about the change of pressure with radius in terms of the change of density with radius. For this pressure, the Lane-Emden equation is solved.

7. Binary stars

A binary system is a pair of stars which are bound to each other under the mutual gravitational attraction. In some binary systems, both the stars may be of comparable masses whereas in others, the mass of one star may be very small in comparison to that of the other. The heavier component is known as the primary, whereas the lighter one as the secondary. Both the components of binary system rotate around their center-of-mass in the Keplerian orbits. The primary makes a small ellipse with focus on the center-of-mass whereas the secondary makes a large ellipse with focus on the same point (center-of-mass).

Both the ellipses lie in a common plane known as the orbital-plane. This orbital-plane is generally inclined to the sky-plane.[3] When the orbital-plane is projected on

[3] A plane whose normal is the line-of-sight is known as the sky-plane.

the sky-plane, the elliptic orbit makes another ellipse, but the focus of the original ellipse (*i.e.*, the center-of-mass) may not be at the focus of the projected ellipse.

For a binary system, it is possible to get the information about the masses and sizes of the component stars. This information about the stars is required as input for the theories of origin and evolution of stars. Some binaries having a compact object, like a white dwarf, a neutron star or a black hole as one component may help for the verification of the General Theory of Relativity. Depending on the method used for its study, the binary may be called a visual binary, a spectroscopic binary or an eclipsing binary. In some cases, a binary may be visual and spectroscopic or eclipsing; it may be spectroscopic and eclipsing.

The center-of-mass of a binary may not be stationary. Since there is no net external force acting on the system, the center-of-mass may move along a straight line and the components of the system trace out serpentine paths. Period of revolution of a binary range from 5-10 years to thousands of years. Hence, it may not be possible for an astronomer to observe a complete orbit of a binary system.

7.1 Visual binary

When a binary system can be seen (resolved) unaided or with the help of a telescope, the system is known as the visual binary. Mizar (ς UMa) has been the first visual binary discovered in 1650.

7.2 Astrometric binary

In a visual binary, it is possible that one of its components is not visible and only the brighter one is visible. The brighter component may be the primary or the secondary of the system. Presence of the faint component can however be inferred from the wiggly motion of the visible component with respect to the background stars. Such a system is known as the astrometric binary.

Sirius (Figure 4) is a famous visual binary in which Sirius A is a main sequence star of spectral type A1 and Sirius B is a white dwarf of spectral type A5 (which can be detected with efforts), orbiting about one another with a period of 49.9 years.

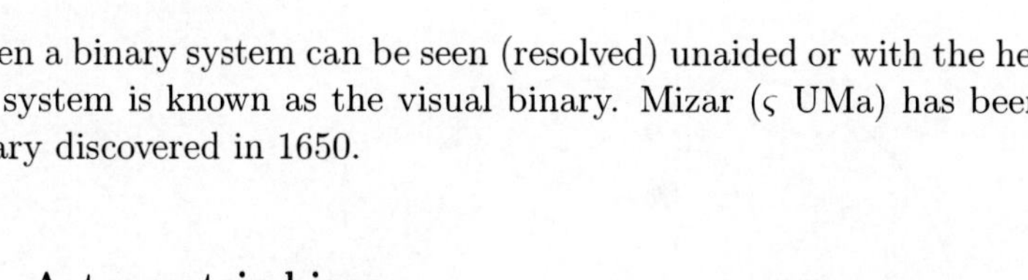

Figure 4: Schematic diagram for the center-of-mass of Sirius (a binary system).

Even with the help of a good telescope, it is not easy to see the white dwarf in the glare of much brighter companion. The faint companion of the Sirius was discovered later on.

7.3 Spectroscopic binary

When the two components of a binary system cannot be resolved or seen even with the help of a telescope, the system may be studied through the oscillations of their spectral lines due to Doppler effect. Such a study is only possible when the angle between the orbital-plane and the sky-plane is reasonably large to have an appreciable component of velocity along the line-of-sight. Spectroscopic binaries are usually too distant and their components cannot be resolved even with the help of a telescope. The first star of this type was ς UMa A, discovered in 1889, the brighter component of the first observed visual binary.

In order to understand the situation, let us consider a simple view (Figure 5) where more bright star A and less bright star B of a binary system are moving in the circular orbits around the center-of-mass, C. Let the angle between the plane of the orbits and the sky-plane is 90°, *i.e.,* the line-of-sight is in the orbital-plane. (This is the best situation from the point of view of the observation of a spectroscopic binary.) Let us consider four positions of the stars A and B denoted by 1, 2, 3 and 4. The directions of motions of stars are shown by the arrows. Here, the primary star A is assumed to be heavier than the secondary star B.

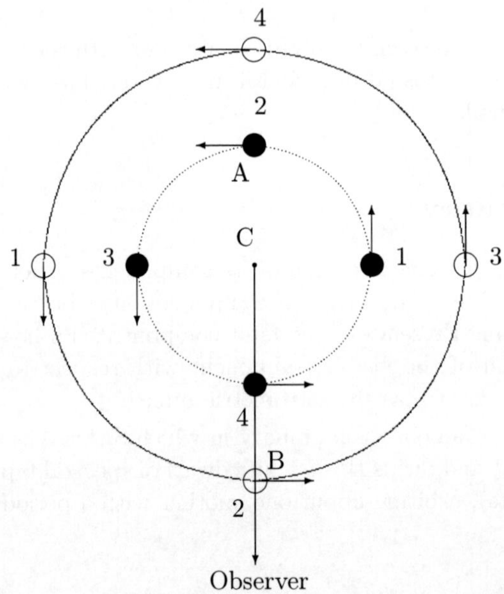

Figure 5: A simple view of a binary system where the primary star A and the secondary star B revolve in the circular orbits around the center-of-mass, C. The line-of-sight of the observer is in the orbital-plane.

When the stars are in the position 2 as well as in the position 4, they are moving across the line-of-sight and the spectral lines from the two stars overlap, as there is no Doppler shift, and we see only one set of the spectral lines. When the stars are in the position 1, the star A moves away from the observer whereas the star B moves towards the observer.

Owing the Doppler effect, the star A gives the red-shifted set of the spectral lines whereas the B gives the blue-shifted set of the spectral lines. Thus, we see two sets of spectral lines of different intensities, corresponding to the brightness of stars. When the stars are in the position 3, the star A moves towards the observer whereas the star B moves away from the observer. Now, the star A gives the blue-shifted set of the spectral lines whereas the B gives the red-shifted set of the spectral lines. Thus, we see two sets of spectral lines of reverse intensities. Therefore, when the stars revolve around the center-of-mass, each spectral line oscillates, as shown in Figure 6.

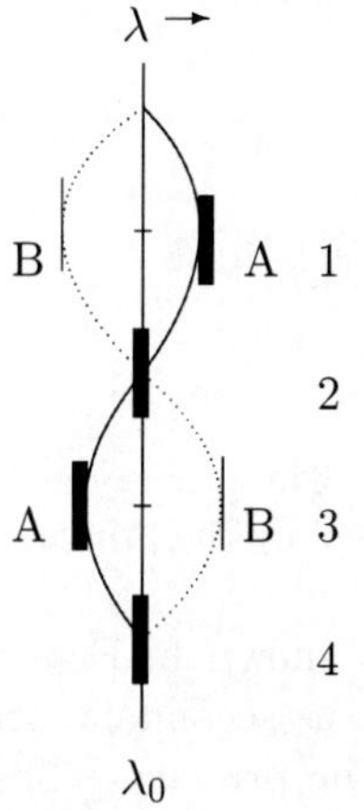

Figure 6: Shift of a spectral line from the rest wavelength λ_0 is shown. Here, 1, 2, 3 and 4 correspond to the positions of the stars A and B, given in Figure 5.

When one of the components of the binary is too faint for its spectral lines, we see only one set of oscillating spectral lines. Such binaries are called the single lined spectroscopic binaries, in contrast to double lined binaries.

7.4 Eclipsing binary

It is the case where the orbital-plane and the sky-plane are inclined in such a manner that the components of the binary system eclipse with each other. Algol was the first eclipsing binary discovered in 1782. The first x-ray eclipsing binary, Hercules-X1, was discovered in 1971. This situation arises when the inclination angle is very large (for ideal situation, it is 90°). This condition is more stringent than that for the spectroscopic binaries. But, the variation of brightness can be measured more easily than the variation of radial velocity. Therefore, a large number of eclipsing binaries, more than 5500, are already known.

When one component eclipses the other, the intensity of a line becomes less than that when the line joining the components is perpendicular to the line-of-sight. Further, the two components have different brightness, therefore, the decrease in the intensity is different when the primary eclipses the secondary from the situation when the secondary eclipses the primary.

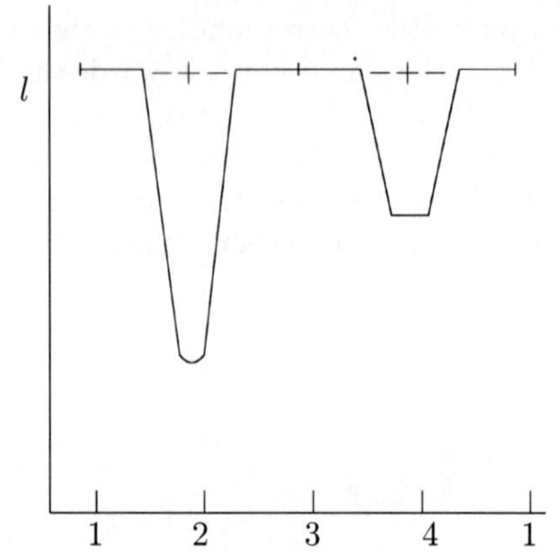

(c)

Figure 7: Light curve of an eclipsing binary. The primary is eclipsed by the secondary at the position 2 whereas the secondary is eclipsed by the primary at the position 4.

Light curve of an eclipsing binary is shown in Figure 7. Assuming that the intensity of primary is more than that of the secondary, schematic light-curve of an eclipsing binary is shown in Figure 7. The primary eclipse occurs at the position 2 (we get light from the secondary only) and the secondary eclipse occurs at the position 4 (we get light from the primary only). These points represent conjunction. The positions 1 and 3 are called quadratures when full light from both the primary and secondary is obtained.

7.5 Determination of masses of components

Consider a binary system having two stars A and B of masses m_1 and m_2, respectively, as shown in Figure 8. The two stars revolve about the center-of-mass, C. Let the distances of the stars A and B from the point C are d_1 and d_2, respectively. If T is the time-period of revolution of each of the stars, the Kepler's third law of planetary motion gives

$$M = \frac{4\pi^2 r^3}{GT^2} \tag{6.1}$$

where $M = (m_1+m_2)$ is the total mass of the system and $r = (d_1+d_2)$ the separation between the components A and B.

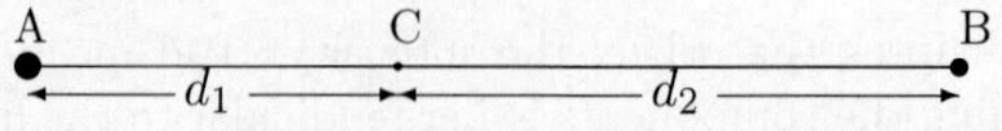

Figure 8: Binary system having two stars A and B.

We have

$$m_1 d_1 = m_2 d_2 \qquad \text{or} \qquad m_1 = \frac{m_2 d_2}{d_1} \tag{6.2}$$

Using equation (6.2) in (6.1), we get

$$\frac{m_2 d_2}{d_1} + m_2 = \frac{4\pi^2 (d_1 + d_2)^3}{GT^2}$$

or

$$m_2 = \frac{4\pi^2 d_1 (d_1 + d_2)^2}{GT^2} \tag{6.3}$$

Using equation (6.3) in (6.2), we get

$$m_1 = \frac{4\pi^2 d_2 (d_1 + d_2)^2}{GT^2} \tag{6.4}$$

Thus, for the known values of d_1, d_2 and T, the masses m_1 and m_2 of individual stars in a binary system can be determined with the help of equations (6.4) and (6.3), respectively.

Note: Here, we have assumed that the orbits of the stars are circular. Generally, they are elliptical. For the elliptical orbits, the analysis is similar, but becomes complicated.

Binary system with limited information

For determination of masses of the stars in a binary system, we often find that the information available is insufficient to apply the method discussed above. Such a situation arises because of the following.

- The orbits of the stars of the binary system may not be mapped exactly on the celestial sphere. It is true when the binary is astrometric where we can see only one star of the binary system and the second star is realized due to the wavy motion of the first star.

- Even when the orbits of both stars in a binary system can be mapped, we have two-dimensional projections on the celestial sphere of the true three-dimensional orbits. Hence, more information is required for the construction of true orbits.

Under such cases, we can determine only the sum of the masses of the stars or we may put limits on the masses rather than to determine the individual masses.

Example 7: The α-Centauri system is 1.338 pc distant with a period of 79.92 years. The A and B components have a mean separation of 23.7 AU. Assuming that the orbits are circular, calculate the total mass of the system.

Solution: We have

$$T = 79.92 \text{ yr} = 79.92 \times 365.25 \times 24 \times 60 \times 60 = 2.522 \times 10^9 \text{ s}$$

$$r = (d_1 + d_2) = 23.7 \text{ AU} = 23.7 \times (1.5 \times 10^{11}) = 3.555 \times 10^{12} \text{ m}$$

Total mass

$$M = \frac{4\pi^2 r^3}{GT^2} = \frac{4 \times 3.14^2 \times (3.555 \times 10^{12})^3}{6.67 \times 10^{-11} \times (2.522 \times 10^9)^2} = 4.18 \times 10^{30} \text{ kg}$$

Example 8: The α-Centauri system is 1.338 pc distant with a period of 79.92 years. The A and B components have a mean separation of 23.7 AU. If the α-Cent A has mean distance of 11.2 AU from the system's barycenter, assuming that the orbits are circular, calculate the masses of the components.

Solution: We have

$$d_1 + d_2 = 23.7 \text{ AU} = 23.7 \times (1.5 \times 10^{11}) = 3.555 \times 10^{12} \text{ m}$$

and

$$d_1 = 11.2 \text{ AU} = 11.2 \times (1.5 \times 10^{11}) = 1.68 \times 10^{12} \text{ m}$$

Therefore,

$$d_2 = 3.555 \times 10^{12} - 1.68 \times 10^{12} = 1.875 \times 10^{12} \text{ m}$$

The time-period is

$$T = 79.92 \text{ yr} = 79.92 \times 365.25 \times 24 \times 60 \times 60 = 2.522 \times 10^9 \text{ s}$$

The masses of the components are

$$m_1 = \frac{4\pi^2 d_2 (d_1 + d_2)^2}{GT^2} = \frac{4 \times 3.14^2 \times (1.875 \times 10^{12}) \times (3.555 \times 10^{12})^2}{6.67 \times 10^{-11} \times (2.522 \times 10^9)^2}$$

$$= 2.197 \times 10^{30} \text{ kg}$$

and

$$m_2 = \frac{4\pi^2 d_1 (d_1 + d_2)^2}{GT^2} = \frac{4 \times 3.14^2 \times (1.68 \times 10^{12}) \times (3.555 \times 10^{12})^2}{6.67 \times 10^{-11} \times (2.522 \times 10^9)^2}$$

$$= 1.969 \times 10^{30} \text{ kg}$$

Example 9: The approximate masses of A and B components of Sirius are about 2.1 and 1.0 solar masses, respectively. If the separation between the Sirius A and Sirius B at a particular time is 20 AU, calculate the distances of the stars from the center-of-mass of the system.

Solution: Suppose, m_A and m_B are the masses, and d_A and d_B the distances from the center-of-mass of the A and B components of Sirius, respectively. We have $m_A = 2.1\ M_\odot$, $m_B = 1.0\ M_\odot$ and $d_A + d_B = 20$ AU. Therefore,

$$\frac{d_A}{d_B} = \frac{m_A}{m_B} = 2.1 \qquad \text{or} \qquad d_A = 2.1 d_B$$

Therefore, we have

$$2.1 d_B + d_B = 20 \qquad \text{or} \qquad d_B = \frac{20}{3.1} = 6.45 \text{ AU}$$

Hence, we have

$$d_A = 20 - 6.45 = 13.55 \text{ AU}$$

Example 10: If the total mass of the Sirius binary is 3.2 solar masses and the period for rotation is 50.1 years, calculate the mean separation between the two stars. Given, mass of Sun = 1.99 $\times 10^{30}$ kg.

Solution: We have the total mass

$$M = 3.2 \times (1.99 \times 10^{30}) = 6.368 \times 10^{30} \text{ kg}$$

Time-period of revolution

$$T = 50.1 \text{ yr} = 50.1 \times 365.25 \times 24 \times 60 \times 60 = 1.58 \times 10^9 \text{ s}$$

The mean distance d between the components of the binary is

$$d = \left(\frac{MGT^2}{4\pi^2}\right)^{1/3} = \left[\frac{6.368 \times 10^{30}(6.67 \times 10^{-11})(1.58 \times 10^9)^2}{4 \times 3.14^2}\right]^{1/3}$$
$$= 2.996 \times 10^{12} \text{ m}$$

Example 11: Observations of Sirius show it to be a binary star. The Sirius is visible with the naked eye and the motion of both stars can be tracked by a telescope. The separation of stars is (on average) 20 AU and the orbital period is 50 years. Observations of Sirius A show it to have an orbital radius of 7 AU. Calculate the masses of Sirius A and B. Given, $G = 6.67 \times 10^{-11}$ m^3/s^2 kg.

Solution: We have time-period

$$T = 50 \text{ yrs} = 50 \times 365 \times 24 \times 60 \times 60 = 1.5768 \times 10^9 \text{ s}$$

$$r = 20 \text{ AU} = 20 \times 1.5 \times 10^{11} = 3 \times 10^{12} \text{ m}$$

$$r_1 = 7 \text{ AU} = 7 \times 1.5 \times 10^{11} = 1.05 \times 10^{12} \text{ m}$$

$$r_2 = 13 \text{ AU} = 13 \times 1.5 \times 10^{11} = 1.95 \times 10^{12} \text{ m}$$

The mass m_1 is

$$m_1 = \frac{4\pi^2 r^2 r_2}{GT^2} = \frac{4(3.14)^2 \times (3 \times 10^{12})^2 \times (1.95 \times 10^{12})}{6.67 \times 10^{-11} \times (1.5768 \times 10^9)^2} = 4.2 \times 10^{30} \text{ kg}$$

The mass m_2 is

$$m_2 = \frac{4\pi^2 r^2 r_1}{GT^2} = \frac{4(3.14)^2 \times (3 \times 10^{12})^2 \times (1.05 \times 10^{12})}{6.67 \times 10^{-11} \times (1.5768 \times 10^9)^2} = 2.2 \times 10^{30} \text{ kg}$$

8. Interstellar medium

The matter between the stars is known as the interstellar medium (ISM) and it has two components: (i) gas, and (ii) dust. It fills the interstellar space and blends smoothly into the surrounding intergalactic space. The energy that occupies the same volume, in the form of electromagnetic radiation, is the interstellar radiation field. By mass, 99% of the ISM is gas in any form, and 1% is dust.

8.1 Interstellar gas

The interstellar gas is in the form of atoms, ions and molecules. The interstellar medium is composed of multiple phases, distinguished by whether the matter is ionic, atomic, or molecular, and the temperature and density of the matter. The interstellar medium is composed primarily by hydrogen followed by helium with trace amounts of carbon, oxygen, and nitrogen comparatively to hydrogen.

In cool, dense regions of the ISM, matter is primarily in molecular form, and reaches number densities of 10^6 molecules/cm^3. In hot, diffuse regions of the ISM, matter is primarily ionized, and the density may be as low as 10^{-4} ions/cm^3. Compare this with a number density of roughly 10^{19} molecules/cm^3 for air, and 10^{10} molecules/cm^3 for a laboratory high-vacuum chamber. Of the gas in the ISM, by number 91% of atoms are hydrogen and 9% are helium, with 0.1% being atoms of elements heavier than hydrogen or helium, known as the metals in the astronomical language. By mass this amounts to 70% hydrogen, 28% helium, and 1.5% heavier elements. The hydrogen and helium are primarily a result of primordial nucleosynthesis, while the heavier elements in the ISM are mostly a result of enrichment in the process of stellar evolution.

8.2 Interstellar dust

The dust found in the space is known as the cosmic dust or as the interstellar dust. Most part of it consists of small dust particles which are of the size from few molecules to 0.1 μm. A smaller fraction of dust consists of larger refractory minerals that condensed as matter left the stars. It is called the star-dust. The dust density in the local interstellar medium of the Local Bubble is approximately $10^{-6}\times$ dust grain/m^3 with each grain having a mass of approximately 10^{-17} kg.

Cosmic dust was once solely an annoyance to astronomers, as it obscures objects they wish to observe. When infrared astronomy began, those dust particles were observed to be significant and vital components of astrophysical processes. Their analysis can reveal information about phenomena like the formation of our Solar System.

9. Review Questions

1. Discuss the process for the formation of a star and the problems encountered in it.

2. Describe the evolution of stars.

3. Describe the Jeans criterion for the formation of stars.

4. What is a white dwarf? Discuss about the mass-radius relationship.

5. What are supernovae? Explain their classification.

6. A star has spherically symmetric density

$$\rho(r) = \rho_c\Big(1 - \frac{r}{R}\Big)$$

where ρ_c is the density at the center of the star and R is the radius of star. For the star, calculate total mass, and pressure and temperature at the center.

7. Discuss various types of binary systems. How are the masses of the components of a visual binary system calculated?

8. Two components A and B of a binary system have masses m_1 and m_2, respectively. If the position vector of B relative to A is $\vec{r}$, show that the position-vectors of A and B, relative to the center-of-mass, respectively, are

$$\vec{r}_1 = -\frac{m_2}{m_1 + m_2}\,\vec{r} \qquad \text{and} \qquad \vec{r}_2 = \frac{m_1}{m_1 + m_2}\,\vec{r}$$

9. Two components A and B of a binary system have masses m_1 and m_2, respectively. If the velocity of B relative to A is $\vec{v}$, show that the position-vectors of A and B, relative to that of the center-of-mass, respectively, are

$$\vec{v}_1 = -\frac{m_2}{m_1 + m_2}\,\vec{v} \qquad \text{and} \qquad \vec{v}_2 = \frac{m_1}{m_1 + m_2}\,\vec{v}$$

10. Comment on the following:

(i) When the angle between the orbital-plane of a binary system and the sky-plane is zero, the system cannot be studied if it is not a visual binary.

(ii) For spectroscopic and eclipsing binary systems, the angle between the plane of the orbit of a binary system and the sky-plane should be as large as possible.

(iii) When a visual binary cannot be a spectroscopic as well as an eclipsing binary.

11. Write notes on the following:

(i) Protostars

(ii) Hayashi lines

(iii) Brown dwarfs

(iv) Main sequence stars

(v) Red-giant-branch phase

(vi) Asymptotic-giant-branch phase

(vii) Post-AGB stars

(viii) Supernova

(ix) White dwarf

(x) Chandrasekhar limit

(xi) Neutron stars

(xii) Pulsars

(xiii) Black holes

(xiv) Jeans criterion

(xv) Conduction

(xvi) Convection

(xvii) Polytropic stellar model

(xviii) Binary systems

(xix) Visual binary

(xx) Astrometric binary

(xxi) Spectroscopic binary

(xxii) Eclipsing binary

(xxiii) Interstellar medium

(xxiv) Interstellar dust

(xxv) Jeans criterion

(xxvi) Jeans mass

(xxvii) Jeans length

(xxviii) Interstellar medium

10. Multiple Choice Questions

1. Which is not a category of binary system?

   A. visual B. gravitational C. spectroscopic D. eclipsing

   Ans. B

2. The Chandrasekhar limit is

   A. 0.8 $M_\odot$ B. 1.4 $M_\odot$ C. 2.2 $M_\odot$ D. 3.1 $M_\odot$

   Ans. B

3. In the spectrum of a supernova, Balmer lines of hydrogen atom are present. The type of the supernova is

   A. I B. II C. I and II both D. neither I nor II

   Ans. B

4. In the spectrum of a supernova, Balmer lines of hydrogen atom are absent. The type of the supernova is

A. I B. II C. I and II both D. neither I nor II

Ans. A

5. In the spectrum of a supernova, Balmer lines of hydrogen atom are absent, strong ionized silicon absorption line is present. The type of the supernova is

A. Ia B. Ib or Ic C. all the Ia, Ib, Ic D. none of Ia, Ib, Ic

Ans. A

6. In the spectrum of a supernova, Balmer lines of hydrogen atom as well as the absorption line of ionized silicon both are absent. The type of the supernova is

A. Ia B. Ib or Ic C. all the Ia, Ib, Ic D. none of Ia, Ib, Ic

Ans. B

7. In the spectrum of a supernova, Balmer lines of hydrogen atom as well as the absorption line of ionized silicon both are absent, the strong lines of neutral helium are present. The type of the supernova is

A. Ib B. Ic C. both Ib and Ic D. neither Ib nor Ic

Ans. A

8. In the spectrum of a supernova, Balmer lines of hydrogen atom, absorption line of ionized silicon, strong lines of neutral helium all are absent. The type of the supernova is

A. Ib B. Ic C. both Ib and Ic D. neither Ib nor Ic

Ans. B

9. Mass of a white dwarf is

A. proportional to its volume B. inversely proportional to its volume
C. proportional to its radius D. inversely proportional to its radius

Ans. B

10. Mass of a neutron star is

A. proportional to its volume B. inversely proportional to its volume
C. proportional to its radius D. inversely proportional to its radius

Ans. B

11. Mass of a neutron star can be

A. 0.8 $M_\odot$ B. 1.2 $M_\odot$ C. 2.2 $M_\odot$ D. 10.4 $M_\odot$

Ans. C

12. For the study of which type of binary, the Doppler effect is used.

A. Visual B. Astrometric C. Spectroscopic D. Eclipsing

Ans. C

13. White dwarfs are generally composed of

A. Hydrogen and Helium B. Carbon and Oxygen

C. Magnesium and Calcium D. Iron and Nickel

Ans. B

14. In which type of binary, both the components can be seen.

A. Eclipsing B. Spectroscopic C. Astrometric D. Visual

Ans. D

15. In which type of binary, only one component can be seen.

A. Eclipsing B. Spectroscopic C. Astrometric D. Visual

Ans. C

16. The best situation for the study of a binary system with the help of Doppler effect is when the angle between the orbital plane and sky plane is

A. 0° B. 45° C. 90° D. 180°

Ans. C

17. The best situation for the study of a binary system with the help of eclipsing method is when the angle between the orbital plane and sky plane is

A. 0° B. 45° C. 90° D. 180°

Ans. C

18. In a degenerate gas, the pressure depends on

A. Volume B. Volume and temperature both

C. Temperature D. Neither volume nor temperature

Ans. B

19. The term 'electron degeneracy pressure' is used in the context of

A. white dwarf B. neutron star C. pulsar D. black hole

Ans. A

20. The term 'neutron degeneracy pressure' is used in the context of

A. white dwarf B. neutron star C. Sun D. black hole

Ans. B

21. Population I stars are rich in

A. Hydrogen B. Helium C. Heavy elements D. Carbon

Ans. C

22. Stars having zero population of heavy elements are known as population

    A. I stars B. II stars C. III stars D. Zero stars

    Ans. C

23. The stars which are rich in heavy elements are known as population

    A. I stars B. II stars i C. III stars D. zero stars

    Ans. A

24. The stars which are poor in heavy elements are known as population

    A. I stars B. II stars i C. III stars D. zero stars

    Ans. B

25. In a black-hole, the relation between the temperature T of event-horizon and mass M of the black-hole is

    A. $T \propto M^{1/2}$ B. $T \propto M$ C. $T \propto M^{-1}$ D. $T \propto M^{-1/2}$

    Ans. C

26. Our universe does not have

    A. white dwarf B. neutron star C. black dwarf D. brown dwarf

    Ans. C

27. Mass of a neutron star cannot be

    A. 1.6 $M_\odot$ B. 2 $M_\odot$ C. 2.4 $M_\odot$ D. 1 $M_\odot$

    Ans. D

28. Which one of the processes for the transport of energy in a star from the deep interior to its photosphere is less probable.

    A. Radiation B. Conduction C. Convection D. All of them

    Ans. B

29. When the energy produced through thermonuclear reaction in the deep interior of a star cannot be transported through radiations at the same rate, the additional mode of the transport of energy becomes

    A. Conduction B. Convection C. Both of them D. None of them

    Ans. B

7. Sun

As stated earlier, the Sun is one of the numerous stars in the universe. Since it is the nearest star to us, it has been studied in detail. The next nearest star is the Proxima Centauri at a distance of 4.2 ly. The Sun is situated at a distance of 27,000 ly from the center of our galaxy Milky Way.

Scientists have obtained good amount of knowledge about the Sun. The Sun is considered to be situated at the center of our solar system. The Sun is a ball of hot gas held together by its gravity. It consists mainly of hydrogen and helium which are in the ionized state, because of very high temperature in the core of the Sun. In addition to the hydrogen and helium, there are other elements like carbon, nitrogen, oxygen, neon, etc., in very small proportions. The mass of the Sun has been estimated as 1.99×10^{30} kg and its average radius as 6.96×10^5 km. The visual magnitude of the Sun is $-$ 26.74 and the absolute magnitude is +4.8. That is, the Sun can be visible to naked eyes from a distance of 10 pc. Some physical parameters of the Sun are summarized in Table 1.

Table 1: Physical parameters of Sun

Age	4.5×10^9 years
Mass ($M_\odot$)	1.99×10^{30} kg
Radius ($R_\odot$)	6.96×10^5 km
Mean density	1.4×10^3 kg/m^3
Mean distance from Earth	1.50×10^8 km
Escape velocity	618 km/s
Luminosity ($L_\odot$)	3.86×10^{26} W
Equatorial rotation period	26 days
Polar rotation period	31 days
Mass loss rate	10^9 kg/s

The Sun rotates about its own axis. However, it does not rotate like a rigid body, but its different layers rotate with different velocities. The layer along its equator rotates with the rotation period of about 26 days. The period of rotation goes on increasing away from the equator and it is about 31 days at the poles. The Sun is a magnetically active star. It supports the strong, changing magnetic field that varies from year to year and reverses its polarity every time in a period of about eleven years. The Sun's magnetic field gives rise to many effects which are collectively known as the solar activities, including the sunspots on the solar surface, solar flares, and variations in the solar wind which carries material moving through the solar system. Since the electron density in the ionosphere around the

Earth varies with the solar activity, the radio communication (where the reflection through the ionosphere is used) depends on the solar activity.

Solar radiations are the primary source of energy for our Earth. The Sun radiates its energy in all directions. The amount of energy received at 1 AU from the Sun (*i.e.*, at the distance of the Earth) is approximately equal to 1.37 kW per square meter. This amount of energy reaching per square meter per second on the surface of the Earth is known as the solar constant. The sunlight arriving the Earth's surface is attenuated by the Earth's atmosphere, so that very little part of the solar energy reaches the Earth's surface.

Although the Sun, being the nearest star to the Earth, has been studied intensively for quite a long time, many questions about it are still waiting for answers. For example, the existence of the Sun; the reversibility of polarity of magnetic field of the Sun. For convenience, the study of the Sun can be divided into two parts: (i) solar interior, and (ii) solar atmosphere.

1. Solar interior

The interior part of the Sun is completely opaque and we cannot look into it. There are no direct observations about the solar interior. However, theoretical models suggest that there are three major regions in the solar interior, known as the core, radiative zone, and convective zone. A schematic view of general structure of the Sun is shown in Figure 1.

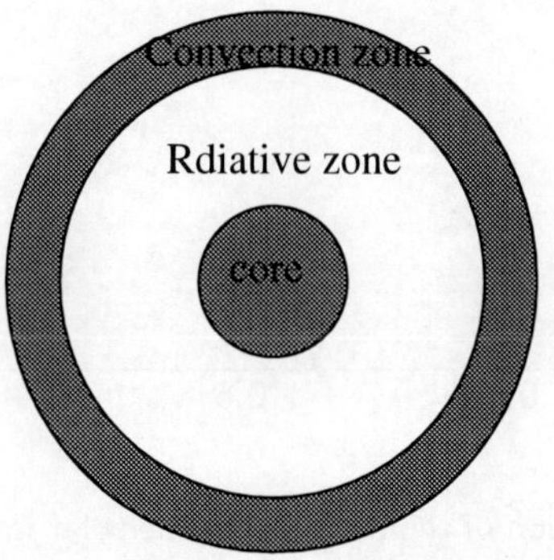

Figure 1: Schematic structure of the Sun.

The Sun does not have a definite boundary. The density of its gases drops approximately exponentially with increasing distance from its center. The solar radius, denoted by $R_{\odot}$, is measured from its center to the edge of the photosphere. At any point in the interior, the pressure associated with the thermal energy of the gas exactly balances the weight of the overlying material, a balance known as the hydrostatic equilibrium. The equations of radiative and hydrostatic equilibrium, together with the perfect gas law, govern the structure of the solar interior from the edge of the core out a distance equal to about one third of the solar radius.

The escape of radiation from the core is not completely blocked. Photons proceed from one collision to the next, diffusing outwards in a manner that looks like as a random walk. A temperature gradient is established between the hot core and the cool outer layers which promotes the outward flow of heat. The mean energy of the photons decreases with the decrease of the temperature. The variation of temperature in the solar interior is shown in Figure 2.

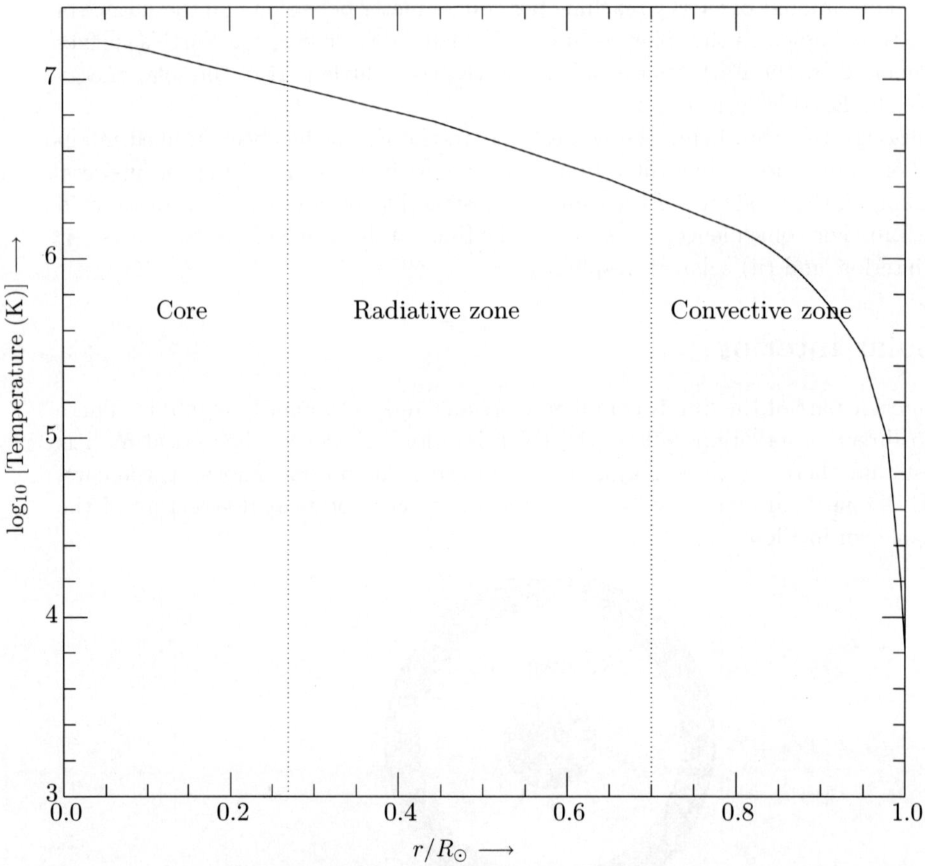

Figure 2: Variation of temperature in the solar interior.

The electromagnetic radiation leaks out slowly, but the interior is very opaque, like a thick fog, so that it undergoes under many reflections. By the time, the radiation reaches the photosphere, most of the photons change through numerous collisions with plasma from x-rays to visible light. If the photons had come straight out, they would have taken two seconds only. But, there are so many reflections that it takes from 10,000 to 170, 000 years to reach the surface of the Sun from the core.

1.1 Core

The core of the Sun is considered to extend from the center to about 0.2 - 0.25 $R_\odot$ and it contains about 34% of total solar mass. It has a density of up to 1.5×10^5 kg/m^3

and a temperature of close to 15.7×10^6 K (by contrast, the surface temperature of Sun is about 5800 K). During most of the solar life, the energy is produced by nuclear fusion through a series of steps called the *p-p* (proton-proton) chain where the hydrogen is converted into helium. The core is the only location in the Sun where the temperature is sufficient to initiate the thermonuclear reactions for production of an appreciable amount of heat. The rest of the star is heated by the energy that is transferred outward from the core. All of the energy produced by the fusion in the core must travel through many successive layers to the solar photosphere before it escapes into the space in the form of sunlight or kinetic energy of the particles.

The rate of nuclear fusion depends strongly on the density and temperature, and therefore the fusion rate in the core is in a self-correcting equilibrium: a slightly higher rate of fusion would cause the core to heat up more and to expand slightly against the weight of the outer layers.

The high-energy photons (gamma rays) released in the fusion reactions are absorbed in only a few millimeters of the solar plasma and then re-emitted. After the final trip through the convective outer layer to the transparent surface of the photosphere, the photons escape in the form of visible light. Each gamma radiation in the Sun's core is converted into several million visible light photons before escaping into space.

Energy generation in Sun

The Sun shines and emits tremendous amount of energy per second. Scientists thought that the self-energy of the Sun could be the source of energy being emitted from the sum. The self energy of Sun is

$$E_s = \frac{3}{5}\frac{GM_\odot^2}{R_\odot}$$

Using the values, we have

$$E_s = \frac{3}{5}\frac{6.67 \times 10^{-11}(1.99 \times 10^{30})^2}{6.96 \times 10^8} = 2.28 \times 10^{41} \text{ J}$$

Luminosity of Sun $L_\odot = 3.85 \times 10^{26}$ J/s. The period for which the self-energy in the Sun is sufficient is

$$T = \frac{2.28 \times 10^{41}}{3.85 \times 10^{26}} = 5.92 \times 10^{14} \text{ s} = 1.88 \times 10^7 \text{ yr}$$

Thus, the self-energy is sufficient for the illumination of sun for a period of 1.88×10^7 years. On the other side the life of the Sun is 4.5×10^9 years. It shows that the self-energy of Sun is not sufficient source for its illumination.

Then, it was thought that the source of such energy could be conversion of its own material (nuclear reaction) into the illumination energy. Nuclear reactions can be initiated at a temperature of the order of 10^6 - 10^7 K. For a celestial body consists of mono-atomic gas with atomic mass μ having radius R and total mass M in hydrostatic equilibrium, the temperature in its core is

$$T = \frac{GM\mu}{5NkR}$$

where $N = 6.023 \times 10^{26}$ is the Avogadro number. Using the parameters for the Sun, we have

$$T = 4.59 \times 10^6 \mu$$

Hence, the temperature of the central core of the Sun is of the order of 10^6 - 10^7 K, which is sufficient for the initiation of nuclear reactions.

When two light nuclei combine to form a single larger nucleus, the process is known as the nuclear fusion. The central temperature and density are so high that the nuclear fusion reaction takes place and the hydrogen is burnt into the helium through the reaction known as the *p-p* chain reaction. It is represented by the following set of reactions. Two protons combine to from a deuteron.

$$_1\mathrm{H}^1 + \,_1\mathrm{H}^1 \rightarrow \,_1\mathrm{H}^2 + \mathrm{e}^+ + \nu \tag{7.1}$$

Here, ν denotes a neutrino particle and e^+ the positron. A positron combines with an electron to produce a pair of photons.

$$\mathrm{e}^+ + \mathrm{e}^- \rightarrow \gamma + \gamma \tag{7.2}$$

The deuteron combines with a proton to form helium isotope.

$$_1\mathrm{H}^2 + \,_1\mathrm{H}^1 \rightarrow \,_2\mathrm{He}^3 + \gamma \tag{7.3}$$

Two helium isotopes obtained from two sets of reactions combine to form a helium nucleus

$$_2\mathrm{He}^3 + \,_2\mathrm{He}^3 \rightarrow \,_2\mathrm{He}^4 + \,_1\mathrm{H}^1 + \,_1\mathrm{H}^1 \tag{7.4}$$

When we consider the combination: 2 (7.1) + 2 (7.2) + 2 (7.3) + (7.4), the net effect is

$$4\,_1\mathrm{H}^1 + 2\mathrm{e}^- \rightarrow \,_2\mathrm{He}^4 + 2\nu + 6\gamma$$

In each reaction here energy is released. Thus, four protons combine to form a $_2\mathrm{He}^4$ nucleus with the release of energy.

In some part of the core of the Sun, the temperature may be sufficient to produce the fusion of three helium nuclei (triple-alpha reaction) to form a carbon nucleus.

$$3\,_2\mathrm{He}^4 \rightarrow \,_6\mathrm{C}^{12}$$

With the formation of carbon, another set of reactions, called the carbon-nitrogen-oxygen (CNO) cycle, has also been suggested for energy generation in the Sun. This set of reactions is as the following. The carbon nucleus combines with a proton to form an isotope of nitrogen.

$$_6\mathrm{C}^{12} + \,_1\mathrm{H}^1 \rightarrow \,_7\mathrm{N}^{13} + \gamma \tag{7.5}$$

This isotope of nitrogen converts into an isotope of carbon.

$$_7\mathrm{N}^{13} \rightarrow \,_6\mathrm{C}^{13} + \mathrm{e}^+ + \nu \tag{7.6}$$

This isotope of carbon combines with a proton to form nitrogen.

$$_6C^{13} + \ _1H^1 \rightarrow \ _7N^{14} + \gamma \tag{7.7}$$

The nitrogen atom combines with a proton to form an isotope of oxygen.

$$_7N^{14} + \ _1H^1 \rightarrow \ _8O^{15} + \gamma \tag{7.8}$$

This isotope of oxygen converts into an isotope of nitrogen.

$$_8O^{15} \rightarrow \ _7N^{15} + \ e^+ + \nu \tag{7.9}$$

This isotope of nitrogen combines with a proton to form carbon nucleus.

$$_7N^{15} + \ _1H^1 \rightarrow \ _6C^{12} + \ _2He^4 \tag{7.10}$$

Combining the reactions (7.5) through (7.10), we get

$$4 \ _1H^1 \rightarrow \ _2He^4 + 2e^+ + 2\nu + 3\gamma$$

In each reaction here energy is released. This is again a net fusion of four protons to form a helium nucleus. It liberates gamma rays, neutrinos and about 25 MeV of thermal energy. In this set of reactions, carbon is not destroyed in the process but it acts only as a catalyst.

For the initiation of this reaction, million kelvin temperature is required. Hence, at the center or in the core, the temperature is about 15 million kelvin. The estimates show that 600 million tons of hydrogen is transformed into helium per second and 5 million tons of material is lost in the form of energy per second. Though this weight appears very large, but in comparison to the solar mass it is insignificant. So far the Sun has consumed only a few hundredth of 1% of the original mass of Sun. The Sun will run out of hydrogen in the next 5 billion years.

Example 1: In the core of Sun, suppose a helium nucleus is formed through fusion of four protons. Calculate the amount of energy released per reaction. Given, mass of proton m_p = 1.0079 amu, mass of helium nucleus m_{He} = 4.0026 amu. 1 amu = 1.66 $\times 10^{-27}$ kg.

Solution: Mass of four protons = 4.0316 amu

Mass defect = mass of four protons − mass of helium nucleus

= 4.0316 - 4.0026 = 0.0290 amu

Mass defect $\Delta m = 0.0290 \times 1.66 \times 10^{-27} = 4.814 \times 10^{-29}$ kg

Energy released = $\Delta mc^2 = 4.814 \times 10^{-29} \ (3 \times 10^8)^2 = 4.3326 \times 10^{-12}$ J

Example 2: In the core of Sun, suppose a helium nucleus is formed through fusion of four protons. Calculate the amount of hydrogen converted into helium per year. Given, mass of proton $m_p = 1.673 \times 10^{-27}$ kg, mass of helium nucleus $m_{He} = 6.664 \times 10^{-27}$ kg, luminosity of Sun $L_\odot = 3.86 \times 10^{26}$ J/s.

Solution: Mass of four protons = $4 \ (1.673 \times 10^{-27}) = 6.692 \times 10^{-27}$ kg.

Mass defect = mass of four protons − mass of helium nucleus

$= 6.692 \times 10^{-27} - 6.664 \times 10^{-27} = 2.8 \times 10^{-29}$ kg

Further,

1 year = $365 \times 24 \times 60 \times 60 = 3.1536 \times 10^7$ s.

The amount of energy emitted in 1 year is

$E = 3.86 \times 10^{26}\ (3.1536 \times 10^7) = 1.217 \times 10^{34}$ J

Mass equivalence of this energy is

$$m = \frac{E}{c^2} = \frac{1.217 \times 10^{34}}{(3 \times 10^8)^2} = 1.35 \times 10^{17} \text{ kg}$$

When the mass defect is 2.8×10^{-29} kg, the amount of hydrogen converted into helium is 6.692×10^{-27} kg.

So the amount of hydrogen converted into helium per year is

$$\frac{1.35 \times 10^{17}(6.692 \times 10^{-27})}{2.8 \times 10^{-29}} = 3.23 \times 10^{19} \text{ kg}$$

Composition of the Sun

The relative abundance (RA) of element (logarithmic base ten) and its first ionization potential (FIP) in eV in the sun are given in Table 2. Here, the abundance of hydrogen is taken as 10^{12}.

Table 2. Relative abundance and first ionization potential

Element	Z	RA	FIP
H	1	12.0	13.6
He	2	10.9	24.6
C	6	8.5	11.3
N	7	8.0	14.5
O	8	8.8	13.6
Ne	10	7.9	21.6
Mg	12	7.4	7.6
Si	14	7.5	8.2
S	16	7.2	10.4
A	18	6.8	15.8
Ca	20	6.3	6.1
Fe	26	7.6	7.9
Ni	28	6.3	7.6

Table 2 shows that there is variety of elements, though the abundance is small, have been found in the Sun. The mass concentrations of hydrogen and helium are generally denoted by X and Y, respectively. The mass concentration of rest of the elements is sometimes denoted by Z or by X_i, where indicates various elements other than hydrogen and helium.

Example 3: Calculate the mean molecular weight of the Sun.

Solution: In the outer layers of Sun, the initial chemical composition has not changed by nuclear reactions. In this case one uses the values $X = 0.71$, $Y = 0.27$, $Z = 0.02$. The mean molecular weight is

$$\mu = \frac{1}{2 \times 0.71 + 0.75 \times 0.27 + 0.5 \times 0.02} = 0.61$$

When the hydrogen is completely exhausted, we have $X = 0$ and $Y = 0.98$, and hence

$$\mu = \frac{1}{0.75 \times 0.98 + 0.5 \times 0.02} \doteq 1.34$$

Example 4: Calculate the temperature of Sun at half the solar radius. The surface value of mean molecular density $\mu = 0.61$.

Solution: The temperature of Sun can be estimated from the perfect gas law as

$$T = \frac{\mu m_H P}{k\rho} = \frac{0.61(1.67 \times 10{-}27)(1.0 \times 10^{14})}{(1.38 \times 10^{-23})(1410)} = 5 \times 10^6 \text{ K}$$

1.2 Radiative zone

The radiative zone inside the Sun is considered to exist from 0.25 $R_\odot$ to 0.7 $R_\odot$. The plasma inside the Sun is sufficiently hot and dense to transfer the intense heat from the core to outward. In this zone, there is no thermal convection. With the increase of the distance, the plasma cools down. The heat is transferred through the radiations, and the ions of hydrogen and helium emit photons, which travel a brief distance before being reabsorbed by other ions. In this manner, the energy makes its way very slowly in the outward direction.

The opacity, is the measure of collisional resistance of gas to the flow of radiant energy. The flow of the energy is measured in terms of the flux, the rate at which the energy crosses a unit area perpendicular to the given direction. When the opacity is large, the radial heat flux due to the flow of radiation τ_r at a distance r from the center of the Sun is described by the radiative conduction equation

$$\tau_r = -\frac{16\sigma T^3}{3\bar{k}} \frac{dT}{dr}$$

where T is the temperature, σ the Stefan-Boltzmann constant and $\bar{k}$ the mean opacity (averaged over all wavelengths). Outside the core, where no further heat is produced by the nuclear reactions, the total heat flux carried out by the photons moving outwards must be constant, so that

$$\frac{T^3}{\bar{k}} \frac{dT}{dr} = \text{constant.}$$

This condition is known as the Radiative Equilibrium. The opacity depends on the density or equivalently on the number of particles per unit volume. The density decreases steadily as we move outwards from the center of the Sun. The variation of density with the distance in the solar interior is shown in Figure 3.

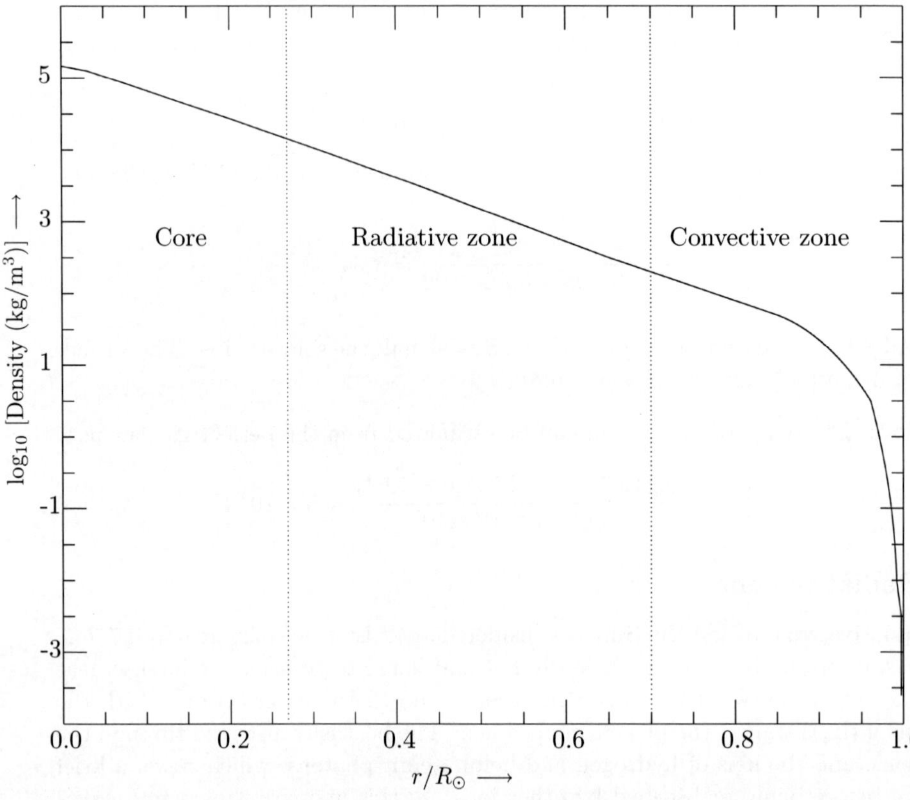

Figure 3: Variation of density in the solar interior.

1.3 Convective zone

The convective zone is the outer layer of the interior of Sun. In the convective zone, the plasma is neither dense enough nor hot enough to transfer the heat energy from the interior to outward through the radiation. The temperature of plasma decreases with the increase of the distance so rapidly that the situation becomes unstable and produces the turbulent convective motions. As a result, the thermal convection takes place, as the thermal columns carry hot material to the solar surface (photosphere). The thermal columns in the convective zone generate imprints on the solar surface, in the form of the solar granulation and super-granulation.

The turbulent convection of this outer part of the solar interior gives rise to a small-scale dynamo that produces the magnetic north and south poles all over the solar surface. The thickness of the layer, about 1.5×10^5 km, corresponds to

$\simeq 0.2R_\odot$. Above this zone, the hydrogen is practically neutral, within the zone, it is partially ionized, and below the zone, it is completely ionized.

2. Solar atmosphere

The region above the photosphere of the Sun is referred to as the solar atmosphere. It can be viewed with the help of various telescopes, operating in various parts of the electromagnetic spectrum. It can be broadly divided into five principal zones, called the photosphere, chromosphere, transition region, corona and heliosphere. The heliosphere, which may be considered the tenuous outer atmosphere of the Sun, extends outward past the orbit of Pluto. The transition region is not easily visible from Earth's surface, but is readily observable from space with the help of the instruments which are sensitive to the far ultraviolet portion of electromagnetic spectrum.

The atmosphere of the Sun is composed of layers of hot gases. These layers, however, are not sharply defined. In fact, each one of them is separated by a transition zone of several kilometers thickness and they greatly differ from each other in the physical characteristics. Starting from the visible solar surface upwards these layers are, the chromosphere and corona, respectively. It transmits the energy, mainly in the form of radiation, coming from below and delivers it out into the space.

The particles of the solar wind move quickly enough to escape from the solar system. Moreover, their speed is such that the disturbances are unable to propagate back to the Sun. The surface layers are unaware of the what is happening out in the wind. For this reason, we do not consider the solar wind to be a part of the solar atmosphere. However, the atmosphere transforms only gradually into the solar wind and it is thus hard to draw an exact boundary between the two regions. The atmosphere may consist of three main regions:

(1) Photosphere

(2) Chromosphere

(3) Corona

The variation of temperature and density of solar atmosphere is shown in Figure 4.

2.1 Photosphere

The photosphere is the top of the convection zone and emits most of the Sun's light. The visible disk of the Sun which we recognize as the solar surface may be considered as the base of the photosphere. This is the innermost layer of the solar atmosphere. The thickness of the photosphere is 0.5 Mm with a density of 10^{23} particles/m^3, and thus, the mass density is $\sim 10^{-4}$ kg/m^3. It is the layer below which the Sun becomes opaque to the visible light. Above the photosphere, the radiation in all the wavelengths are free to propagate into the space, and the entire energy of the Sun escapes outward. The change in the opacity is due to the decreasing amount of H^- ions, which can absorb the visible light easily.

The photosphere is actually 10 - 100 km thick, being slightly less opaque than the air on the Earth. Because the upper part of the photosphere is cooler than the lower one, an image of the Sun appears brighter in the center than on the edge or limb of the solar disk, and this phenomenon is known as the **limb darkening**. Sunlight has approximately a black-body spectrum that indicates its kinetic temperature of about 5800 K.

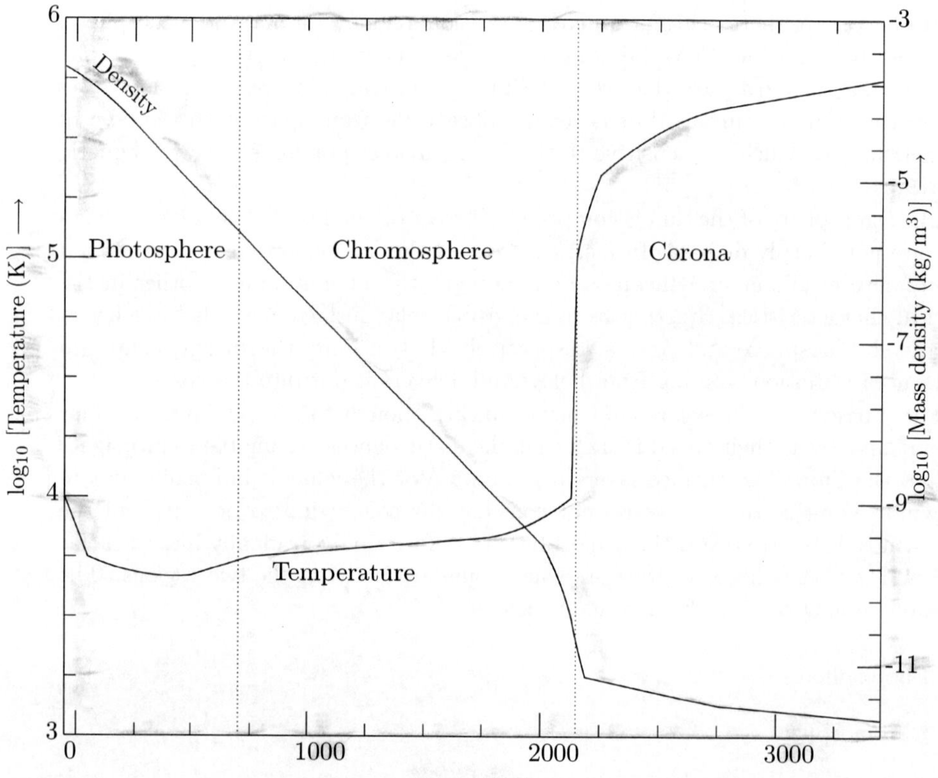

Figure 4: Variation of temperature and density in the solar atmosphere.

2.2 Sunspots

The clearest visible sign of a solar activity is the sunspot (Figure 5). The existence of sunspots is known for long time, as the largest ones can be seen with the naked eye by looking at the Sun through a suitably dense layer of fog. More precise observations became available beginning in the 17th century, when Galileo started to use a telescope for astronomical observations.

A sunspot looks like a ragged hole in the solar surface. In the interior of the spot, there is a dark region, called the umbra, surrounded by a less dark region, called the penumbra. The surface temperature in a sunspot is about 1500 K less than that of its surroundings, and is responsible for the dark appearance of spots. The diameter of a typical sunspot is about 10,000 km and their lifetime is in the range from a few days to several months, depending on its size. The larger spots are more likely to be long-lived. Sunspots often occur in pairs or in the groups of several spots. By following the motions of the sunspots, the period of rotation of the Sun can be determined.

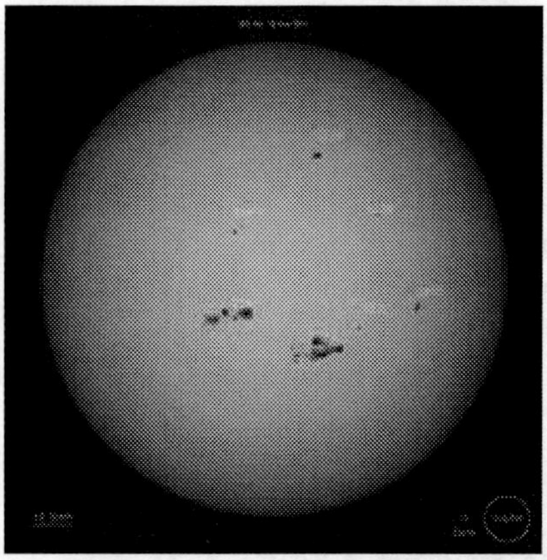

Figure 5: Sunspots. Courtesy to Google website.

The variations in the number of sunspots have been monitored for almost 250 years. The frequency of spots is described by the Zurich sunspot number Z, also known as the Wolf number, expressed as $Z = C(S + 10G)$, where S the number of individual spots and G the number of spot-groups visible at a particular time. The constant C depends on the observer and the conditions of the observation. Sunspots often occur in pairs where the components have opposite polarity. The structure of such bipolar groups can be understood if the field rises into a loop above the solar surface, connecting the components of the pair. If gas is streaming along such a loop, it becomes visible as a loop prominence. The periodic variation in the number of sunspots gives a variation in the general solar magnetic field.

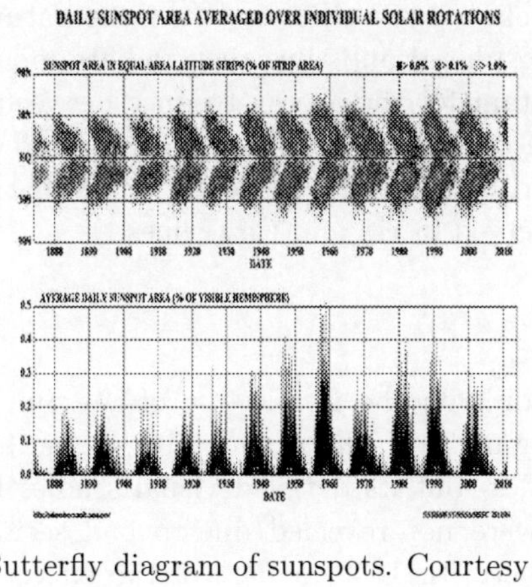

Figure 6: Butterfly diagram of sunspots. Courtesy to Google website.

At the beginning of a new activity cycle, spots first begin to appear at the latitudes of about 40°. The characteristic pattern in which the spots appear, is known as the butterfly diagram (Figure 6). Spots of the next cycle begin to appear while those of the previous one are still present near the equator. Spots belonging to the new cycle have a polarity opposite to that of the old ones. (Spots in the opposite hemispheres also have opposite polarity.) Since the field is reversed between consecutive 11 years cycle, the complete period of solar magnetic activity is 22 years.

2.3 Faculae

A bright area around the sunspots and covering a fraction of solar disk is known as the faculae (Figure 7). These are phenomena belonging to the i transition region between the photosphere and chromosphere. Photospheric faculae are definitely associated with the sunspots, and they are found to appear before the spots and live for several months.

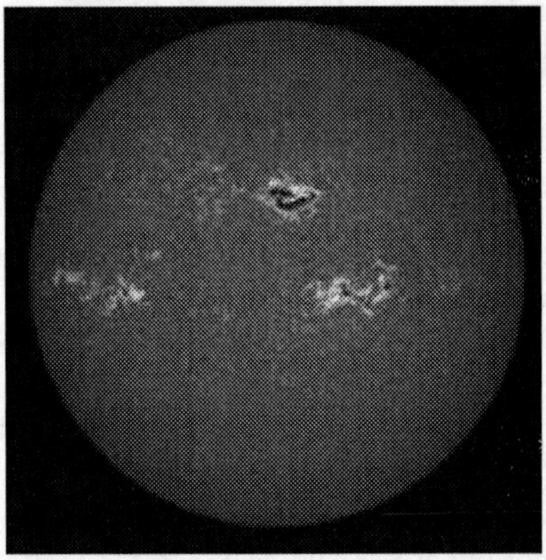

Figure 7: Faculae. Courtesy to Google website.

The average lifetime of photospheric faculae is 90 days. They are brighter than their surroundings, and are believed to be hotter (the temperature in the faculae is higher than that of the photosphere) and thus they radiate more energy. For this reason the faculae are, important for the energy balance between the sunspots and the photosphere. There are also faculae which are not associated with the sunspots, but they are invariably smaller, fainter and relatively short-lived. The chromospheric faculae have been mostly observed in H_α and Ca II lines.

2.4 Granulation

Typical grain-like fine structures on the visible disc of the Sun are known as the photospheric granulation (Figure 8). The granulation of the Sun was known to the observers for many decades, but its detailed visual studies had been difficult, because the fine structures were not revealed due to bad seeing, because of the Earth's atmospheric turbulence. In 1959, a part of the effect of bad seeing was overcome by taking photographs of the Sun from a balloon sent to a height of 5486 m above the Earth's surface.

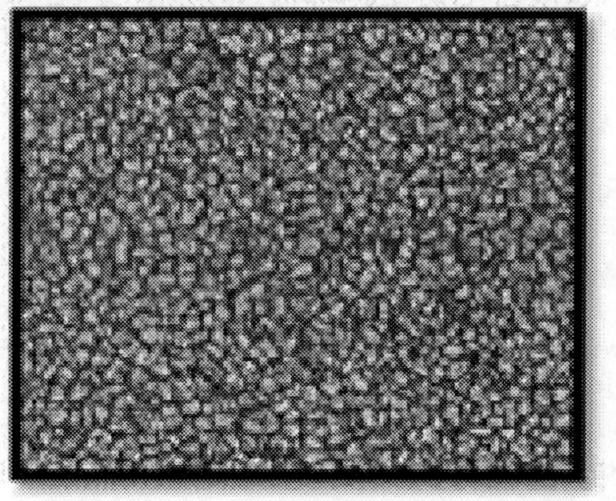

Figure 8: Granulation. Courtesy to Google website.

In the envelope of stars, like Sun, the radiative energy transport has to be supplemented by the convective transport of energy. Convection cells of the gaseous mass carrying higher energy from the hotter deeper layers of the envelope rise through the upper cooler layers, transfer there the excess energy and subsequently sink down again to the deeper layers. The process is repeated incessantly.

The diameters of the granules vary from 300 km to 18,000 km, with a mean value of about 700 km. Their shapes resemble with the irregular polygons or cells which are separated from one another by a rather narrow and dark filament-like regions. Granules are found to vary in both size and number with the 11 years solar cycle of magnet.

2.5 Chromosphere

Above the photosphere lies the second major layer of the solar atmosphere, known as the chromosphere. It is named the chromosphere because of its appearance as a pink flash during the total solar eclipse. The chromosphere word is derived from the Greek word 'chroma' meaning colour, because the chromosphere is visible as a coloured flash.

The chromosphere is the region between the temperature minimum and the corona. The layer extends to nearly 20,000 km above the photosphere with gas-density gradually decreasing with the distance, but the temperature rising rapidly upwards. The atmosphere becomes transparent in the continuum, so the chromospheric continuum can be observed at the limb during the total solar eclipse.

The chromosphere is rarer and more transparent with a particle density of 10^{17} m^{-3}, and thus the mass density is about 10^{-10} kg/m^3. Minimum temperature of $\sim$ 4000 K exists in the chromosphere. This part of solar atmosphere is cool enough to support the existence of simple molecules, such as CO and H_2O, which can be identified through their absorption spectra.

The temperature in the chromosphere increases gradually with the altitude, ranging up to around 10^5 K near the top. When we study the chromosphere in H_α, we often find small transient features. These include surges and eruptions, small brightening called micro-flares, transient small filaments separating small bipolar elements.

2.6 Spicules

During a total solar eclipse or with the use of a coronagraph, spiky structures of hot, tenuous gas are seen to rise from the lower chromosphere and extended up to 10,000 km, or more, sometimes even up to 20,000 km. The diameters of these spikes are, on an average between 500 to 600 km with larger ones having diameter about 2000 to 5000 km. These projected chromospheric streamers are known as the spicules (Figure 9). The velocity of their upward movement is about 30 km/s, and their average lifetime is about 10 minutes. Spicules are observed as dark mottles in H_α spectroheliograms to cover a good fraction of solar disc.

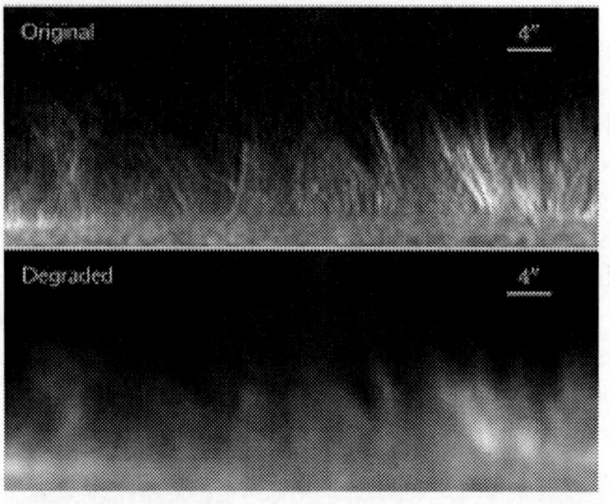

Figure 9: 'Spicules. Courtesy to Google website.

When viewed near the solar limb, a large number of such structures are seen together in the projection as to give the appearance of a 'forest of spikes'. They appear to be a moving disturbance through the solar chromosphere driven by the enhanced magnetic field and probably, they play an important role in transferring the mass from the lower chromosphere to the corona.

2.7 Plages and filaments

Large bright and dark clouds were revealed in the chromosphere when the spectroheliograms in the monochromatic light of H_α and Ca II K were examined. These are, respectively, called the plages (Figure 10) and the filaments (Figure 11). In the spectroheliograms, they appear as the clouds of hydrogen and calcium, but in fact they are just the regions of chromospheric gas which radiate more intensely at the wavelengths of H_α and Ca II K line.

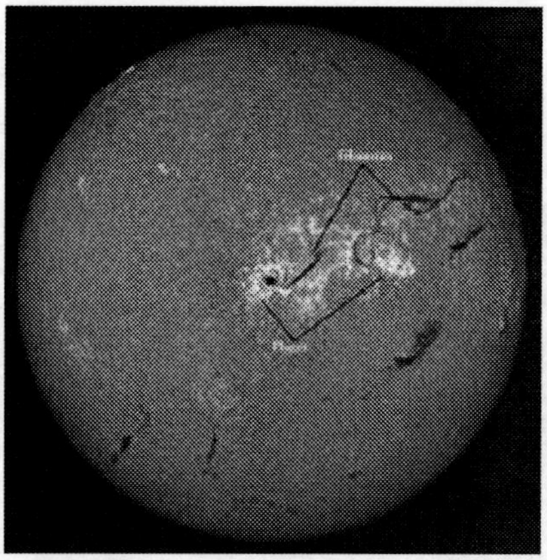

Figure 10: Plages. Courtesy to Google website.

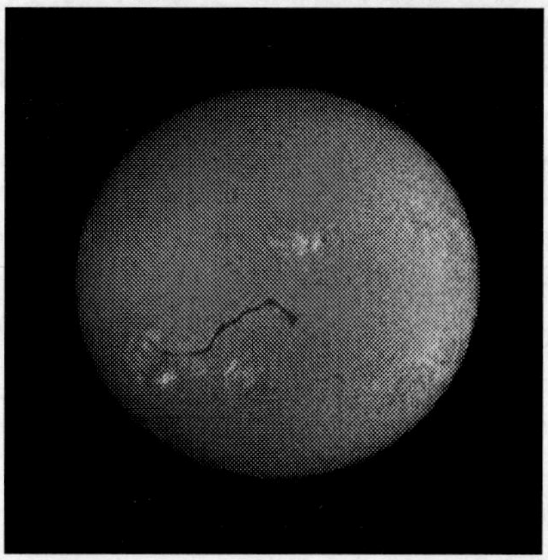

Figure 11: Filaments. Courtesy to Google website.

Like faculae and spicules, plages are also associated with the enhanced magnetic field of Sun. They are mostly seen in the regions of sunspots but sometime are also seen where there are not visible sunspots.

2.8 Corona

The corona is a Latin word, meaning 'crown'. It is an atmosphere around the Sun, containing plasma and extending millions of kilometers into the space. It can be easily seen during a total solar eclipse as shown in Figure 12, and can also be observed with the help of an instrument, called the coronagraph, which creates an artificial solar eclipse.

Two spectral lines at wavelengths 5303 Å and 6374 Å having very high intensity were found in the spectrum of the corona in the year 1869. Scientist attempted to assign them to some elements or their ions. Since the temperature of photosphere is 5800 K, the scientists had opinion that the temperature in the atmosphere of Sun should not be more than 5800 K. They produced spectrum of elements and their singly ionized ions in their laboratories, but could not find these two lines in their spectra. Then, some scientists suggested that these two lines were emitted by some unknown element, named, 'coronium'. Now, the point was to describe the details of the coronium.

Figure 12: Structure of corona (white region) during the total solar eclipse. The dark sphere is the Sun which is covered by the Moon. Courtesy to Google website.

Some scientists were not satisfied with this concept, and they went on producing spectra of highly ionized ions in their laboratories. B. Edlen in the year 1940 found that the lines of wavelength 5303 Å and 6374 Å were produced by Fe X and Fe XIV, respectively.[1] That is, it took 70 years to find the source for these two lines. Some other lines which have been found in the corona are assigned to various elements as given in Table 3.

Table 3. Wavelengths and elements

Wavelength (Å)	Element	Wavelength (Å)	Element
3328	Ca XII	5303	Fe XIV
3388	Fe XIII	6374	Fe X
3601	Ni XVI	6702	Ni XV
3643	Ni XIII	7060	Fe XV
3987	Fe XI	7892	Fe XI
4086	Ca XIII	8024	Ni XV
4231	Ni XII	10747	Fe XIII
5116	Ni XIII	10798	Fe XIII

These highly ionized elements are produced at a temperature of several million kelvin.

[1]As per convention, for example, neutral iron element is written as Fe I; singly ionized calcium is written as Ca II. Thus, Fe X is nine times ionized iron, *i.e.*, Fe^{+9} and Fe XIV is thirteen times ionized iron, *i.e.*, Fe^{+13}.

These spectral features have since been traced to highly ionized Iron (Fe XIV) which indicates a plasma temperature more than 10^6 K. The corona is observed at the time of eclipse as a faint halo of very low density of 10^{15} particles/m^3, and thus the mass density is about 10^{-12} kg/m^3, it extends to the Earth's orbit where density is 10^7 m^{-3} (*i.e.*, mass density is about 10^{-20} kg/m^3) and beyond.

The temperature of the corona is several million kelvin. In spite of large number of attempts, no complete theory yet proposed to explain the temperature of corona; at least some of its heat is known to be from the magnetic reconnection.

Temperature of corona

For a quite long time, some lines in the spectra of Sun could not be assigned to any known atom or ion. Among these lines, the two at $\lambda = 5303$ Å and $\lambda = 6374$ Å were quite prominent. In absence of any species to which these lines could be assigned, it was thought that there could be an element, named *coronium*, in the corona, which is emitting these lines. However, no one was in aposition to tell about the physical characteristics of the coronium. Later on, B. Edlen found that these lines could be generated by the highly ionized ions. For example, the lines of 5303 Å and 6374 Å could be generated by the Fe XIV and Fe X ions, respectively.

As these highly ionized ions can be produced at a temperature of million kelvin, in the beginning it had been unbelievable that the temperature in the solar corona could be of million kelvin. Later on, it was found that there was no other option, but to accept it. Though the temperature of the solar atmosphere is 10^6 K, still the spectrum for the Sun corresponds to a temperature of about 6000 K, which is the temperature of the photosphere of the Sun. It is because the density of the corona is very low.

The temperature of solar corona generally lies in the range of 1 - 5 million kelvin. Such a hot plasma is subject to lose energy through the radiation in all directions and the thermal conduction to the transition region below. As the coronal temperature is maintained at several million kelvin, these energy losses must obviously be compensated by an energy source. The explanation of mechanism through which the energy is supplied to the corona is a great challenge to the scientists. Many solutions have been proposed, but none of them has yet achieved a consensus. The corona can be divided into three segments as the following.

1. K-corona (K stands for continuum)
2. E-corona (E stands for emission)
3. F-corona (F stands for Fraunhofer)

K-corona

The name K-corona is derived from the German word 'Kontinum', means the continuum. The inner corona which is sometimes known as the real corona extends from 1.03 $R_\odot$ to 2.5 $R_\odot$. This part of corona, known as the K-corona, imitates the continuous spectrum of the photosphere, but the Fraunhofer lines are absent. This may be explained by considering that the free electrons in the highly ionized coronal gas scatter photospheric light to produce the observed continuum. It has been suggested that the Thomson scattering is operative in the process.

E-corona

The E-corona is due to the spectral emission lines produced by the ions, present in the coronal plasma. These emission lines include the highly ionized atoms of iron, calcium, and some other elements. It may be observed in the broad or hot spectral emission lines and is the main source of information about the corona's composition. The E-corona is much fainter than the K and F coronae and therefore its observation is very difficult.

F-corona

The F-corona which lies at $R > 2.5R_\odot$ displays the solar spectrum with Fraunhofer lines superimposed on the continuum. This is sometimes known as the 'false' part of the corona and the prefix F stands for Fraunhofer. The spectrum of the F-corona is produced by the Sun's light scattered by the tiny dust particles in the interplanetary space. The lines are assumed to be formed by the forward scattering produced by the particles with sizes somewhat larger than the wavelengths scattered.

The F-corona together with the K-corona constitute what is known as the white corona, which merges into the interplanetary space with decreasing brightness.

2.9 Prominences

A quiescent prominence is an immense, dense, cool vertical sheet of plasma in the corona with a density $n \sim (0.5-1.0)\times 10^{17}$ particles/m^3, about 500 times more than the ambient coronal value and a temperature (5000 – 10,000 K) about 500 times less. It changes slowly in shape and is long-lived. The main observational features of the prominences are as the following:

1. The geometry of a prominence is a thin vertical sheet. The length of the sheet lies between 60 and 600 Mm (typically 200 Mm), while its height ranges between 10 and 100 Mm (typically 50 Mm) and its width between 4 and 15 Mm (typically 6 Mm).

2. The particle density is between 10^{16} and 10^{17} particles/m^3, and thus the mass density is about 10^{-11} – 10^{-10} kg/m^3. The temperature is usually between 5000 K and 8000 K. Both the density and temperature tends to decrease with height.

3. The magnetic field has a magnitude between 3 and 30 Gauss (typidally 5 G) and usually increases with the increase of height.

4. Prominences are long-lived, lasting for 1 - 300 days, with high-latitude ones enduring typically 140 days.

5. Prominences tend to form where there are converging flows and shearing flows.

3. Solar neutrinos

We have seen that in the thermonuclear reactions occurring in the core of the Sun, neutrinos are produced. Raymond Davis began measuring the neutrino flux from the Sun using a detector located about 1.5 km below the Earth's surface in the Homestake Gold Mine in Lead, South Dakota. He found that the number of neutrinos detected was almost one-third of the expected value. Because of very low cross section of neutrino interactions with other matter, the neutrinos can easily travel through all the matters, and even through the Earth, while the other particles being originated from the space cannot. Thus, the underground detector assured of measuring the neutrinos created 500 seconds earlier in the core of the Sun.

The Davis neutrino detector had 100,000 gallons of C_2Cl_4 (perchloroethylene). One isotope of chlorine ${}_{17}Cl^{37}$ is capable of interacting with the neutrinos of sufficient energy to produce a radioactive isotope of argon, having a half-life of 35 days. The reaction is

$$ {}_{17}Cl^{37} + \nu_e \rightarrow {}_{18}A^{37} + e^- $$

The threshold energy for this reaction, 0.814 MeV, is less than the energies of the neutrinos produced in each step of the p-p chain except for the reaction

$$ {}_1H^1 + {}_1H^1 \rightarrow {}_1H^2 + e^+ + \nu_e $$

However, the reactions that account for 77% of the neutrinos detected in the Davis experiment is due to the decay of ${}_5B^8$

$$ {}_5B^8 \rightarrow {}_4Be^8 + e^+ + \nu_e $$

Unfortunately this reaction is very rare. Once every few months, the atoms of the argon produced were counted. The counting rate is measured in terms of the solar neutrino unit (SNU). 1 SNU is defined as 10^{-38} reactions per target atom per second. With approximately 2.2 $\times 10^{30}$ atoms of ${}_{17}Cl^{37}$ atoms in the tank, when only one argon atom is produced each day, this rate would correspond to 5.35 SNU.

Standard solar models predict that the outcome of the experiment should get a capture rate of 7.9 SNU whereas the actual data give 2.23 $\pm$ 0.26 SNU. That is, only one argon atom is produced every two to three days. This discrepancy in the production and detection rates of solar neutrinos is known as the **solar neutrino puzzle**.

Other neutrino experiments, fundamentally different from the ${}_{17}Cl^{37}$ experiment, confirmed the existence of solar neutrino puzzle.

Scientists continued attempts to resolve the discrepancy between the experiment and theory ever since the problem first appeared. Given the small number of neutrinos that are detected, one obvious solution is that there is some thing wrong with the experiments or with the analysis of the resulting data. However, all attempts to find the errors that could produce symmetrically low counts failed.

Can the standard model be fundamentally flawed in some way? Do we understand the solar interior well enough to predict neutrino fluxes accurately?

The search for a theoretical resolution to the problems of solar neutrinos and oscillations has considered two general approaches: Either some fundamental physical process operating in the standard models is incorrect, or something happens to the neutrinos on their way from the Sun's core to the Earth. The first of these possibilities resulted in the re-examination of a host of features of the solar model. One point was to enhance the helium abundance in the core. It improved the agreement between the theory and observations with regard to the solar oscillations, but it made the solar neutrino problem even worse.

An elegant solution to the solar neutrino problem that has been proposed suggests that the standard solar models are essentially correct but that the neutrinos produced in the Sun's core actually change before they reach the Earth. The Mikheyev-Smirnov-Wolfenstein (or MSW) effect involves the transformation of neutrinos from one type to another.

The neutrinos produced in various branches of the p-p chain are all electron neutrinos (ν_e). Two other type of neutrinos also exist - the muon neutrinos (ν_μ) and the tau neutrinos (ν_τ). The MSW effect suggests that neutrinos oscillate between the electron neutrinos, muon neutrinos, or tau neutrinos during their passage through the Sun. The neutrino oscillations are caused by the interactions with the electrons as the neutrinos travel towards the solar surface.

One testable consequence of the MSW effect is that neutrinos would necessarily have mass, with $m(\nu_e) << m(\nu_\mu) << m(\nu_\tau)$. The required mass of the electron neutrino is well within the upper mass limit of approximately 7.2 eV established by the experiment.

4. End state of Sun

The age of Sun is about 5 billion years. In the next about 5 billion years, the Sun will cool down and will expand outward to a size having diameter several times larger than the current value (becoming a red giant), before blowing out its outer layers in the form of planetary nebula, and leaving behind a solid body, known the white dwarf. This white dwarf will cool down and its colour will be brown; called the brown dwarf. Finally, it will exhaust its all energy and will become a part of the dark matter in the universe.

5. Magnetohydrodynamics

The magnetohydrodynamics (MHD) is a physical-mathematical framework about the dynamics of magnetic fields in electrically conducting fluids, e.g., in plasmas and liquid metals. The word magnetohydrodynamics is comprised of the words magneto- meaning magnetic, hydro- meaning water (or liquid) and -dynamics referring to the movement of an object under forces.

The central point of the MHD theory is that conductive fluids can support magnetic fields. The presence of magnetic fields leads to the forces which in turn act on the fluid (typically a plasma), and therefor changing the geometry (or topology) and strength of the magnetic fields themselves. A key issue for a particular conducting fluid is the relative strength of the advecting motions in the fluid, compared to the diffusive effects caused by the electrical resistivity.

The standard non-relativistic form of the MHD equations are the basic conservation laws, which in the SI units are as the following.

Continuity equation

$$\frac{\partial \rho}{\partial t} + \nabla.(\rho\, \vec{u}) = 0$$

where ρ is the mass density and $\vec{u}$ the fluid bulk velocity.

Momentum equation

$$\rho\Big(\frac{\partial}{\partial t} + \vec{u}.\nabla\Big)\vec{u} \;=\; \vec{j} \times \vec{B} - \nabla p$$

where p is the gas pressure, $\vec{B}$ the magnetic field (the magnetic flux density), $\vec{j} = \nabla \times \vec{B}/\mu_0$ the current density, and μ_0 the vacuum permeability.

Ideal Ohm's law

$$\vec{E} + \vec{u} \times \vec{B} = 0$$

Faraday's law

$$\frac{\partial \vec{B}}{\partial t} = -\nabla \times \vec{E}$$

Conservation of magnetic field

$$\nabla.\vec{B} = 0$$

Energy equation

$$\frac{\mathrm{d}}{\mathrm{d}t}\Big(\frac{p}{\rho^{\gamma}}\Big) = 0$$

where $\gamma = 5/3$ is the ratio of specific heats for an adiabatic equation of state.

These MHD equations are to be solved simultaneously, either analytically or numerically. An important dimensionless number in the MHD theory is the plasma beta β, defined as the ratio of gas pressure p to the magnetic pressure

$$\beta = \frac{p}{B^2/2\mu_0}$$

When $\beta << 1$, the magnetic field dominates in the fluid, and the fluid is forced to move along with the field. When $\beta >> 1$, the magnetic field is weak and the field is swirled along along by the fluid.

In the ideal MHD, the topology of the magnetic field (the connectivity of the field lines) does not change. If a set of magnetic field lines is twisted into a knot, they remain unchanged. This means that it is very difficult to change the magnetic topology of some field configuration by the MHD reconnection, if the resistivity is very low. It is therefore possible to store a large amount of energy in the magnetic fields by twisting them into a complex topology by the fluid motions. This energy can subsequently be released into the heat and kinetic energy, if the scales of the magnetic field become very small leading to a large magnetic gradients or if the resistivity increases. In this case the ideal MHD breaks down allowing the magnetic reconnection to occur.

5.1 Applications of MHD

Magnetism is found throughout the Universe. Magnetic fields are known to exist in the planets, stars, accretion disks, the interstellar medium, galaxies and in active galactic nuclei. Often these magnetic fields are generated and maintained by the magnetohydrodynamics dynamo action (*e.g.*, the solar dynamo or planetary dynamos, which are described by hydromagnetic dynamo theory), and in many cases the magnetic field is dynamically dominant, determining the evolution of the object. On the Sun and other stars, these magnetic structures may take the form of cool spots (sunspots) and magnetic bright points.

Magnetic structures can store enormous amounts of energy that may be released through reconnection of magnetic field lines (MHD reconnection) leading to solar flares and stellar or even galactic eruptions. It is in this way that the Sun causes storms in the heliospherical magnetic field and ultimately aurorae and space weather.

6. Helioseismology

In the helioseismology, we study of the propagation of wave oscillations, in particular the acoustic pressure waves (p-waves), in the Sun. Contrary to the seismic waves on the Earth, the solar waves have practically no shear component (s-waves). The solar pressure waves are considered to be produced through the turbulence in the convection zone near the solar surface. Certain frequencies are amplified by constructive interference. The acoustic waves are transmitted to the outer photosphere of the Sun. These oscillations are detectable on almost any time series of solar images, but are best observed by measuring the Doppler shift of photospheric absorption lines. Changes in the propagation of oscillation waves through the Sun reveal inner structures and help the astrophysicists to develop extremely detailed profiles for the interior conditions in the Sun.

Helioseismology can be used to image the far side of the Sun from the Earth, including sunspots. In simple terms, sunspots absorb helioseismic waves. This sunspot absorption causes a seismic deficit.

The analysis of data from the SOHO spacecraft shows the seismic images of the central portion of the solar far side being produced nearly continuously since late 2000.

6.1 Solar oscillations

Individual oscillations in the Sun are damped and they die out within a few periods. However, the interference between these localized waves produces global standing waves, also known as the normal modes. Analysis of these overlapping modes are studied under the discipline of global helioseismology. These modes are essentially divided up into three categories, according to the restoring force that drives them: acoustic, gravity, and surface-gravity wave modes.

• For the p-mode or acoustic waves, the pressure is their restoring force, and they are called the p-mode. Their dynamics are determined by the variation of the speed of sound inside the Sun. P-mode oscillations have frequencies > 1 mHz and are very strong in the 2-4 mHz range, where they are often referred to as *5-minute oscillations.*[2] The p-modes at the solar surface have amplitudes of hundreds of kilometers and can be detected easily with Doppler imaging or sensitive spectral line intensity imaging.

• For the g-mode or gravity waves, the gravity is their restoring force, and they are called the g-mode. The g-mode oscillations are low frequency waves (0-0.4 mHz) and they are confined to the interior of the Sun below the convection zone (which extends from 0.7-1.0 solar radius), and practically cannot reach the surface. Since the 1980s, there have been several claims of g-mode detection, none of which have been confirmed.

• f-mode or surface gravity waves are also gravity waves, but occur at or near the photosphere, where the temperature gradient drops below the adiabatic lapse rate. Some f-modes of moderate and high degree, between $l = 117$ and $l = 300$, have been observed by the Michelson Doppler Imager (MDI).

6.2 Analysis of oscillation data

The data from time-series of solar spectra show all the oscillations overlapping. Thousands of modes have been detected (with the true number perhaps being in the millions). The mathematical technique of Fourier analysis is used to recover information about individual modes from this mass of data.

The simplest modes to analyze are the radial ones; however most solar modes are non-radial. A nonradial mode is characterized by three wave numbers: the spherical-harmonic degree l and azimuthal order m which determine the behaviour of the mode over the surface of the star and the radial order n which reflects the properties in the radial direction (see the diagram on the top right for an example).

In general the frequencies ω_{nlm} of stellar oscillations depend on all three wave numbers. It is convenient, however, to separate the frequency into the multiplet frequency ω_{nl}, obtained as a suitable average over azimuthal order m and corresponding to the spherically symmetric structure of the star, and the frequency splitting $\delta\omega_{nlm} = \omega_{nlm} - \omega_{nl}$.

Analysis of oscillation data must attempt to separate these different frequency components. In the case of the Sun the oscillations can be observed directly as functions of position on the solar disk as well as time. Thus here it is possible to analyze their spatial properties. This is done by means of a generalized 2-dimensional Fourier transform in position on the solar surface, to isolate particular values of l and m. This is followed by a Fourier transform in time which isolates the frequencies of the modes of that type. In fact, the average over the stellar surface implicit in observations of stellar oscillations can be thought of as one example of such a spatial Fourier transform.

7. Review Questions

1. Provide a brief description about the Sun.
2. Describe the interior of Sun.

[2]Note: 5 minutes per cycle is 1/300 cycles per second = 3.33 mHz.

3. Describe the atmosphere of Sun.
4. Explain how the temperature of corona was decided.
5. What is the source of energy in the core of the Sun.
6. Discuss the solar neutrinos.
7. In the core of Sun, suppose a helium nucleus is formed through fusion of protons. Calculate the amount of energy released per reaction. Given, mass of proton is 1.0079 amu, mass of helium nucleus is 4.0026 amu. 1 amu = 1.66×10^{-27} kg.
8. In the core of Sun, suppose a helium nucleus is formed through the fusion of protons. Calculate the amount of hydrogen converted into helium per year. Given, mass of proton = 1.673×10^{-27} kg, mass of helium nucleus = 6.664×10^{-27} kg, luminosity of Sun $L_\odot = 3.86 \times 10^{26}$ J/s.
9. Describe various modes of helioseismology.
10. Write notes on the following:

    (i) Solar interior
    (ii) Energy generation in Sun
    (iii) p-p chain
    (iv) CNO cycle
    (v) Photosphere
    (vi) Sunspots
    (vii) Faculae
    (viii) Granulation
    (ix) Chromosphere
    (x) Spicules
    (xi) Plages and filaments
    (xii) Corona
    (xiii) Temperature of corona
    (xiv) Prominences
    (xv) Solar neutrinos
    (xvi) End state of Sun
    (xvii) Magnetohydrodynamics
    (xviii) Helioseismology
    (xix) Solar oscillations

8. Multiple Choice Questions

1. A photon produced in the core of the Sun reaches the photosphere in about

   A. 2 sec B. 8 m 20 sec C. 1 year D. 10 million years

   Ans. D

2. Which is not the interior part of the Sun?

   A. core B. convective zone C. corona D. chromosphere

   Ans. D

3. Which is not the part of solar atmosphere?

   A. chromosphere B. transition region C. corona D. convective zone

   Ans. D

4. Which is not the part of solar atmosphere?

   A. chromosphere B. transition region C. corona D. radiation zone

   Ans. D

5. Umbra and penumbra are the parts of

   A. corona B. chromosphere C. solar wind D. sunspot

   Ans. D

6. Suppose, the Sun is taken to a distance of 10 pc from the Earth. Can the Sun be seen by unaided eyes?

   A. No B. only in the morning C. Yes D. only at the noon

   Ans. C

7. The sunspots lie up to the latitudes of about

   A. 10° B. 40° C. 60° D. 90°

   Ans. B

8. The time-period of solar magnetic activity for getting the same polarity is

   A. 5 yr B. 11 yr C. 22 yr D. 44 yr

   Ans. C

9. The time-interval for having consecutive maximum number of sunspots is

   A. 5 yr B. 11 yr C. 22 yr D. 44 yr

   Ans. B

8. Solar System

According to the helio-centric system, Sun is the main object around which all the planets, asteroids (small planets) and other smaller bodies are revolving. Hence, each cosmic body revolving around a sun is not planet. A planet must have at least specific mass (size). Because of this reason, the Pluto has been removed from the list of planets. A body revolving around the Sun may have its satellites revolving around it. A simple exercise of numbers gave the idea of missing planet, which lead to the discovery of asteroids. All these bodies are directly or indirectly controlled by the gravitational field of the Sun. Besides them, there are some bodies which reach to the region between the planets, such as comets and meteors. All these bodies together constitute a group, called the **solar system**. The solar system may be considered as consists of:

(i) planets

(ii) asteroids (minor planets)

(iii) other minor planets

(iv) comets

(v) meteors

In this chapter, we have discussed about the solar system. Students in the schools are given information about the solar system.

1. Planets

In the language of a common person, a planet is a natural body which revolves around the Sun. This is incomplete definition, as a specific mass (or size) of the revolving body is required in order to call it a planet. A body having more than the specific mass (or size) and revolving around the Sun is known as a planet. We know that there are eight planets:[1] Mercury, Venus, Earth, Mars, Jupiter, Saturn, Uranus, and Neptune. These planets revolve around the Sun in distinct elliptical orbits, with the Sun at one (common) focus of all the ellipses. Besides the revolution around the Sun, a planet spins about its own axis. Let us first look into some physical parameters of planets.

[1]We have later on discussed why the Pluto no more belongs to the list of planets. Before 24 August 2006, there have been 9 planets: Mercury, Venus, Earth, Mars, Jupiter, Saturn, Uranus, Neptune and Pluto.

1.1 Mass

Planets are physical bodies, and thus each of them has a mass. Estimated values of the masses of planets are given in Table 1. The Jupiter is 318 times heavier than the Earth. That is, out of the material of Jupiter, 318 Earth-like bodies can be constructed. Table 1 shows that the heaviest planet is the Jupiter and the lightest planet is the Mercury.

1.2 Density

A planet has its finite volume and finite mass. Though the distribution of mass in a planet, in general, is not uniform, we can however calculate average density of a planet. Estimated values of mean densities of planets are given in Table 1. The Earth has the largest density and the lowest density is of the Saturn. Interesting to note that the density of Saturn is smaller than that of water. That is, all the planets could be put in the water, only the Saturn would float.

Table 1. Some parameters for planets

Planet	Mass (kg)	Density (10^3 kg/m^3)	Distance (AU)	Orbital period
Mercury	3.28×10^{23}	5.43	0.39	87.97 days
Venus	4.87×10^{24}	5.24	0.72	224.71 days
Earth	5.97×10^{24}	5.52	1.00	365.25 days
Mars	6.39×10^{23}	3.94	1.52	1.88 yr
Jupiter	1.90×10^{27}	1.33	5.20	11.86 yr
Saturn	5.68×10^{26}	0.70	9.54	29.46 yr
Uranus	8.68×10^{25}	1.19	19.18	84.02 yr
Neptune	1.02×10^{26}	1.66	30.06	164.77 yr

1 AU = 1.496×10^8 km

1.3 Mean distance from the Sun

Each of the planets revolves around the Sun in an elliptical path. The Sun is situated at one of the foci of ellipse of the path. Each planet has its own elliptical path. Thus, there is a common point, where one focus of each elliptical orbit coincides. This common point is the position of the Sun. Values of the mean distances of planets from the Sun is reported in Table 1 in AU (Astronomical Unit). The distance of a planet is according to its velocity of revolution and vice versa. Suppose, a planet of mass m revolves around the Sun of mass M with velocity v, the equilibrium is governed by the gravitational force and the centrifugal force.

$$\frac{GMm}{r^2} = \frac{mv^2}{r} \qquad \text{or} \qquad v = \sqrt{\frac{GM}{r}}$$

where r is the distance between the Sun and planet. Notice that the velocity does not depend on the mass of the planet. It shows that a planet closer to the Sun revolves with higher speed. For convenience, the orbit of planet is generally taken as circular.

1.4 Orbital period

The time taken by a planet to make one revolution around the Sun is known as the orbital period. The orbital period T can be obtained as

$$T = \frac{2\pi r}{v} = \frac{2\pi r^{3/2}}{\sqrt{GM}} \quad \text{or} \quad T^2 = \frac{4\pi^2 r^3}{GM}$$

This relation $T^2 \propto r^3$ shows the Kepler's third law of planetary motion. It shows that a planet at larger distance from the Sun has larger period of revolution. The periods of revolution of planets are given in Table 1. It is interesting to note that the Neptune takes 164.77 years to complete one revolution round the Sun.

1.5 Orbital eccentricity

Suppose, semi-major axis and semi-minor axis of the elliptical path of a planet revolving around the Sun are a and b, respectively. Then, the eccentricity of the orbit is expressed as

$$e = \sqrt{1 - b^2/a^2}$$

For a circular orbit $a = b$ and the eccentricity is zero. Values of orbital eccentricities of the paths of planets are given in Table 2. It shows that the path of any planet is not circular in shape.

Table 2. Some parameters for planets

Planet	Orbital eccentricity	Rotation period	Equatorial radius (km)	Polar oblateness	Escape velocity (km/s)
Mercury	0.2056	58.6462 days	2439.99	0.0	4.2
Venus	0.0068	243.026† days	6051.813	0.0	10.4
Earth	0.0167	$23^h56^m4\overset{s}{.}1$	6378.137	0.003352	11.2
Mars	0.0934	$24^h37^m22\overset{s}{.}1$	3397.515	0.005186	5.0
Jupiter	0.0485	$9^h55^m29\overset{s}{.}1$	71398	0.064809	59.6
Saturn	0.0557	$10^h39^m24^s$	60330	0.107621	35.4
Uranus	0.0472	$17^h14^m24^s$	25559	0.024	21.3
Neptune	0.0086	16^h3^m	24750	0.0259	23.4

† denotes the retrograde motion

1.6 Rotation period

Besides the revolution around the Sun, a planet spins about an axis passing through its center of gravity. This spin axis intersects the surface of the planet at two points, called the poles. Values of the rotation period of planets are given in Table 2.

Members of solar system, in general, move from west to east. According to the convention, the motion from west to east is taken as positive. When the direction of motion is from east to west, it is known as the retrograde motion.

1.7 Equatorial radius

A plane perpendicular to the spin axis and passing through the center of gravity of the planet intersects its surface in a circle, called the equator. Owing to the spin motion, the planet is oblate spheroid and the radius of equator (called the equatorial radius) is in general larger than the distance of a pole from its center of gravity. Values of equatorial radius of planets are given in Table 2.

1.8 Polar oblateness

Suppose, the equatorial and polar radii of a planet are, respectively, R_e and R_p. The polar oblateness (or flattening ratio) of the planet is expressed as

$$\text{polar oblateness} = \frac{R_e - R_p}{R_p}$$

Values of polar oblateness of planets are given in Table 2. Zero value of polar oblateness for a planet (Mercury and Venus) shows that the planet is spherical in shape, and the equatorial and polar radii of the planet are equal.

1.9 Escape velocity

Escape velocity is defined as the velocity that an object requires to escape from the surface of a heavenly body such as a planet, satellite. Let us consider a spherical body of mass M and radius R. Suppose, an object of mass m is thrown with velocity v from the surface of the body. When the object is at a distance r from the center of the body, the force of attraction between the body and the object is

$$F = -G\,\frac{Mm}{r^2}$$

The potential energy V is expressed as

$$F = -\frac{\mathrm{d}V}{\mathrm{d}r} \qquad \text{or} \qquad V = -\int F\,\mathrm{d}r + C$$

where C is a constant of integration. Using the expression for F, we get

$$V = -G\,\frac{Mm}{r} + C$$

Since $V = 0$ when $r = \infty$, we have $C = 0$ and therefore

$$V = -G\,\frac{Mm}{r}$$

When the object is on the surface of the body, the potential energy is

$$V = -G\,\frac{Mm}{R}$$

Thus, in order to send the object out of the gravitational field of the body at least this much amount of kinetic energy be given to the object. The minimum velocity

v_e given to the object so that it goes away from the body's gravitational field is known as the escape velocity. Thus, we have

$$\frac{1}{2}\, m v_e^2 = G\,\frac{Mm}{R} \qquad \text{or} \qquad v_e = \sqrt{\frac{2GM}{R}} \tag{8.1}$$

It is interesting to note that the escape velocity does not depend on the mass of the object. If g denotes the acceleration due to gravity of the body, we have

$$g = \frac{GM}{R^2} \tag{8.2}$$

From equations (8.1) and (8.2), we have

$$v_e = \sqrt{2gR}$$

Values of escape velocity from the planets are given in Table 2.

Example 1: Mass and radius of Moon are 7.34×10^{22} kg and 1.74×10^6 m, respectively. Calculate the escape velocity for the Moon. (Given, $G = 6.67 \times 10^{-11}$ N m^2/kg.)

Solution: The escape velocity is

$$v_e = \sqrt{\frac{2GM}{R}} = \sqrt{\frac{2 \times 6.67 \times 10^{-11} \times 7.34 \times 10^{22}}{1.74 \times 10^6}} = 2372 \text{ m/s}$$

The escape velocity of the Moon is 2.4 km/s. It is much smaller than that for the Earth. Because of this low value of escape velocity, the Moon does not have any atmosphere, as all the material from the surface of the Moon escaped away.

Existence of atmosphere around a celestial body

For a heavenly body of mass M radius R and acceleration due to gravity g, the escape velocity is

$$v_e = \sqrt{\frac{2GM}{R}} = \sqrt{2gR}$$

That is, an object having velocity more than v_e escapes from the surface of the heavenly body. Suppose, T is the temperature of atmosphere around the body. The gas of molecular mass m in the atmosphere has velocity expressed as

$$\frac{1}{2}\, m v^2 = \frac{3}{2}\, kT \qquad \text{or} \qquad v = \sqrt{\frac{3kT}{m}}$$

Thus, if $v > v_e$, the molecules escape from the atmosphere of the body. Consequently, the body does not have atmosphere. For example, the satellite Moon has no atmosphere.

However, at a given temperature T, all the molecules do not have the same velocity v. Some molecules may have large velocity and some may have small velocity. The molecules follow the Maxwellian distribution of velocity and the average velocity is

$$\overline{v} = \sqrt{\frac{8kT}{\pi m}}$$

Following the theory of J.H. Jeans, if a gas is to remain on the body for a period of 1 billion years, the mean speed $\overline{v}$ should be at least $v_e/5$. Thus, for the existence of atmosphere around a celestial body, we have a condition

$$\sqrt{\frac{8kT}{\pi m}} \leq \frac{v_e}{5} \qquad \text{or} \qquad \frac{8kT}{\pi m} \leq \frac{2GM}{25R}$$

Thus, the minimum molecular mass m that can remain in the atmosphere of a celestial body is

$$m = \frac{100kTR}{\pi GM}$$

For $T = 300$ K on the Earth, we have

$$m = \frac{100 \times 1.38 \times 10^{-23} \times 300 \times 6.4 \times 10^6}{3.14 \times 6.67 \times 10^{-11} \times 6 \times 10^{24}} = 2.10 \times 10^{-27} \text{ kg}$$

Since 1 amu $= 1.66 \times 10^{-27}$ kg, we have $m = 1.27$ amu. Thus, there is no possibility of hydrogen ($m = 1$) to exist in the Earth's atmosphere. However, in the upper regions of atmosphere, the temperature is very low and therefore, atomic hydrogen can exist in the Earth's atmosphere.

Let us look for the Moon with atmospheric temperature $T = 400$ K. We have

$$m = \frac{100 \times 1.38 \times 10^{-23} \times 400 \times 1.7 \times 10^6}{3.14 \times 6.67 \times 10^{-11} \times 7.3 \times 10^{22}}$$

$$= 6.14 \times 10^{-26} \text{ kg} = 37 \text{ amu}$$

It shows that the gases nitrogen (28), oxygen (32), which are the predominant constituents of Earth's atmosphere cannot be present on the Moon.

1.10 Equatorial gravity

As the equatorial radius R_e is larger than the polar radius R_p, the gravity at the equator is smaller than that at a pole. Table 3 has the values of equatorial gravity of a planet.

1.11 Surface temperature

With the spin motion of a planet, surface temperature at a point on the planet changes with time. This change is quite appreciable in case of inner planets (Mercury, Venus, Earth and Mars).[2] The surface temperature is given in Table 3. Some planets have thick atmosphere, and therefore, the temperature of the outer layer of cloud around the planet is reported, instead of the surface temperature.

Table 3. Some parameters for planets

Planet	Equatorial gravity (m/s^2)	Surface temp (K)	Albedo	Obliquity of orbit	Obliquity of rotation
Mercury	3.704	90 - 700	0.13	$7°0'15''$	$0.°0$
Venus	8.870	740 - 250*	0.72	$3°23'40''$	$-2.°7$
Earth	9.780	223 - 328	0.45	0°	$23°27'$
Mars	3.694	136 - 243	0.16	$1°51'$	$25°11'$
Jupiter	22.65	130*	0.43	$1°18'17''$	$3°7'$
Saturn	8.792	94*	0.61	$2°29'22''$	$26°43'$
Uranus	8.699	52*	0.35	$0°46'23''$	$-82°8'$
Neptune	11.33	53*	0.40	$1°46'22''$	$29°34'$

⋆ denotes the temperature at the top of the atmosphere

Example 2: The power output of Sun is 4×10^{26} W. Calculate temperature of Venus situated at a distance of 110 million km from the Sun. Radius of Venus = 6×10^6 m.

Solution: Power received by Venus per unit area is

$$\text{Power per unit area} = \frac{\text{Total power}}{\text{Total area}} = \frac{4 \times 10^{26}}{4\pi(1.1 \times 10^{11})^2} = 2600\ \text{W/m}^2$$

The surface area that Venus presents to the Sun is that of a disk with a radius equal to that of Venus. This area is

$$\pi \times (6 \times 10^6)^2 = 1.1 \times 10^{14}\ \text{m}^2$$

so that total power received by Venus is the product of power per unit area and the presented area is

$$2600 \times 1.1 \times 10^{14} = 2.9 \times 10^{17}\ \text{W}.$$

The albedo of Venus is 0.72. That is only 28% of the incident radiation reaches the planet. The energy reaching the planet is

$$0.28 \times 2.9 \times 10^{17} = 8 \times 10^{16}\ \text{W}.$$

For Venus to have a steady state temperature, it must radiate this same amount of energy from its surface. For the unknown temperature, we have

$$P = \sigma T^4 \times 4\pi r^2 = (5.67 \times 10^{-8})T^4 \times 4\pi(6 \times 10^6)^2 = 2.6 \times 10^7 T^4$$

[2]Planets may be classified into two categories: (i) inner planets, and (ii) outer planets. The orbit of missing planet (asteroids) is the boundary between the planets of two categories.

The power received is equal to the power radiated. Thus, we have

$$8 \times 10^{16} = 2.6 \times 10^7 T^4 \qquad \text{and} \qquad T = 240 \text{ K}$$

This value of 240 K is a bit lower than a more sophisticated calculation. It may be noted that internal heating also plays a role in increasing the ambient temperature.

1.12 Albedo

Solar energy (sunlight) continuously falls on the surface of a planet. A part of that energy is reflected back, whereas the rest is absorbed by the planet. The Albedo of a planet is defined as the ratio of the amount of energy reflected by the planet to the amount of energy incident on the planet.

$$\text{Albedo} = \frac{\text{energy reflected}}{\text{energy incident}}$$

Albedo of planets are reported in Table 3. The albedo of Venus is the largest and therefore it appears as the most beautiful planet.

1.13 Obliquity of orbit

The elliptical path of the Earth around the Sun is known as the ecliptic. The plane in which the ecliptic lies is known as the ecliptic-plane. Each of the planets has a plane of its orbit, which is not parallel to the ecliptic-plane. The angle between the plane of the orbit of planet and the ecliptic-plane is known as the obliquity of the orbit of planet. Table 3 displays also the values of the obliquity of the orbit of planets.

1.14 Obliquity of rotation

Rotation (spin) axis (and thus the equatorial plane) of a planet is inclined with the ecliptic plane. The angle between the equatorial plane of planet and the ecliptic plane is known as the obliquity of rotation. Values of the obliquity of rotation of planets are displayed in Table 3.

2. Mercury

The Mercury (Figure 1) is the innermost and smallest planet in our solar system, orbiting around the Sun in 87.969 days. It is almost spherical in shape. Since the Mercury is very close to the Sun, it is quite difficult to observe it from the Earth. The maximum angular separation of Mercury relative to the Sun is 28.3°. So, Mercury is seen only for about one hour and a half under the favourable conditions, either before the sunrise, or after the sunset.

When viewed from the Earth, the apparent magnitude of Mercury varies from 2.3 to 5.7. Since the orbit of Mercury lies inside the orbit of the Earth, it shows Moon-like phases. It does not contain any atmosphere, because of its lower surface gravity. Mercury is similar to Moon in appearance. Its surface is full of craters produced either by volcanoes, or by impacts of small foreign debris. Mercury has a large iron core and is found to have a bipolar magnetic field of strength about 200 - 300 nT (1 nT = 10^{-9} Tesla). It has no natural satellites. Surface temperature of Mercury varies from about 90 to 700 K.

Figure 1: Mercury. Courtesy to Google website.

3. Venus

The Venus (Figure 2) is the second planet in the solar system revolving around the Sun in 224.7 days. It is almost spherical in shape. Venus is in many respect an Earth-like planet, particularly in its size, mass and retention of atmosphere. Because of this, it is sometimes called the Earth's sister-planet. Owing to its exquisite brilliance and appearance in the morning and in the evening sky alternately, Venus is imagined as a goddess of love and beauty.

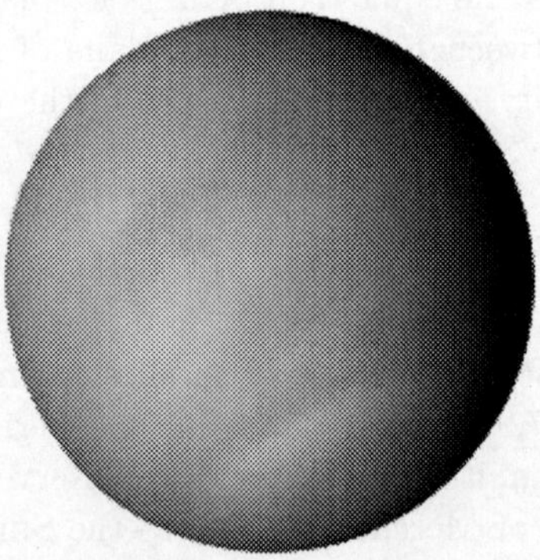

Figure 2: Venus. Courtesy to Google website.

As seen from the Earth, Venus is the most brilliant object next to the Moon in the night sky. In fact, this is the only planet that can be seen in the broad day light with the naked eyes. Its apparent magnitude reaches up to -4.6. Its maximum angular separation from the Sun is 47.8°. Venus can be seen up to three hours under favourable circumstances either before the sunrise or after the sunset. It reaches its maximum brightness shortly before the sunrise or shortly after the sunset. For this reason, it is often called the Morning Star or the Evening Star.

The telescopic images of Venus are rather dull and featureless, although its appearance in different Moon-like phases are quite attractive. Incidentally, it was the characteristic phase pattern of Venus that led Galileo Galilee to produce a convincing argument against the geocentric concept of the solar system.

Venus is totally surrounded by its own cloud system. Clouds reflect light quite efficiently. In the case of Venus, more than 70% of incident sunlight is reflected by its topmost thin layer of clouds. Since its surface cannot be seen from outside, its rotation period cannot be ascertained directly.

Venus has a thick atmosphere containing ~96.6% of CO_2, 3.2% of N_2 and no water vapour. Possibly all water molecules have been dissociated into H_2 and O_2, and escaped from the atmosphere of Venus. CO_2 has a property that it allows visible light to pass through but absorbs the infrared radiations that the surface returns or re-emits. So, the sunlight can get in, and after being thermalized, it is radiated in the infrared radiations which are efficiently trapped due to the presence of CO_2 in the atmosphere. This effect, known as the **green house effect**, is responsible for raising the surface temperature of Venus to a value as high as 650 K. The atmospheric pressure at the planet's surface is believed to be 92 times that of the Earth.

Venus contains a lot of sulfur which finally forms sulfuric acid and the clouds of Venus are made up of the droplets of sulfuric acid. Venus also has no natural satellites.

4. Earth

The Earth (Figure 3) is the densest planet in our solar system. It makes one round around the Sun in 365.25 days and is sometimes referred to as the World, the Blue Planet, or by its Latin name, Terra. As we are housed on the surface of Earth, we must therefore be knowing a lot about it. The most reliable method that is used today for its age determination is the method of radioactive chronology. The Earth is estimated to be formed 4.54 billion years ago, and life appeared on its surface within a billion years. Millions of species, including human, live on its surface. Earth is currently the only place in the universe where life is known to exist.

The most important thing about the Earth's atmosphere is that it contains a lot of oxygen and nitrogen which are usually not found on other planets. They contain either too much H_2, He, CO_2 or methane, ammonia, etc. It is also important that the Earth's atmosphere contains ozone in sufficient amount which readily absorbs the ultra-violet radiations coming from the outer atmosphere and protects the life on the Earth's surface.

The Earth's magnetic field also shunts out most of the lethal solar cosmic rays from the equatorial regions which otherwise would have been extremely harmful to the life. Interaction of Earth's magnetic field with the solar cosmic rays makes the latter channeled to the polar regions of the Earth. There they can penetrate sufficiently deep into the atmosphere and produce grand display of the auroral lights as a result of excitation and spontaneous de-excitation of the atmospheric nitrogen and oxygen.

Figure 3: Earth. Courtesy to Google website.

The Earth has a sufficiently strong magnetic field of 32 μT (1 μT= 10^{-6} Tesla). It means that some part of the Earth's interior must be in the molten state. Detailed seismological studies have revealed that at a depth of about 2900 km from the surface, the matter changes its state, from solid to liquid state. At a depth about 5100 km from the surface there is another transition from liquid to solid state again. The central density and temperature of Earth reaches as high as 1.5 $\times 10^4$ kg/m^3 and 6900 K, respectively.

71% of the Earth's surface is covered with water with an average depth of about 3 km. Earth's axis of rotation is tilted 23.4^o away from the perpendicular to its orbital plane, producing seasonal variations on the Earth's surface.

Most part of the electromagnetic radiations from the outer space is blocked by the Earth's atmosphere. Radiations reaching on the surface of Earth, however, mainly lie in two bands, known as the optical window and the radio window. The optical window ranging from 3000 Å to 10,000 Å has been exploited for several centuries in order to get information/knowledge about the cosmic objects as well as about our solar system. Since the radiations of mm and cm wavelengths cannot be seen by human eyes directly, therefore, they could not be noticed until 1932, when a Karl G. Jansky accidentally discovered the radio window, which ranges from a few mm to a few decameters in wavelength.

The Earth has only one natural satellite, called the Moon.

5. Moon

The Earth has only one natural satellite called the Moon (Figure 4). Its radius is about 27% that of the Earth, mass being 1.23% that of the Earth, and density about 3.34 $\times 10^3$ kg/m^3. Besides its revolution around the Earth, the Moon spins about its own axis. The time-period of the revolution is exactly equal to the time-period of its spin motion. It is 27.32166 days. So, the Moon keeps its one face permanently turned towards the Earth. That is why we always see the same part of the surface of Moon. On the surface of Moon, there are craters and mountains.

Figure 4: Moon. Courtesy to Google website.

The Moon surface is found to be highly cratered, interspersed by darkish plains, called Maria. Craters are mostly impact type with some of them having distinct ray patterns. Since the Moon has very little magnetic field, its interior must be geologically dead, but the surface must retain craters of all ages, as it is practically devoid of atmosphere (about 10^{12} atoms/m^3). The heights of lunar mountains can be measured by measuring the length of shadows they cast during the morning or the late afternoon's low-lying Sun.

Lunar rocks that have been brought to the Earth by the Apollo astronauts have provided many interesting clues regarding the formation of Moon, which is now believed to have either taken place simultaneously with the Earth or been captured in the very early times.

6. Solar eclipse

As seen from the Earth, a solar eclipse occurs when the Moon passes between the Sun and Earth, and the Moon blocks the Sun fully (Figure 5) or partially. This can happen only at the new Moon day. For a total eclipse, the disk of the Sun is fully obscured by the Moon. In a partial or annular eclipse, only a part of the Sun is obscured.

Since the duration of a total solar eclipse is very small, it shows that the angular sizes of both the Moon and Sun are nearly equal, when seen from the Earth.

If the Moon were in a perfectly circular orbit and in the same orbital plane, there have been total solar eclipses every month. However, the Moon's orbit is inclined (tilted) at more than 5° to the Earth's orbit around the Sun, so its shadow at new Moon usually misses the Earth.

Some dates when total solar eclipse will take place in future are: 2 July 2019,

14 December 2020, 4 December 2021, 8 April 2024, 12 August 2026, 2 August 2027 and 22 July 2028.

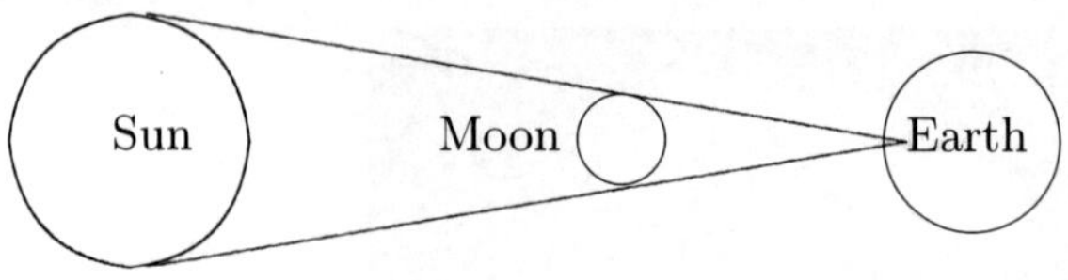

Figure 5: During the total solar eclipse, Moon covers the Sun.

Some dates when total solar eclipse occurred in the recent past are: 21 August 2017, 9 March 2016, 20 March 2015, 13 November 2012, 11 July 2010, 22 July 2009, 1 August 2008, 29 March 2006, 23 March 2003, 4 December 2002, 21 June 2001.

7. Lunar eclipse

A lunar eclipse occurs when the Moon passes directly behind the Earth into its shadow (Figure 6). This can occur only when the Sun, Earth and Moon are aligned exactly, or very closely so, with the Earth in the middle. Hence, a lunar eclipse can occur only at a night of full Moon.

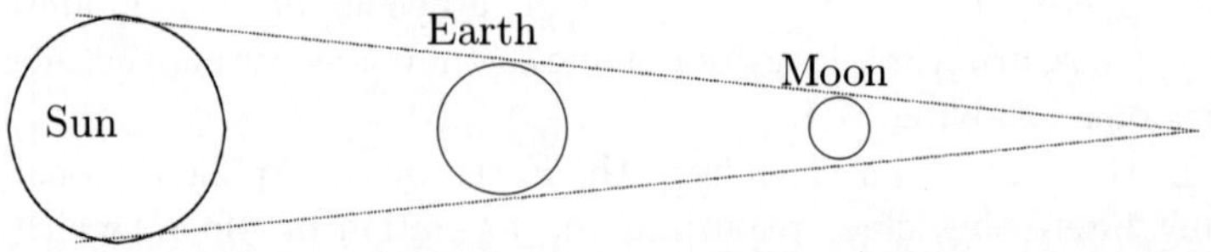

Figure 6: During the total lunar eclipse, Moon moves through the shadow of Earth.

The type and length of a lunar eclipse depends upon the Moon's location relative to its orbital nodes.[3]

Some dates when lunar eclipse will take place in future are: 21 January 2019, 26 May 2021, 16 May 2022, 8 November 2022, 14 March 2025, 7 September 2025, 3 March 2026 and 31 December 2028.

Some dates when total lunar eclipse occurred in the recent past are: 27 July 2018, 31 January 2018, 28 September 2015, 4 August 2015, 8 October 2014, 15 August 2014, 20 December 2011, 15 June 2011, 21 December 2010, 21 February 2008, 28 August 2007, 3 March 2007, 28 October 2004, 4 Mat 2004, 9 November 2003, 16 May 2003, 9 January 2001.

[3]The lunar orbital nodes are the points where the orbit of the Moon crosses the orbit of the Earth. The ascending or north node is where the Moon crosses from the south of the ecliptic to the north. In Hindu mythology, it is known as the *Rahu.* The descending or south node is where it crosses from the north of the ecliptic to the south. In Hindu mythology, it is known as the *Ketu.*

Example 3: Suppose, a solar eclipse takes place when the Sun and Moon are separated from the Earth by 1.53×10^8 km and 3.79×10^5 km, respectively. Calculate the width of strip on the Earth's surface from which a total solar eclipse is observed. The diameters of Sun, Earth and Moon are 1.392×10^6 km, 3480 km, 12800 km, respectively.

Solution: For the solar eclipse, the Moon is in between the Sun and Earth. It is easier to work with half shadow which is cast by half of the Moon obscuring half of the Sun and produces a shadow of length A″C″ on the surface of Earth (as the A″C″ is small, the surface can taken as plane.) The triangles ABC, A′BC′, A″BC″ are all right-angled and enlargements of each other so that they all subtend the same angle, θ, at B. The position of B is initially unknown but can be calculated from the information available from triangles ABC and A′BC′:

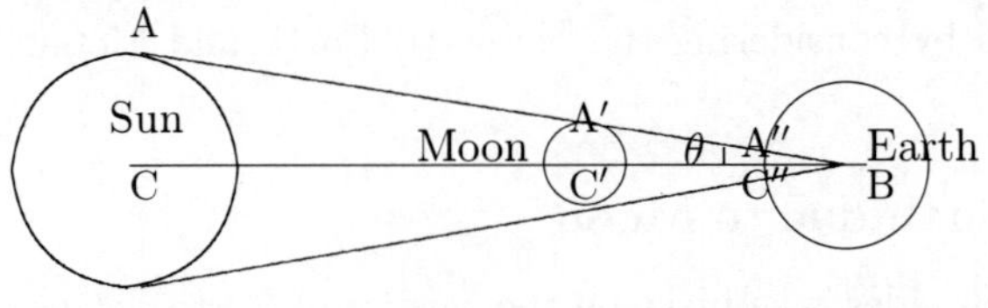

AC = BC $\tan\theta$ and A′C′ = BC′ $\tan\theta$

AC is the Sun's radius = $1.392 \times 10^6/2 = 6.96 \times 10^5$ km

A′C′ is the Moon's radius = 3480/2 = 1740 km

CC″ is the Earth-Sun separation = 1.53×10^8 km

Earth-Moon distance = 3.79×10^5 km

Thus, CC′ = $1.53 \times 10^8 - 3.79 \times 10^5 = 1.526 \times 10^8$ km

The distance CC′ is equal to BC - BC′. Three equation are now available:

$6.96 \times 10^5 = \text{BC} \tan\theta \quad 1740 = \text{BC}' \tan\theta \quad \text{BC} - \text{BC}' = 1.526 \times 10^8$

As there are only three unknowns, the equations can be solved to give BC′ = 3.83×10^5 km and $\tan\theta = 4.55 \times 10^{-3}$ (BC is not required). This means that B is within the Earth's volume, slightly further away from the Moon than the Earth's center. its separation from the Earth's center is ($3.83 \times 10^5 - 3.79 \times 10^5 =$) 4000 km. The distance BC″ is therefore 4000 km plus the Earth's radius (6400 km) which is 10,400 km.

The size of half-shadow on the Earth's surface is

A″C″ = BC″ $\tan\theta = 10400 \times (4.55 \times 10^{-3}) = 47$ km

The path of the total eclipse is thus only about 100 km across. It should be noted that this calculation is subject to quite large rounding errors as the scales involved vary from thousands of km to hundreds of millions of km.

8. Tides

We are familiar with the relation

$$F = -G \frac{m_1 m_2}{r^2} \tag{8.3}$$

expressing the gravitational force between two point-masses m_1 and m_2, separated by a distance r. Here, G is the universal gravitational constant and the negative sign shows that the force is attractive. The relation (8.3) is valid for two point-masses or for two spherical bodies where the separation r between their centers is much larger than their radii (sizes). When the separation r between two spherical bodies is comparable to the radii of bodies, the shapes of the bodies are deformed and they are elongated along the line joining their centers. The elongation is symmetrical on both sides of the bodies.

These effects are clearly observable on the Earth's surface in the form of tides, which may be understood by considering the pairs: (i) Earth and Moon, and (ii) Earth and Sun.

8.1 Tidal force on Earth due to Moon

In order to understand how tides take place on the surface of Earth, let us consider the Earth with its center at O and the moon with its center at O′, separated by a distance r (Figure 7). For simplicity, we assume that both the Earth and Moon are not spinning. Let R be the radius of Earth which is about 3.7 times that of the Moon. The mass of the Earth is about 81.3 times that of the Moon. Since we are interested in the tidal effects observed on the surface of earth, it is convenient to consider that the mass of the Moon is situated at its center O′.

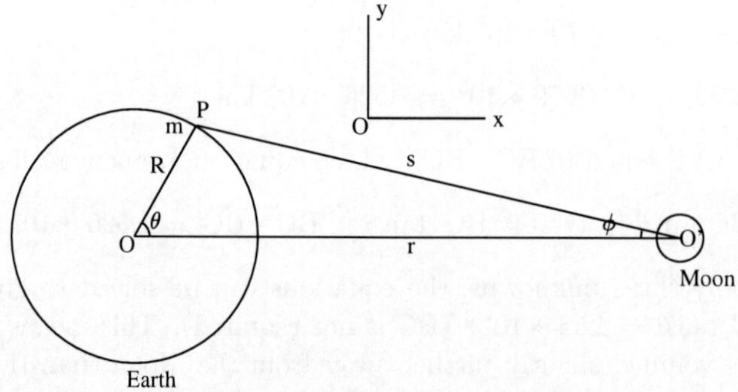

Figure 7: Earth and Moon are separated by a distance r.

Let M be the mass of the Moon. (In fact various mass-elements of Moon attract different mass-elements of Earth. It leads to a complicated situation. In order to get reasonably reliable results, we have considered that the mass of Moon is situated at its center.) The average separation between the centers of Earth and Moon is about 3.82 $\times 10^8$ m, which is about 60 times of the radius of Earth. Therefore, different elements of the body of Earth experience different gravitational force due to the Moon, as the separation between an element of Earth and the Moon varies. Further, the effects would be symmetrical about the line joining the centers O and O′. For convenience, we consider the forces in the xy-plane (lying in the plane of the paper) only, where the x-axis is parallel to OO′. Suppose, a mass m is situated at O. The x and y components for the force between the mass m at O and the Moon are

$$F_{Ox} = -G\,\frac{Mm}{r^2}, \qquad F_{Oy} = 0$$

Now, suppose the mass m is situated at a point P on the surface of Earth, which is at a distance s from O′. The x and y components for the force at P are

$$F_{Px} = -G\,\frac{Mm}{s^2}\,\cos\phi, \qquad F_{Py} = G\,\frac{Mm}{s^2}\,\sin\phi$$

The difference between the forces at P and at O is

$$\Delta F = (F_{Px}\hat{\imath} + F_{Py}\hat{\jmath}) - (F_{Ox}\hat{\imath} + F_{Oy}\hat{\jmath})$$

$$= GMm\Big(-\frac{\cos\phi}{s^2} + \frac{1}{r^2}\Big)\hat{\imath} + \frac{GMm}{s^2}\,\sin\phi\hat{\jmath} \tag{8.4}$$

From Figure 7, we can write

$$s^2 = (r - R\cos\theta)^2 + (R\sin\theta)^2 = r^2\Big[1 + \frac{R^2}{r^2} - \frac{2R\cos\theta}{r}\Big] \approx r^2\Big[1 - \frac{2R\cos\theta}{r}\Big] \tag{8.5}$$

where the term $R^2/r^2 << 1$ is neglected. On using equation (8.5) in (8.4), we get

$$\Delta F = GMm\Big[-\frac{\cos\phi}{(r^2 - 2Rr\cos\theta)} + \frac{1}{r^2}\Big]\hat{\imath} + \frac{GMm\sin\phi}{(r^2 - 2Rr\cos\theta)}\hat{\jmath} \tag{8.6}$$

On neglecting the terms having second and higher orders of R/r in equation (8.6), we get

$$\Delta F = \frac{GMm}{r^2}\Big[-\cos\phi\Big(1 + \frac{2R\cos\theta}{r}\Big) + 1\Big]\hat{\imath} + \frac{GMm\sin\phi}{r^2}\Big[1 + \frac{2R\cos\theta}{r}\Big]\hat{\jmath} \tag{8.7}$$

On using the first order relations, $\cos\phi \approx 1$ and $\sin\phi \approx (R\sin\theta/r)$, in equation (8.7), and neglecting the terms containing R^2/r^2, we get

$$\Delta F = -\frac{GMmR}{r^3}(2\cos\theta\hat{\imath} - \sin\theta\hat{\jmath}) \tag{8.8}$$

It shows a differential force on the surface of the Earth relative to that at O.

Though all the forces on various elements of the earth are directed towards the center of the moon, the forces relative to the force at O (equation 8.8) are directed outward along the x-axis and inward along the y-axis. The factor 2 in equation (8.8) shows that the magnitude of outward force is two times larger than that of the inward one. Consequently, the Earth is elongated along the line OO′, and hence the tidal bulges are pronounced along the Earth-Moon line. The bulges are on both sides of the Earth, as the shape of the Earth is oblate spheroidal with larger radius along the line joining the Earth and Moon. (Similar effects, about 20 times more pronounced than those on the Earth, would be observable on the surface of the Moon. However, here, we are interested in the effects observed on the surface of Earth only.)

Owing to the revolution of Moon around the Earth, the position of the tidal bulge changes smoothly on the surface of the Earth. This phenomenon is periodic and one observes two high tides in a period of 24 h 53 m. This period known as lunar day is the average time interval between two successive transits of Moon through the observer's meridian.

8.2 Accounting the rotational motion of Earth

Let us see the effect of accounting for the rotational motion of the Earth. (Since Moon is being considered to be a point mass, we are not accounting for the rotational motion of Moon.) If the time-period of Earth's rotation about its own axis is equal to the period of revolution of Moon around the Earth, then the position of the tidal bulges would have remained fixed on surface of Earth. But, in fact the time-period of Earth's rotation is shorter than that of the Moon's revolution around the earth by about 53 m. Therefore, the frictional forces on the surface of Earth drag the bulge axis ahead of the Earth-Moon line. The bulge is pulling the Moon forward, speeding it up and causing it to move farther out. Since a frictional force is always dissipative, Earth's rotational kinetic energy is being lost regularly, and hence, the Earth's spin rate is decreasing regularly. So, the Earth's rotational period is increasing regularly. Estimates show that the Earth's rotational period is, at present, increasing at a rate of 0.0016 s/century. The angular momentum of the Earth is obviously decreasing with time.

Following the conservation of angular momentum for the Earth-Moon system, the angular momentum lost by the Earth is being transferred to the orbital motion of the Moon, and therefore, the angular momentum of Moon is increasing regularly. With the increase of angular momentum of Moon, the distance of Moon from the Earth is increasing regularly. Estimates show that at present, the Moon is drifting away from the Earth at the rate of 3 to 4 cm/year.

8.3 Tidal force on Earth due to Sun

Another celestial body, after the Moon, which produces observable tidal effects on the Earth's surface is the Sun. Similar calculations can be performed for the Earth-Sun system. The ratio of the effects on the Earth's surface due to Moon and due to Sun may be written as (using equation 8.8)

$$\frac{(\Delta F)_{\text{Moon}}}{(\Delta F)_{\text{Sun}}} = \frac{(M/r^2)_{\text{Moon}}}{(M/r^2)_{\text{Sun}}} = \frac{7.35 \times 10^{22}/(3.844 \times 10^8)^3}{1.989 \times 10^{30}/(1.496 \times 10^{11})^3} = 2.2$$

Obviously, the tidal effect of Moon is 2.2 times larger than that of the Sun. The tidal effect of Sun also is periodic. There are two high tides due to Sun (other than those due to the Moon) in a period of one solar day, which is equal to 24 h. Hence we have two sets of periodic tidal bulges, one due to Moon and other due to Sun. The time periods of the two sets are different.

8.4 Superposition of tidal forces

These two sets of tidal bulges show a phenomenon of superposition. When the Sun, Earth and Moon are aligned along a straight line, at the time of full Moon or a new Moon, the tidal effects are added to create the largest possible bulges on the Earth's surface. These situations are known as the **spring tides**. The spring tide occurs twice each lunar month all year long without regard to the season.

At the first and third quarters of a lunar month, the Sun, Earth and Moon form a right-angle. In this situation, the tides produced by Moon and by Sun reduce the effects of each other to the minimum possible. These situations are known as the **neap tides**.

8.5 Future scenario

Equation (8.8) shows that with the increase of r, the tidal effects would decrease rather rapidly, as the exponent of r is 3. Since the Moon is receding away from the Earth, the tidal effects due to Moon would not be observable on the surface of Earth in far future. Since the distance between Sun and Earth is not changing, the Earth would always observe tidal effects due to Sun. Hence, in the far future, the tidal effect would not depend on the Moon calendar, and the intensity of the effect would remain about one-third of the observed maximum effect at present.

8.6 Tidal forces in the context of other planets

Equation (8.8) shows that the tidal effects become more pronounced with the decrease of the distance r from the planet, generating the tidal effect. The effect may be so severe that a body coming into the range of the tidal force may break down into small pieces. This is an explanation for the material found in the ring(s)around some planets. The phenomenon would obviously be more intense in case of massive planets, for example, Jupiter and Saturn. Jupiter is the most massive planet in our solar system. Thus, if some comet comes close to the surface of Jupiter (*i.e.*, within the Roche limit), it would break into pieces. Such event has been witnessed by a large section of people that the nucleus of the comet Shoemaker-Levy-9 was broken into 22 fragments when it passed through perijove (nearest point in orbit of Jupiter) on 8 July 1992.

9. Mars

The planet Mars (Figure 7) appears markedly reddish, and therefore, it is sometimes called the red planet. Mars has been the first planet whose actual distance from the Earth was measured using the methods of trigonometric parallax by Cassini in 1673. However, in 1659, Huygens noticed a dark triangular mark on the disc of Mars, which he found to rotate regularly once every 24.5 hours, giving the first measurement of rotation period of Mars. In 1781, Hershel showed that the axis of rotation of Mars was inclined by almost the same amount as that of the Earth. So, the seasons on the Mars must follow the terrestrial rules, namely, alternately opposite seasons in the opposite Martian hemispheres.

Mars has a thin atmosphere with a pressure about 1/200-th that of the Earth, and with the composition of 95.3% CO_2, 2.7% N_2, 1.6% Ar, 0.3% O_2, 0.07% CO, 0.03% H_2O and very little Ne, Kr, Xe and O_3. Mars atmosphere is very dry because all the water is frozen into ice, and the maximum day-time temperature is about -30° C. Even the CO_2 is frozen into the dry ice for most of the time.

Figure 8: Mars. Courtesy to Google website.

The planet has polar caps composed of both water ice and dry ice. The extent of the northern cap is found to vary with the season, but not the southern cap. The surface as seen through a telescope is reddish, and this colour is due to the presence of iron oxide in the Marian limonites.

The question of existence of life on the Mars has been discussed from time to time. But, no concrete evidence for it has been established so far.

Mars has been found to have two natural satellites, named Phobos and Deimos. The Phobos is nearer as compared to the Deimos. The Phobos revolves with the time-period of 0.318 day, and the Deimos revolves with the time-period of 1.262 days. Interesting to note that both the satellites are rotating synchronously with their respective periods of revolution. That is, they keep always their same face turned towards the Mars. They are believed to have been captured from outside.

It is found that Mars has very weak magnetic field of about 60 nT; a very weak ionosphere containing plasma of singly ionized oxygen duly captured from the solar wind.

10. Jupiter

The Jupiter (Figure 8) is the largest planet of our solar system. It occupies $\sim$ 0.1% of the mass of Sun and a radius $\sim$ 1/10th of the Sun. Its average density is almost equal to that of the Sun. Had it been 8 times more massive, it could have become a star by itself. From the Earth, it appears as the second brightest planet, next to Venus (however not always, because the Mars at its closest position appears somewhat brighter than Jupiter).

With the help of a big telescope, it appears as a nebulous object with a sharp elliptical boundary. It is because the Jupiter rotates very fast about its own axis, and has therefore assumed the shape of an oblate spheroid with its polar diameter, $\sim$ 6% smaller. A point on its equator has to move with a speed of about 45,000 km/hr in order to complete a full axial rotation in only 9.85 hr. Jupiter displays a number of dark and light colourful bands round the body. They are seen to be perpendicular to the rotation axis. Since it is a giant gaseous planet, we can see only the atmosphere and the cloud system from outside.

Figure 9: Jupiter. Courtesy to Google website.

There is a permanent oval-shaped spot in the southern hemisphere of the planet, which is visible since its discovery in 1664. It was dramatically described as the Great Red Spot in 1878. The Great Red Spot is $\sim$ 14,000 km wide and 30,000 - 40,000 km long, and deep brownish red in colour. It swifts very slowly with time as the rotational period of the Jupiter is $9^h\ 55^m\ 37.6^s$. There are however other smaller, red and white spots with lifetime of several years. Also seen on its surface are the belts and zones which are stable, coloured formation, though the width and colour do vary with time. The belts are reddish and brownish in colour. The motion of gas inside a belt is downward.

The Jovian atmosphere is primarily made up of H and He in almost the same proportion as found in the interstellar clouds or the Sun.

Like the Sun, Jupiter shows differential rotation. Its equatorial region rotates with a period of $9^h\ 50^m\ 30^s$. So, it is very difficult to determine the actual period of rotation of the rigid part of the planet. Fortunately, Jupiter has a strong bipolar magnetic field which suggests that its inner core must be in a fluid state, supporting the electric currents. The differential rotation is also responsible for the stratification of the atmosphere parallel to the rotation axis.

The solid and liquid parts of Jupiter are buried inside its thick atmosphere of depth about 1000 km. It is believed that possibly it has a solid iron, nickel and silicate core of few tens of the Earth masses, surrounded by a thick layer of metallic liquid hydrogen and overlying thick layer of liquid molecular hydrogen. The central temperature of Jupiter exceeds about 3×10^4 K with a central pressure $\sim 3 \times 10^6$ times the pressure on the surface of Earth.

One of the greatest puzzling discoveries made by the Pioneer probes in 1979 was that the Jupiter emits twice the radiation that it receives from the Sun, indicating that the Jupiter has an internal source of heat. Evolutionary calculations show that the Jupiter is still cooling; He^4 is still gradually settling down towards the center, releasing tremendous amount of gravitational energy. This heat is being transported outwards by convection currents inducing flows in the metallic hydrogen, responsible for the origin of Jovian magnetic field.

Jupiter has the largest number of satellites. At least 67 satellites of Jupiter are known. The four prime satellites of Jupiter are the Io, Europa, Ganymede and Callisto, discovered in 1610 by Galileo. The motion of inner three satellites, namely, Io, Europa and Ganymede is not independent of each other; in fact their orbital periods around the Jupiter are locked in the perfect tidal resonances with a ratio of the orbital periods 1 : 2 : 4. The satellite Io, a little bigger than our Moon, has its surface spotted with numerous volcanoes but without any mountains. There is also no trace of impact craters on the surface of Io, and no trace of water either.

Since the time of Galileo, the tally of Jupiter satellites has been steadily increasing from 4 to at least 67. Rings of Jupiter have also been discovered in 1979, which are sparse and narrow having thickness of smaller than 1 km and being composed of small particles, few micrometer (μm) in size.

Europa, slightly smaller than our Moon, appears appreciably brighter than the Io because of its high reflectivity. The surface is extremely smooth and covered with ice bearing no sign of impact craters at all. Europa's surface is renewed continuously by small fountains of fresh water that comes through the icy crust.

Ganymede, the biggest of all the satellites in the solar system, is about 1.5 times bigger than our Moon. It is also bigger than the Pluto and Mercury. Ganymede is composed of water or ice and silicate rocks in almost equal proportion. Its surface bears the marks of impact craters.

Callisto, the outermost of the Galilean satellites has a rather dark complexion and is still quite big in size, only a little smaller than Ganymede. From the texture of surface, it seems that Callisto is made mostly of stone and ice, and bears marks of impact craters.

The other satellites of Jupiter are considerably small in size.

11. Saturn

The Saturn (Figure 9) is the second largest planet of our solar system, after Jupiter, having beautiful ring system. In the beginning, the planet appeared to have ears. These ears later on turned out to be a beautiful system of rings surrounding the planet. Saturn is the only planet in the solar system whose average mass-density is lower than that of water. One can say that the Saturn can float on water like a piece of wood.

Like Jupiter, Saturn is also found to radiate about 1.8 times the energy it receives from the Sun. One reason why the Saturn is so featureless is that the velocity of wind in the equatorial band is about 1800 km/hr. So, every feature is finely striated parallel to the Saturn's equator. The outflow of Saturn's own internal energy becomes the driving force behind achieving such high velocities of wind.

Figure 10: Saturn. Courtesy to Google website.

Like Jupiter, Saturn has also got a very strong magnetic field of its own (about 40 μT near its equator). Saturn has a rocky core of about 3 Earth masses and a temperature of 1.5 $\times 10^4$ K. Beneath the clouds, there is a thick layer of molecular liquid hydrogen and metallic liquid hydrogen, the latter starting at a depth of 30,000 km. Although the conducting mantle of Saturn rotates with a period of 10^h 39^m 24^s, its equatorial atmosphere rotates somewhat slowly with a period of 10^h 10^m.

The rings of Saturn are made up of normal water-ice, having particle sizes widely varying from μm to meter. The thickness of the rings is extremely small, $\sim$ 100 m at the most, whereas width of the major ring system is $\sim$ 1.3 $\times 10^5$ km. The reason why Galileo missed the identification of the ears of Saturn as its ring was that he could no longer see those ears in 1612 which he saw in 1610-1611.

In fact the Saturn's ring system disappears, when viewed from the Earth, once every 15 years, and in the intervening period one alternately sees the northern and the southern faces of the ring with a maximum opening angle of about 27°. That the ring is not continuous across its breadth was first noted by Cassini in 1675. He thought that there had been two rings separated by a dark portion, now known as the Cassini's division.

The outer ring has been designated as A and the inner B. In 1850, yet a smaller region inner than B was reported and was duly designated as C. Thanks to the wonderful space missions, today we know that there are about 7 prominent rings around the Saturn, named alphabetically D, C, B, A, F, G and arranged sequentially from the innermost to the outermost.

Roche demonstrated that any satellite in orbit within $\sim$ 2.44 times the radius of the planet finally breaks up into small pieces and forms a ring.

More than sixty-two known moons orbit the planet of which Titan is the largest. In size, Titan is slightly bigger than the Moon. It is the only satellite in the solar system that has an atmosphere. Its atmosphere contains 99% of nitrogen, the temperature is about 90 K, and the atmospheric pressure is 1.522 bars. Other prime satellites of Saturn are the Mimas, Enceladus, Tethys, Dione, Rhea, Iapetus.

Except for Phoebe, the first 11 satellites of Saturn are synchronous, that is, axial periods of rotation are the same as those of their orbital periods of revolution around Saturn. Compared to Titan, Tethys, Dione and Rhea are the three next bigger satellites composed of ice and rock. All the other satellites are just small irregular fragments of ice and rock.

12. Uranus

The Voyager 2 passed by the Uranus on 24 January 1986. Before that very little was known about the Uranus as its angular size seen from the Earth never exceeds 4″. The Uranus (Figure 10) was the first planet discovered using a telescope by Herschel in 1781 (the planets up to the Saturn are naked eye planets), and its orbit lies beyond that of Saturn. It was later on named after the god of the sky and the father of Saturn. When the planet comes closest to the Earth, that is, during an opposition, it can barely be seen with the naked eye and its total visual magnitude becomes +5.5.

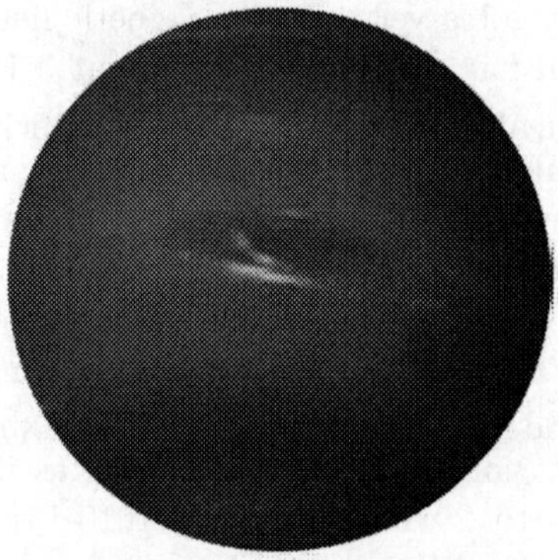

Figure 11: Uranus. Courtesy to Google website.

Uranus is surrounded by a system of eleven rings of which five were discovered in 1977 by observing the occultation of a star by the disc of Uranus. The last two rings were discovered by Voyager 2. The Uranian system of rings are found to be confined within the Roche limits of Uranus.

Since the axis of rotation of Uranus is tilted by 98° with respect to the pole of the ecliptic, Uranus revolves around the Sun keeping its axis turned to a side. Even though Uranus rotates very fast about its own axis, its outer surface shows very little oblateness. Thus, it is expected that the Uranus should have a rocky core about the size of the Earth. The envelope is a gaseous mixture of H, He, H_2O, NH_3 and CH_4, instead of a liquid mantle.

Due to a large tilt of its rotation axis, Uranus' north pole faces the Sun for 42 years (one half of the period of its orbital revolution) and during the next 42 years the south pole faces the Sun. So, the poles get alternately warmed up during the planet's complete revolution around the Sun.

The atmosphere is composed mostly of hydrogen (85% by mass) with about 12% of helium by mass and very small amounts of methane and other hydrocarbons.

The Uranus appears blue in colour because its atmospheric methane differentially absorbs red light from the Sun. The visible surface is a layer of haze containing C_2H_2 and C_2H_6, formed due to the photo-dissociation of CH_4.

The average temperature at the cloud top is 52 K. Uranus reflects ~50% of the incident sunlight from its cloud top. There are intense radiation belts and radio emissions because it has a sufficiently strong magnetic field (23 μT on its magnetic equator). The axis of rotation and that of the magnetic field are however not aligned. They subtends an angle of 60° with each other.

The narrow rings of Uranus appear to be very dark and are possibly composed of meter-sized icy chunks. The thickness of the rings are less than 12 km, except the epsilon ring, which is 22 - 93 km.

More than 27 moons of Uranus revolve synchronously with the axial rotation of Uranus. The biggest ones, the Titania, Oberon, Ariel, Umbriel and Miranda, can be spotted as tiny bright dots with the help of a large telescope. They are made up of dark gray ice and rock. The Ariel appears brightest containing many features and extensive flows of icy material.

13. Neptune

The Neptune (Figure 11) is the eighth and the farthest planet in our solar system. It is slightly smaller in size, yet more massive than Uranus. It is the first planet of the solar system whose existence was predicted by the theory of gravitation, and which was subsequently discovered in 1846. After the discovery of Uranus in 1781, it was found that Uranus was not exactly following its predicted orbit. So, the problem was taken up and the conclusion drawn was that some other yet unknown planet must be perturbing the orbit of the Uranus. The discovery of the Neptune marked the greatest triumph of Newton's theory of gravitation.

Figure 12: Neptune. Courtesy to Google website.

Neptune has not so far completed one complete revolution around the Sun since its discovery. It appears as a point of light when seen through a pair of binoculars and has a magnitude at opposition about 7.8. With the help of the latest telescopes,

Neptune appears as a tiny bluish disc of size 2.3″. The bluish colour is once again due to the presence of methane in its atmosphere.

A ring system around the Neptune was reported from stellar occultation observations in 1985. Then the Voyager 2 flew past Neptune on August 25, 1989, which provided the wealth of information about the planet, its ring system and satellites.

Neptune has got two complete thin rings at the distances of 62.9 and 52.3 thousand km from its center. There are also evidences of 3 sheets or incomplete rings. Rings are once again made up of micron-sized dust particles.

Neptune appears in many respects similar to Uranus and Jupiter. There is a Great Dark Spot on Neptune almost at the same latitude of 22° S as the Great Red Spot on Jupiter. Even though Neptune's period of axial rotation is $16^h\ 3^m$, its cloud system is slightly slower in the same direction with a period of $\sim$ 18 hr, whereas on Jupiter, Saturn and Uranus, the clouds revolve around the planet faster than the planets' axial rotation.

The temperature at the cloud top near the equator as well as at the poles is about 60 K, whereas at the mid-latitudes, it is only 50 K. Neptune has a rocky core surrounded by a waterly mantle and its thick atmosphere is primarily composed of H and He with an admixture of methane. Ionized water molecules in the mantle conduct electric current thereby giving the planet its magnetic field strength 13 μT. The magnetic axis of Neptune is tilted by 47° with respect to its rotation axis and the center of the magnetic dipole seems to be greatly displaced from the center of the Neptune towards south. The core contains rocky material and is believed to be cold and inert.

At least 14 satellites of Neptune are known. However, only one satellite, Triton, is prime. Triton is slightly bigger than our Moon and revolves synchronously with axial rotation, but in a retrograde sense with a period of 5.88 days. The Triton, having also a highly inclined orbit, was perhaps captured by the Neptune. The pinkish eyes that constitute Triton's south polar cap, is probably a layer of frozen methane and nitrogen, and nitrogen is believed to have been a product of ammonia. The other satellites of Neptune are quite small in size.

14. Asteroids

As stated earlier that every body revolving around the Sun is not a planet. For a body to be called a planet, it must have size (mass) more than the value specified by the International Astronomical Union (IAU). This value can be modified. For example, the Pluto was in the list of planets up to 24 August 2006, and after that it has been removed from the list, as the value of the specified mass was modified. There are large number of bodies revolving around the Sun having small size. They are generally called the minor planets. The Asteroids (Figure 12) or minor planets are the bodies orbiting around the Sun, mainly in between the Mars and Jupiter.

14.1 Titius-Bode law

The Titius-Bode law first predicted a missing planet between the Mars and Jupiter. In order to have the idea about the Titius-Bode law, take the following series:

0, 1, 2, 4, 8, 16, 32, ...

Now, follow the following steps on each of the terms of the series:

(i) multiply each term by 3, to get

0, 3, 6, 12, 24, 48, 96, ...

(ii) add 4 to each of the terms, to get

4, 7, 10, 16, 28, 52, 100, ...

(iii) divide each term by 10, obtaining

0.4, 0.7, 1.0, 1.6, 2.8, 5.2, 10.0, ...

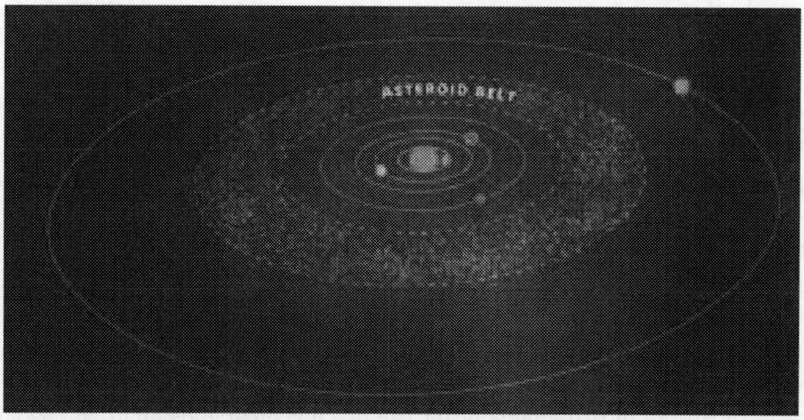

Figure 13: Asteroids. Courtesy to Google website.

These final numbers when taken in AU, they correspond to the distances of planets from the Sun: 0.4 AU corresponds to Mercury, 0.7 AU to Venus, 1.0 AU to Earth, 1.6 AU to Mars, 5.2 AU to Jupiter, 10.0 AU to Saturn and so on. It obviously shows that no known planet has been assigned to 2.8 AU. Thus, the Titius-Bode law predicted a missing planet at a distance of 2.8 AU from the Sun.

Table 4: Parameters of planets

Planet	Distance (AU)	Bode No.
Mercury	0.39	0.4
Venus	0.72	0.7
Earth	1.00	1.0
Mars	1.52	1.6
Missing planet		2.8
Jupiter	5.20	5.2
Saturn	9.54	10.0

14.2 Search for missing planet

Investigations for the search of missing planet were made during the last decades of 18th century. On 1 January 1801, an object having magnitude 8 (very dim) was noticed moving across the sky. The same was found on the successive nights. The body was recognized as a new planet and it was named the Ceres. Very soon in 1802, another similar object was discovered and named the Pallas. Then Juno was discovered in 1804 and Vesta in 1807. Information about these four bodies is summarized in Table 5.

Table 5: Discovery of asteroids

Year	Asteroid	Diameter (km)
1801	Ceres	960
1802	Pallas	570
1804	Juno	233
1807	Vesta	530

By that time, Uranus was discovered. After the discovery of Vesta, from 1807 to 1845, the planetary system has been as the following.

Mercury, Venus, Earth, Mars, Vesta, Juno, Ceres, Pallas, Jupiter, Saturn, Uranus

All these asteroids are regarded as the minor planets. They were also asteroids. Because of their faintness and smallness, they were found to twinkle like the stars.

The orbital characteristics of these newly discovered minor planets seemed to fill the gap. In the Titius-Bode relation for the mean distance from the Sun about 2.8 AU, placing them between the orbits of Mars and Jupiter. As on 1 September 2017, there are more than 200 asteroids having diameter more than 100 km; between 1.1 and 1.9 million satellites having diameter larger than 1 km.

The asteroids were mostly named after the mythological gods and demons which were soon exhausted, and the list continued with the names of wives, friends, cities, colleges of discoverers. However, officially the minor planets, also called the asteroids, are designated by serial numbers in addition to their common names, if any. For example, the first masculine name for an asteroid was given to 433 Eros. Examples for these, named after countries are 136 Austria, 1125 China, 1279 Uganda and so on.

The asteroid number 4130 has been discovered by astronomers working in Indian Institute of Astrophysics, Bangalore, India. It is named as 'Ramanujan', in the honour of great mathematician from India.

The total mass of all the known asteroids does not exceed one-tenth of the mass of our Moon. The orbits of the asteroids were found to form two distinct groups: one group having length of semi-major axis between 2.3 and 2.8 AU and the other group with the length of semi-major axis between 3 and 3.2 AU. There are practically no asteroids having the length of semi-major axis 2.52 AU, 2.82 AU or 2.90 AU. These gaps are collectively known as the Kirkwood gaps, first predicted by Daniel Kirkwood in 1857 and confirmed in 1866.

Because of the large orbital eccentricities, some of the asteroids are found to come regularly close to the Earth. The shape of an asteroid can be spherical to irregular depending on its size. In fact the biggest few asteroids namely, the Ceres, Pallas, Vesta and Hygiea are quite spherical in shape. Most of the other asteroids remain, however, extremely irregular. By observing the light curve of the asteroids, the rotation period as well as the shape can be figured out.

Most asteroids have a rotation period of only a few hours. For example, the period of Eros is $\sim 5^h\ 17^m$, Juno has a period of $7^h\ 13^m$, and so on. Asteroids can have their satellites too. The first such satellite was discovered in 1973. The asteroid 532 Herculina is found to have a tiny satellite. This asteroid has a diameter of 217 km, whereas the satellite has a size of 50 km only.

Note: It is now clarified that every body revolving around the Sun is not a planet. Up to the year 1800, a planet was defined as an object having a fairly circular orbit around the Sun and that did not show any cometary activity such as a coma or a tail. After discovery of the Ceres, Pallas, Vesta, etc., a point was raised to put a limit on the size of the body in order to distinguish between a planet and an asteroid. However, no clear demarcation was decided.

15. Small planets away from Neptune

Up to 24 August 2006, Pluto has been in the list of planets in our solar system. Since the size of Pluto is about 18% of the Earth, calling it as a planet had been objected by some astronomers. On the other side, search for additional planets in the Kuiper belt (away from Neptune) is progressing. There are about 800 known Kuiper belt objects; most of them are only 100 km in diameter. The Kuiper belt objects of diameter larger than 1200 km are given in Table 6. Here, we have included Pluto also in order to compare it with other objects.

An object, tentatively named as 2003 UB_{313}, bigger than Pluto was found in the Kuiper belt. Like Pluto, it has a satellite. With the discovery of 2003 UB_{313}, the point under consideration was that if the Pluto was continued to be recognized as a planet, then 2003 UB_{313} was to be recognized as the tenth planet. In case the 2003 UB_{313} to be recognized as a small planet, then Pluto also was to be recognized as a small planet. On 24 August 2006, the International Astronomical Union (IAU) decided to exclude Pluto from the list of planets. Likewise Quaor, 2004 DW, Sedna, the Pluto became a body revolving around the Sun beyond the Neptune.

The Pluto is found to have at least 5 satellites. The prime satellite of Pluto is the Charon.

Table 6: Parameters of bodies.

Body	Distance (AU)	Diameter (km)	Year of discovery
Pluto	39.53	2302	1930
Quaor	42	1250	2002
2004 DW	55	1600	2004
Sedna	90	1800	2004
2003 UB_{313}	97	3100	2005

16. Satellites

A body revolving around a planet is known as a satellite. Some manmade bodies have been put for revolving around the Earth, for example. Thus, the satellites are of two types: (i) natural satellites, and (ii) artificial satellites. Here, we are concerned with the natural satellites.

As we discussed, the Moon is the only natural satellite of Earth. Other planets also have natural satellites. However, the Mercury and Venus do not have any satellites. Here, we do not consider for the smaller bodies revolving around a planet. Even then, some planets, *e.g.*, Jupiter and Saturn, have large number of satellites. However, some of them are prime ones. The prime satellites of the planets are given in Table 7.

Table 7: Information about prime satellites.

Satellite	Mean radius (km)	Semi-major axis (km)	Discovery year	Planet
Moon	1737.10	384,399	perhistoric	Earth
Phobos	11.1 ± 0.15	9,380	1877	Mars
Deimos	6.2 ± 0.18	23,460	1877	Mars
Io	1818.1 ± 0.1	421,800	1610	Jupiter
Europa	1560.7 ± 0.7	671,100	1610	Jupiter
Ganymede	2634.1 ± 0.3	1,070,400	1610	Jupiter
Callisto	2408.4 ± 0.3	1,882,700	1610	Jupiter
Mimas	198.2 ± 0.4	185,540	1789	Saturn
Enceladus	252.3 ± 0.6	238,040	1789	Saturn
Tethys	536.3 ± 1.5	294,670	1684	Saturn
Dione	562.5 ± 1.5	377,420	1684	Saturn
Rhea	764.5 ± 2.0	527,070	1672	Saturn
Titan	2575.5 ± 2.0	1,221,870	1655	Saturn
Iapetus	734.5 ± 4.0	3,560,840	1671	Saturn
Ariel	578.9 ± 0.6	190,900	1851	Uranus
Umbriel	584.7 ± 2.8	266,000	1851	Uranus
Titania	788.9 ± 1.8	436,300	1787	Uranus
Oberon	761.4 ± 2.6	583,500	1787	Uranus
Miranda	235.8 ± 0.7	129,900	1948	Uranus
Triton	1353.4 ± 0.9	354,800	1846	Neptune
Charon	603.6 ± 1.4	17,536	1978	Pluto*

*Pluto is no more a planet.

The planet Mars has two satellites, Phobos and Deimos. Jupiter has the largest number of satellites. At least 67 satellites of Jupiter are known. The four prime satellites of Jupiter are the Io, Europa, Ganymede and Callisto. Interesting to note that these four satellites were discovered in 1610. The planet Saturn is the next planet to have large number of satellites. At least 62 satellites of Saturn are known. The seven prime satellites of Saturn are the Mimas, Enceladus, Tethys, Dione, Rhea, Titan, Iapetus.

The planet Uranus has at least 27 satellites. The five prime satellites of Uranus are the Ariel, Umbriel, Titania, Oberon, Miranda. At least 14 satellites of Neptune are known. However, only one satellite, Triton is prime. The Pluto is though not a satellite after August 2006, but it has at least 5 satellites and the prime satellite is the Charon.

17. Comets

Unlike a planet which moves across the sky with great regularity, there are some strangers which visit the inner part of solar system quite unpredictably. As in general, they have hairy-like tail and therefore have been named comets (Figure 13) (the word 'comet' has been derived from the coma-like appearance).

Figure 14: Halley comet

It appears that all the comets are members of the solar system, having size up to a few kilometres. However, when they come very close to the Sun, a tail is developed which might extend up to millions of kilometer. Some comets, for example, the Halley's comet, have been known to the mankind for more than 2000 years. Occasionally, once or twice in a decade some comets become so bright as can be seen with the naked eyes.

When a comet is away from the Sun, its size is of a few kilometers. But, when it approaches the Sun, it becomes brighter and develops a tenuous halo around it, known as the coma of the comet. The actual cometary body, now called the nucleus of the comet, remains deeply buried inside the head, consists of the nucleus and coma.

The tails of the comet are generally of two types - one is straight, long and somewhat longitudinal. This is usually made up of ionized atoms and molecules, and is known as the ion tail or the tail of type I. The other type of tail, the so called tail of type II, are more diffused, more thick, shorter and curved. They were probably composed of dust particles and hence known also as the dust tail. Occasionally, a Sun-ward small spike-like tail is also seen in some comets. It is known as the anomalous anti-tail.

Almost all comets are endowed with tails as and when they approach the Sun inward of Mars' orbit. The comets are distinguished from the asteroids by the characteristic developments of coma and tail. Obviously, the comets must be composed of some volatile substances, at least partially, not rock entirely.

In fact, the comets are mostly composed of water-ice, ammonia and methane-ice before they come to the vicinity of the Sun. Dust grains of various sizes are embedded in them, making them the dirty-ice-balls. At a large distance from the Sun this cometary body serves as the bare nucleus of the comet. It shines by the reflected sunlight. Therefore, its brightness is a function of its distance from the Sun and the Earth.

The spectrum of a comet is quite complex. There is a continuum superposed by emission lines and molecular bands. The continuum results from the direct reflection of sunlight from the grains. From the emission lines and bands, it is found that coma is composed of CN, C_2, C_3, CH, NH, NH_2, OH and others, which are mostly the dissociation products of H_2O, CO, CH_4, CO_2, H_2CO and NH_3. The spectrum of a comet obviously varies with its distance from the Sun. When a comet occasionally comes within 0.1 AU of the Sun, the lines of Fe, Cr and Ni are also seen. The size of coma with the diluted gas and ions evaporated from the nucleus by the action of the sunlight assumes a dimension as big as 10^5 km.

The tail of a comet usually begins to form when the comet comes within 1.5 to 2 AU of the Sun and it stems out from the coma in a direction opposite to the Sun. The speed of the particles in the solar wind is about 430 km/s. With their high flux of momentum, they are able to streamline the flow of the ions and to duct them suitably along the magnetic field lines of Sun that they must follow. The interaction of the magnetic field with the ambient plasma of the cometary ion tail is extremely complex.

The type II tails are mostly made up of dust particles of micron size. Their motions are affected most conspicuously by the pressure of the solar radiation. The dust grains absorb the sunlight preferentially from the direction of Sun. In this process, the momentum of the absorbed photons produces a recoil on the dust grain in a direction opposite to that of the Sun. However, when the dust grains re-radiate in the infrared, they do so isotropically in all directions, producing no net recoil on the dust grain. This is how the radiation pressure of the Sun is able to drive the dust grains and channelize them in a direction opposite to that of the Sun.

Since the magnitudes of these accelerations are not very large as compared to those produced by the solar wind on the gas molecules and ions, the dust tails are produced at a slower rate. Since the comet follows a curved path, the dust tail appears to be curved, as the distant-most part of the dust tail remembers the direction of Sun with respect to the former location of comet and the nearer portion of the tail follows the latest direction of Sun at the pressure position of the comet.

A comet must lose mass because of evaporation, though at a variable rate depending on the distance from the Sun. The rate of evaporation from a bright comet becomes about 10^{29} - 10^{30} molecules/s, when it is inside the Earth's orbit. This is how both the coma and tail are sustained for a long time. During a fly-by, a comet can lose $\sim$1/1000-th of its mass. The process of ejection of gas and dust particles from the comet's surface takes place preferentially from that part of its surface facing the Sun. Thus, the ejected gas also produces a recoil on the comet which tries to slow it down during incoming and accelerates while outgoing from the Sun. A comet is also found to rotate about its own axis.

Every year, a dozen or two of comets are observed, of which 4 or 5 may be new. All the observed comets are usually named after the discoverers as well as after the specific year and their alphabetical order discovery/recovery in a given calendar year. Only fraction of comets are found to be periodic and they are further designated by the letter P. The periodic comets are classified as short and long periodic ones, the dividing line standing at $P = 200$ yrs. So, the Halley having an average period of 76 years is regarded as a short periodic comet. Some extremely short periodic ones have periodicity less than 10 years.

18. Meteors, meteorites and meteoroids

The meteors (Figure 14) are often known as the shooting stars. Almost every 5 - 10 min one finds a shooting star popping up from somewhere in the night sky. They appear as a bright streak of light often covering 10° - 100° across the sky. It is this burning state of matter that they are referred to as the meteors. The objects which become meteors due to friction with the atmosphere are usually small particles coming from somewhere in the solar system. These particles are generally classified as meteoroids.

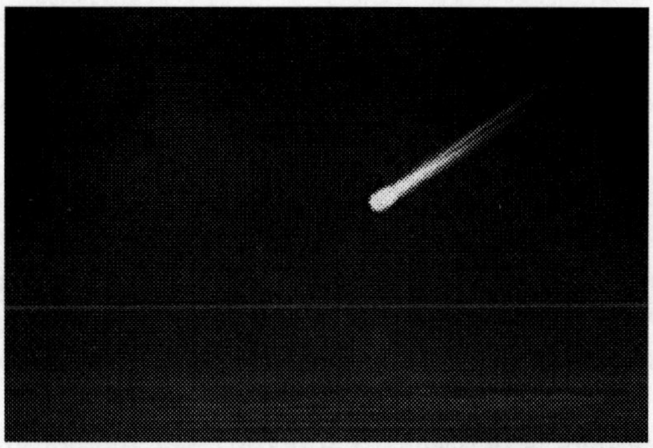

Figure 15: Meteor. Courtesy to Google website.

Occasionally, a meteoroid can be so big that even after becoming meteors they are not completely burnt out and a sizable remnant hits the surface of Earth. The extraterrestrial matter recovered this way is referred to as a meteorite (Figure 15).

When the tracks of a given meteor can be simultaneously observed from two different stations on the Earth with a proper record of the tracks in terms of the celestial coordinates, the actual track through the atmosphere can easily be produced. When the time record of a flight is kept, the speed of the meteor can be calculated. It is found that the speed before a comet encounters the atmosphere is usually quite high ranging from 11 km/s to 72 km/s.

Fortunately, meteors can also be recorded during the day time. This is possible because the meteors leave a hot and ionized column of atmosphere behind their velocity track. In this column, the atoms of air are collisionally excited and even ionized. Before they recombine, they form a column of high density electron plasma in the atmosphere along the track.

Figure 16: Meteorite. Courtesy to Google website.

An average meteor that is seen with the naked eye in a clear night has its track originating usually at a height of 100 km above the ground, the original size of the particle being a few mm only. Brighter meteors are caused by the cm-sized particles. The most bright ones called the 'fire balls', can be anything between a pebble and a cobble. The density of these objects also varies from 3 $\times 10^2$ kg/m^3 to 3 $\times 10^3$ kg/m^3. The naked eye meteors have masses in the range form 10^{-4} kg to 10^{-2} kg with the density of about 3 $\times 10^2$ kg/m^3.

The meteors which are recorded photographically but not seen with the naked eye, have masses of about 10^{-5} kg and have similar densities. The meteors detected by the radars are usually much smaller and less massive, having mass in the range from 10^{-9} kg to 10^{-7} kg and they have the mean density of about 8×10^2 kg/m^3.

There are occasionally bursts of meteors, called the meteor showers, apparently found to originate in a constellation and seen for over few hours to few days. The hourly rate of such meteors is between 15 and 100. The meteor showers are usually named after the constellation in which they seem to originate. For example, the meteor shower named Perseids originates in the constellation Perseus.

The fire-balls which finally hit the surface of the Earth are called the meteorites.

19. Zodiacal light

The Zodiacal light (Figure 16) is a faint, roughly triangular, whitish glow seen in the night sky which appears to extend up from the vicinity of the Sun along the ecliptic or zodiac. This phenomenon was first investigated by the Cassini in 1683 and first explained by the Fatio de Duillier in 1684. When the plane of the ecliptic passes through the zenith of a local observer, the ecliptic intercepts with the horizon at a right angle. Under such configuration, small particles that are concentrated mainly along the ecliptic scatter the sunlight in the forward direction.

Usually, about 3/2 hour after the local sunset during January and February or about 3/2 hour before the sunrise in September, the ecliptic passes through the zenith of an observer located in the latitude belt of 15° N - 25° N. A conical glow of sunlight is visible for about 1/2 hour along the plane of ecliptic projected away from the Sun, and extending up to the zenith. Such glows are known as the Zodiacal light. Thousands of years ago, Egyptians had made a record of these lights. It is called the zodiacal light because the glow extends along the ecliptic which runs through the zodiacal constellations.

Figure 17: Zodiacal. Courtesy to Google website.

20. Oort cloud

The Oort cloud (Figure 17) is a hypothesized spherical cloud of comets which may lie roughly 50,000 AU, or nearly a light year away from the Sun. In 1950, Jan Oort made an interesting investigation of the orbits of very long period comets. Some of them having the semi-major axis longer than 2 $\times 10^4$ AU. It meant, according to Oort, that all the comets are not only bound to the solar system but many of them are strongly populated at a distance of about 2 $\times 10^4$ AU from the Sun. He therefore hypothesized the existence of a huge population of comets, later on called the Oort cloud, encircling the solar system at a distance of about 2 $\times 10^4$ AU - 10 $\times 10^4$ AU.

The Oort cloud is thought to have two separate regions: a spherical outer Oort cloud and a disc-shaped inner Oort cloud. Objects in the Oort cloud are largely made up of ices, such as water, ammonia, and methane. The outer Oort cloud is loosely bound to the solar system, and this is easily affected by the gravitational pull both of passing stars and of the Milky Way galaxy itself. It may divert a large number of comets from their orbits within the cloud and send them towards the Sun.

A question arises how could the Oort cloud stay in its original position for over 4.5 billion years. In order to remove this problem, Gerard Kuiper suggested the existence of a belt of comets at a distance far closer than the Oort cloud. The Kuiper belt is hypothesized in the form of a flattened belt of comets beyond the orbit of Neptune but within 150 AU of the Sun, with a pool of comets numbering 10^8 - 10^{10}.

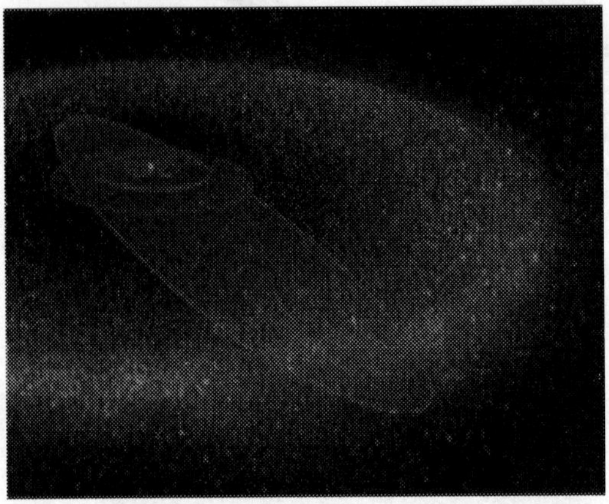

Figure 18: Oort cloud. Courtesy to Google website.

The orbital characteristics of most of the short period comets are consistent with their presumed origin rather in the Kuiper belt than in the Oort cloud. Also hypothesized is the existence of an inner Oort cloud at a distance of 2 $\times 10^4$ AU, having a bigger pool of comets, numbering about 6 $\times 10^{12}$. Kuiper belt and the farther scattered disc have been observed and mapped by the scientists.

21. Origin of solar system

The formation and evolution of the solar system is estimated to have begun 4.55 to 4.56 billion years ago with the gravitational collapse of a small part of a giant molecular cloud. There are a large number of different theories about the origin of the solar system, none of which seems to explain all the known peculiarities of the system. Nevertheless, it is one of the most important issues that relates to the origin of the Earth, Moon, planets and Sun, and therefore of the stars, in general, in the galaxy.

The solar system is believed to have evolved considerably since the formation of planets and their respective satellites. Some satellites might have been formed independently and later on been captured by their planets. Some satellites, as such the Earth's satellite Moon, might have been the result of giant collisions between the bodies.

21.1 Early theories of solar system formation

Emmanuel Kant, towards the middle of the 18th century, proposed his famous theory about the origin of the solar system. He thought that the Sun was a product of self gravitational condensation of a cosmic nebula, made up of gas and dust particles. The nebula was initially rotating and because of its rotation, it was flattened. The planets were the nuclei of matter in this flattened nebula, where the concentrations took place. Thus, he explained why the planets lie in the equatorial plane of the Sun.

Simon Laplace, towards the end of the 18th century, expanded the idea of Kant. He imagined that the spinning gas cloud, out of which the solar system was born, was gradually shrinking due to cooling. Because of the contraction, the size and hence the moments of inertia of the cloud about its rotation axis kept on decreasing, and as a result of that the axial spinning was accelerated. As the angular speed of rotation increased, the gas cloud became more flattened and assumed a lenticular shape with the sharp edge. At some stage, the centrifugal force at the rim balanced the gravitational force of attraction due to the inner parts of the cloud.

Perhaps this is how the rim got separated into a number of equatorial rings. The series of rings thus released, finally condensed into the planets. Perhaps, the planets themselves also were initially in the form of rotating clouds or cloud-lets which, on further contraction, led similarly to the formation of satellites around them. In this model, the Keplerian distribution of angular velocity among the planets and satellites were duly explained. Some of the modern theories of the solar system formation still support the basic ideas of Kant and Laplace.

The major criticism about the Kant-Laplace model is its inability to explain the distribution of the sizes of planets. It is found that the largest planet Jupiter occupies a central position in the system of planets.

Comte de Buffon was the first to suggest in 1745 (that is even before Kant) that the planets were formed through the impact of a comet which collided with the Sun on its way through the space. The idea was revived in first two decades of twentieth century where the collision of comet was replaced by that of another star with the Sun. The collision of Sun with a foreign star led to the ejection of matter from both the stars. These ejected materials perhaps formed a gaseous disc around one of these stars, which finally fragmented into planets.

Very soon it became clear that collision of two stars in the sky has practically zero probability. Stars are literally distributed in otherwise free space like any two marbles at an average separation of about 10 km. Any two such objects moving however irregularly, would find very little chance of meeting in a head-on collision; not even once in 10 billion years. Therefore, it was suggested that instead of a head-on collision, two wandering stars might have interacted gravitationally at perhaps a close enough range, a range which may be as small as a few hundred to thousand times their radii.

For such a gravitational interaction of any two objects of finite size at a close range of separation, both the stars are attracted slightly more strongly as compared to their rear parts, making either or both of the stars assume the shape of a prolate spheroid, the elongated sides being often referred to as the tidal bulges. For a close enough interaction, the tidal bulges can be appreciably large and in the extreme cases can even separate from the respective parent bodies. Such a tidally ejected trail of hot gas was believed to have finally cooled down to form small asteroid-like objects, called the planetesimals, which finally conglomerated to form the body of the planets and the satellites.

Sir James Jeans (1916) believed in the tidal theory for the formation of the solar system, but deviated from the idea of forming planetesimals from a tidally ejected gas. He thought and in fact showed that the ejection would take place in the form of a curved cigar because of the relative angular momentum of two stars about each other. This cigar-like ejection finally splits up into big knots, and the rest of the material fell back on the Sun. Since the originally ejected material had already developed a sense of rotation, this fell-back material provided the Sun a rotation in the same sense as of the planetary revolutions. This theory of Jeans became quite popular in the twenties of twentieth century and remained so till the end of the second world war.

Meanwhile, Raymond Lyttleton in 1940 and Fred Hoyle in 1944 suggested another theory for the solar system formation. Lyttleton thought that the Sun was originally a part of a double star system. In the course of time, the pair interacted with a third star, which due to its tidal interaction with the pair took away one star of the pair and left the other, namely our Sun, behind with an envelope of hot gas cloud. This cloud finally condensed into the planets. Hoyle added a grain of salt to this whole hypothesis by saying that the original double star system had one component which was slightly more massive than the other, and exploded as a supernova leaving our Sun behind. So, he did not have to invoke the interaction of the third star for the want of ejecta, which he could obtain from the supernova outburst of the companion. The gaseous ejecta were captured by the solar system.

Later on, the theories of Carl von Weizäcker and Gerard Kuiper revived the idea of Kant and Laplace, suggesting that the initial disc of the gas developed semi-permanent eddies and the local condensations were caused by the turbulence.

Let us discuss the specific features of the solar system that must be addressed to be any reasonable theory of the solar system formation.

21.2 Specific features of the solar system related to its origin

Here, we mention some outstanding features of the solar system that must be explained by a reasonably good theory of its origin.

(i) The total mass of planets, asteroids and comets constitutes less than 0.2% of the mass of Sun or hence of the solar system.

(ii) The total angular momentum of all planets due to their revolution around the Sun constitutes about 98% of the total angular momentum of the solar system. Thus, the material containing about 0.2% of the total mass of the solar system shares about 98% of the total angular momentum of solar system.

(iii) All planets revolve around the Sun in the same sense as the rotation of the Sun.

(iv) Planets are mostly confined within about 25° of the equatorial plane of the Sun.

(v) Chemical composition of the giant planets, particularly the Jupiter and Saturn, is almost identical to that of the Sun itself. The chemical composition of the Sun, other nearby stars and the interstellar medium out of which the stars are formed are also almost identical. This way, the local interstellar medium, the

Sun and the gaseous giant planets all share the same average cosmic abundance of various elements.

(vi) The inner and the outer planets have widely different chemical compositions. The inner planets are absolutely devoid of volatiles and of cosmically most abundant elements, namely, hydrogen and helium.

(vii) The chemical composition of the asteroids and the meteorites are quite similar. The meteorites are the oldest known objects in the solar system with an estimated age of 4.55 to 4.60 billion years.

(viii) The Moon and the Mercury bear an irrefutable evidence of heavy meteooitic bombardment, that took place about 3.8 billion years ago. The age of the crust of the inner planets and the Moon seems to be about 3.8 - 4.3 billion years.

(ix) The crystalline structure of Ni and Fe chondrules inside chondritic meteorites suggests that they must have been formed at high temperatures with a cooling rate not exceeding one or two degree Celsius per million years. Such a low cooling rate is possible only when these meteorites were crystallized inside an object of the size not less than 200 km. This is often regarded as the evidence for the existence of planetesimals, before the formation of inner planets.

Based on these well established genesis related features of the solar system, the following facts about the origin of the solar system have substantially been agreed upon by about 1972.

1. The Sun and the nebula from which the solar system was born, called the solar nebula, were identical in the composition. This conclusion is based on the similarity in the composition of the solar photosphere and carbonaceous chondrites for the relative abundances of all the heavy elements. It is noted that the mantle of the Sun has lost substantial amount of Li^7 and Be^9 due to the convective mixing up to a certain depth at which the temperature becomes just too high to induce the nuclear reactions responsible for the destruction of Li and Be. The chondrites on the other hand, have lost all the volatiles namely, hydrogen and helium.

2. The solar nebula was essentially homogeneous in composition and hence had been well mixed. This is based on the similarity of all the non-radiogenic isotope ratios in the terrestrial, the lunar and the meteoritic rocks, subject to their availability.

3. The nucleosynthesis of the solar material seems to have taken place about 10^8 years prior to the onset of the condensation in the solar nebula. That is, any local supernova explosion that triggered the early condensation and brought about the change in the ratios of the radio isotopes had occurred within about 10^8 years before the condensation of the solar nebula began. This conclusion is based on the products of I^{129} and PU^{244} decay among various radiogenic isotopes of Xe.

4. The Earth, Moon and stony meteoritic parent bodies became separate entities at least 4.5 billion years ago. This is based on the radio-chronological studies of Rb/Sr and Pb/Pb isotopes systematics.

22. Review Questions

1. For the solar planets, describe the following terms:

   (i) Mass, (ii) Density, (iii) Mean distance from the Sun, (iv) Orbital period, (v) Orbital eccentricity, (vi) Rotation period, (vii) Polar obliquity, (viii) Polar oblateness (ix) Equatorial gravity, (x) Surface temperature, (xi) Albedo, (xii) Obliquity of the orbit, and (xiii) Obliquity of the rotation axis

2. Show that a planet closer to the Sun revolves with higher speed.
3. Show that a planet closer to the Sun has smaller period of revolution.
4. Two planets A and B, at the distances r_1 and r_2 from the Sun are revolving with velocities v_1 and v_2, respectively. If $r_1 < r_2$, show that $v_1 > v_2$.
5. Describe the solar eclipse.
6. Describe the lunar eclipse.
7. Describe the tides found on the Earth.
8. What is Titius-Bode relation? What has been the outcome of that.
9. Show that the speed of revolution of Earth is faster than that of Jupiter.
10. What is orbital period for a planet revolving around the Sun? Derive a relation between the speed of revolution and the distance of planet from the Sun.
11. Write down specific features of solar system which should be addressed by a theory for its origin.
12. Describe the Laplace theory for formation of rings around a planet.
13. Explain why Pluto is no more a planet.
14. Write notes on the following:

    (i) Escape velocity
    (ii) Mercury
    (iii) Venus
    (iv) Earth
    (v) Moon
    (vi) Solar eclipse
    (vii) Lunar eclipse

(viii) Tides
(ix) Mars
(x) Jupiter
(xi) Saturn
(xii) Uranus
(xiii) Neptune
(xiv) Asteroids
(xv) Titius-Bode law
(xvi) Pluto
(xvii) Satellites
(xviii) Comets
(xix) Meteors, meteorites and meteoroids
(xx) Zodiacal light
(xxi) Oort cloud

23. Multiple Choice Questions

1. Conventionally, the retrograde motion is from

   A. west to east B. east to west C. north to south D. south to north

   Ans. A

2. Solar planet having the largest mass is

   A. Earth B. Mars C. Jupiter D. Saturn

   Ans. C

3. Solar planet closest to the Sun is

   A. Mercury B. Venus C. Earth D. Mars

   Ans. A

4. Solar planet having smallest period of revolution is

   A. Mercury B. Venus C. Earth D. Neptune

   Ans. A

5. Solar planet having largest speed of revolution is

   A. Mercury B. Venus C. Earth D. Neptune

   Ans. A

6. Solar planet having lowest orbital eccentricity is

   A. Mercury B. Venus C. Earth D. Neptune

   Ans. B

7. Solar planet having retrograde rotation about its own axis is

   A. Mercury B. Venus C. Earth D. Jupiter

   Ans. B

8. Which of the following is not a planet?

   A. Mercury B. Venus C. Earth D. Moon

   Ans. D

9. Mean distance of asteroids from the Sun is

   A. 1 AU B. 0.4 AU C. 2.8 AU D. 5.2 AU

   Ans. C

10. From the Earth's surface, the same face of the Moon is seen always because

    A. The Earth revolves around the Sun in 1 year.

    B. The Earth spins about its own axis in 1 day.

    C. The Moon revolves around the Earth.

    D. The rotation and revolution periods of Moon are equal.

    Ans. D

11. The number of satellites of Jupiter discovered by Galileo in 1610 is

    A. 2 B. 3 C. 4 D. 8

    Ans. C

12. The Roche limit for a ring system around a planet is how many times the radius of the planet.

    A. 1.00 B. 1.44 C. 2.4 D. 4.2

    Ans. C

13. Titan is the satellite of

    A. Jupiter B. Saturn C. Uranus D. Neptune

    Ans. B

14. Which of the following satellites of Jupiter has the largest size.

    A. Io B. Europa C. Ganymede D. Callisto

    Ans. C

15. Imagine that four planets, Earth, Mars, Jupiter and Saturn are put in a big pond having water. Which one would float in the water.

    A. Earth B. Mars C. Jupiter D. Saturn

    Ans. D

16. Which of the following radiates more energy than it receives?

    A. Earth B. Mercury C. Mars D. Jupiter

    Ans. D

17. Which of the following radiates more energy than it receives?

    A. Earth B. Mars C. Saturn D. Uranus

    Ans. C

18. Which of the following of Saturn has the largest size?

    A. Iapetus B. Enceladus C. Dione D. Titan

    Ans. D

19. Which of the following has the largest size?

    A. Ariel B. Umbriel C. Titania D. Miranda

    Ans. C

20. Which of the following appears blue?

    A. Mars B. Jupiter C. Uranus D. Saturn

    Ans. C

21. Which of the following appears red?

    A. Mars B. Jupiter C. Uranus D. Saturn

    Ans. A

22. Which of the following appears blue?

    A. Mars B. Jupiter C. Neptune D. Saturn

    Ans. C

23. Who explained that the planets lie in the equatorial plane of the Sun.

    A. Newton B. Galileo C. Kant D. Einstein

    Ans. C

24. In the past, which of the following was a planet?

    A. Ceres B. Pluto C. Pallas D. Vesta

    Ans. B

25. A planet revolves around the sun in a circular orbit. Its orbital velocity does not depend on

    A. mass of sun B. distance from sun

    C. mass of planet D. gravitational constant

    Ans. C

26. A planet revolves around the sun in a circular orbit. Its time-period of revolution does not depend on

    A. mass of sun B. distance from sun
    C. mass of planet D. gravitational constant

    Ans. C

27. The orbits of three satellites, Io, Europa and Ganymede of Jupiter are locked in the perfect tidal resonances with a ratio of the orbital periods, respectively,

    A. 1 : 2 : 3 B. 1 : 3 : 5 C. 1 : 2 : 4 D. 1 : 4 : 9

    Ans. C

28. In water which of the planets would float?

    A. Earth B. Mercury C. Saturn D. Neptune

    Ans. C

29. Saturn's ring system disappears, when viewed from the Earth, once every

    A. 11 years B. 15 years C. 22 years D. 42 years

    Ans. B

30. Each of the north and south poles of Uranus face the sun for how many years.

    A. 11 years B. 15 years C. 22 years D. 42 years

    Ans. D

31. Since its discovery, which of the following has not yet completed one revolution around the Sun.

    A. Saturn B. Neptune C. Jupiter D. Uranus

    Ans. B

9. Nucleosynthesis

The baryonic matter, where the elements are made out of protons, neutrons and electrons, is a small part of the content of the universe. The cosmological models support the idea that the total visible mass in the form of stars, planets, satellites, living beings, etc. (visible baryonic matter) plus the mass contributed by the energy, is only 4.6%. The rest of the mass is considered to be in the form of dark matter (23%) and dark energy. However, there are no direct observational evidences for these forms of matter and energy, and is a subject of research.

The matter found in stars and interstellar clouds is in the form of atoms or ions (plasma). In some regions, such as the high densities inside the white dwarfs and neutron stars, the matter may be found in the degenerate forms.

In the entire known universe, the hydrogen (atomic number 1) is found to be the most abundant element; the helium (atomic number 2) is found second. However, later on, the rank of abundance does not follow the correspondence with the atomic number. For example, the oxygen (atomic number 8) has abundance rank 3, and carbon (atomic number 6) has abundance rank 4. The nuclear reactions are responsible for the abundance of various elements. Answer to a question, how carbon, iron and other heavy elements are found inside the Earth, may be that the Earth is made out of the recycled material thrown out during death of stars. In the present chapter, we have discussed various elements and the nuclear reactions prevailing in side the stars.

1. Stellar nucleosynthesis

Stellar nucleosynthesis is a process through which the natural abundance of the elements changes within a star due to nuclear fusion reactions in the cores and overlying mantles of the star. The stars are said to evolve (age) with the changes in the abundance of the elements within. The fusion process in the core increases the atomic weight of its gaseous elements, causing pressure loss and contraction accompanied by increase of temperature. Structural changes of the star (evolution) become necessary to stabilize it. Stars lose most of their mass when it is ejected late in their stellar lifetimes, thereby increasing the abundance of elements heavier than helium in the interstellar medium.

The fusion of nuclei in a star, starting from its initial hydrogen and helium abundance, provides that energy and synthesizes new nuclei as a byproduct of that fusion process.

2. Hydrogen fusion

As discussed earlier, the hydrogen burning is a process of nuclear fusion of four protons to form a nucleus of $_2He^4$. There are two predominant processes through which the hydrogen burning in a star takes place.

As discussed earlier, in the low mass stars, such as the Sun, the dominant process of hydrogen burning is the proton-proton chain reaction (pp-chain reaction), whereas in the high mass stars, the dominant process is the CNO cycle. Which type of hydrogen burning process dominates inside a star is decided on the basis of temperature therein. The pp-chain reaction starts at temperatures around 4×10^6 K, making it the dominant mechanism in the smaller stars. The CNO chain requires a higher temperature of approximately 15×10^6 K, but thereafter it increases more rapidly in efficiency than the pp-chain reaction as the temperature grows. Above approximately 17×10^6, the CNO cycle becomes the dominant source of energy.

This temperature is achieved in the cores of main sequence stars with at least 1.3 times the mass of the Sun. The Sun itself has a core temperature of around 15.7×10^6 K and only 0.8% of the energy being produced in the Sun comes from the CNO cycle.

2.1 Deuterium burning

Deuterium $_1D^2$ is a stable isotope of hydrogen with a mass approximately double that of a proton. It consists of one proton and one neutron. Deuterium burning is a nuclear fusion reaction that occurs in stars and some sub-stellar objects, in which a deuterium nucleus and a proton combine to form a $_2He^3$ nucleus. It takes place as the second stage of the proton-proton chain reaction, in which a deuterium nucleus formed from two protons fuses with a further proton. .

Stellar models suggest that if there were no deuterium burning, then there should be no stars with masses more than about two or three times the mass of the Sun in the pre-main-sequence phase because hydrogen burning would occur while the object was still accreting matter.

2.2 Proton-proton chain reaction

The proton-proton chain reaction is one of two nuclear fusion reactions, along with the CNO cycle, by which stars convert hydrogen to helium and which dominates in stars the size of the Sun or smaller.

In general, proton-proton fusion can occur only if the temperature (*i.e.*, kinetic energy) of the protons is high enough to overcome their mutual electrostatic or Coulomb repulsion. The first step involves the fusion of two $_1H^1$ nuclei (protons) into deuterium, releasing a positron and a neutrino as one proton changes into a neutron.

$$_1H^1 + \ _1H^1 \rightarrow \ _1D^2 + e^+ + \nu_e + \gamma$$

This first step is extremely slow. The half-life for a successful p-p fusion in the core of the Sun is estimated to be a billion years, even at the extreme density and temperatures found there. So produced positron is almost immediately annihilated with an electron, and their mass-energy and kinetic energy, is carried off by two gamma ray photons.

$$e^- + e^+ \rightarrow 2\,\gamma$$

The deuterium produced in the first stage can fuse with another proton to produce a isotope of helium, ${}_2\mathrm{He}^3$:

$${}_1\mathrm{D}^2 + {}_1\mathrm{H}^1 \rightarrow {}_2\mathrm{He}^3 + \gamma$$

This process is extremely fast in comparison to the first step. It is estimated that, under the conditions in the Sun's core, a newly-created deuterium nucleus exists for only about 4 seconds before it is converted to He^3. In the Sun, the He^3 produced in these reactions exists for only about 400 years before it is converted into He^4. From here there are four possible paths, pp I, pp II, pp III and pp IV to generate ${}_2\mathrm{He}^4$. The frequencies of the first three paths are, respectively, 86%, 14% and 0.11%. There is also an extremely rare pp IV branch. The pp I branch is dominant at temperatures of 10 to 14 MK and is as the following.

$${}_2\mathrm{He}^3 +_2 \mathrm{He}^3 \rightarrow {}_2\mathrm{He}^4 + 2\,{}_1\mathrm{H}^1$$

The pp II branch is dominant at temperatures of 14 to 23 MK and is as the following.

$${}_2\mathrm{He}^3 +_2 \mathrm{He}^4 \rightarrow {}_4\mathrm{Be}^7 + \gamma \qquad {}_4\mathrm{Be}^7 + e^- \rightarrow {}_3\mathrm{Li}^7 + \nu_e$$

$${}_3\mathrm{Li}^7 + 1\mathrm{H}^1 \rightarrow 2\,{}_2\mathrm{He}^4$$

The pp III branch is dominant if the temperature exceeds 23 MK and is as the following.

$${}_2\mathrm{He}^3 +_2 \mathrm{He}^4 \rightarrow {}_4\mathrm{Be}^7 + \gamma \qquad {}_4\mathrm{Be}^7 +_1 \mathrm{H}^1 \rightarrow {}_5\mathrm{B}^8 + \gamma$$

$${}_5\mathrm{B}^8 +_1 \mathrm{H}^1 \rightarrow {}_4\mathrm{Be}^8 + e^+ + \nu_e \qquad {}_4\mathrm{Be}^8 \rightarrow 2\,{}_2\mathrm{He}^4$$

The pp III chain is not a major source of energy in the Sun (only 0.11%), but was very important in the solar neutrino problem because it generates very high energy neutrinos (up to 14.06 MeV). The pp IV branch, as the following, is predicted but has never been observed due to its rarity.

$${}_2\mathrm{He}^3 +_1 \mathrm{H}^1 \rightarrow {}_2\mathrm{He}^4 + e^+ + \nu_e$$

Comparing the mass of the final ${}^4_2\mathrm{He}$ atom with the masses of the four protons reveals that 0.007 or 0.7% of the mass of the original protons has been lost. This mass has been converted into energy, in the form of gamma rays and neutrinos released during each of the individual reactions.

Energy released as gamma rays interact with electrons and protons and heat the interior of the Sun. Neutrinos do not interact significantly with any matter and therefore do not help to support the Sun against the gravitational collapse. Their energy is lost and the neutrinos in the pp I, pp II and pp III branches carry away 2.0%, 4.0%, and 28.3% of the energy in those reactions, respectively.

2.3 CNO cycle

Besides the proton-proton chain reaction, the CNO cycle (for carbon-nitrogen-oxygen) is another set of fusion reactions by which the stars convert hydrogen to helium. Unlike the pp-chain, the CNO cycle is a catalytic cycle. Theoretical models suggest that the CNO cycle is the dominant source of energy in stars more massive than about 1.3 times the mass of the Sun. The proton-proton chain is more important in the stars having mass equal to that of the Sun or less. This difference stems from temperature dependency differences between the two reactions. The pp-chain reactions start occurring at temperatures around 4×10^6 K, making it the dominant energy source in the smaller stars. A CNO cycle starts occurring at approximately 15×10^6 K, but its energy output rises much more rapidly with increasing temperatures. At approximately 17×10^6 K, the CNO cycle starts becoming the dominant source of energy. The Sun has a core temperature of around 15.7×10^6 K, and only 1.7% of He^4 nuclei are produced in the Sun through the CNO cycle. Various CNO cycles are proposed by the scientists. The CNO-I proposed by Carl von Weizscker and Hans Bethe is as the following.

$${}_6C^{12} + {}_1H^1 \rightarrow {}_7N^{13} + \gamma \qquad {}_7N^{13} \rightarrow {}_6C^{13} + e^+ + \nu_e$$

$${}_6C^{13} + {}_1H^1 \rightarrow {}_7N^{14} + \gamma \qquad {}_7N^{14} + {}_1H^1 \rightarrow {}_8O^{15} + \gamma$$

$${}_8O^{15} \rightarrow {}_7N^{15} + e^+ + \nu_e \qquad {}_7N^{15} + {}_1H^1 \rightarrow {}_6C^{12} + {}_2He^4$$

The positrons almost instantly annihilate with electrons, releasing energy in the form of two gamma radiations. The neutrinos escape from the star carrying away some energy. One nucleus goes to become carbon, nitrogen, and oxygen isotopes through a number of transformations in an endless loop.

In the CNO-II cycle, instead of the production of a ${}_6C^{12}$ and an alpha particle, a ${}_8O^{16}$ and a photon are produced. Like the carbon, nitrogen, and oxygen involved in the main branch (CNO-I), the fluorine produced in the minor branch is merely an intermediate product in the CNO-II, and the sequence of reactions is as the following.

$${}_7N^{15} + {}_1H^1 \rightarrow {}_8O^{16} + \gamma \qquad {}_8O^{16} + {}_1H^1 \rightarrow {}_9F^{17} + \gamma$$

$${}_9F^{17} \rightarrow {}_8O^{17} + e^+ + \nu_e \qquad {}_8O^{17} + {}_1H^1 \rightarrow {}_7N^{14} + {}_2He^4$$

$${}_7N^{14} + {}_1H^1 \rightarrow {}_8O^{15} + \gamma \qquad {}_8O^{15} \rightarrow {}_7N^{15} + e^+ + \nu_e$$

The CNO-III cycle is a subdominant branch only for massive stars where the reactions are started when one of the reactions in CNO-II results in a ${}_9F^{18}$ and a gamma radiation instead of a ${}_7N^{14}$ and an alpha particle, and the sequence of reactions is as the following.

$${}_8O^{17} + {}_1H^1 \rightarrow {}_9F^{18} + \gamma \qquad {}_9F^{18} \rightarrow {}_8O^{18} + e^+ + \nu_e$$

$${}_8\mathrm{O}^{18} + {}_1\mathrm{H}^1 \rightarrow {}_7\mathrm{N}^{15} + {}_2\mathrm{He}^4 \qquad {}_7\mathrm{N}^{15} + {}_1\mathrm{H}^1 \rightarrow {}_8\mathrm{O}^{16} + \gamma$$

$${}_8\mathrm{O}^{16} + {}_1\mathrm{H}^1 \rightarrow {}_9\mathrm{F}^{17} + \gamma \qquad {}_9\mathrm{F}^{17} \rightarrow {}_8\mathrm{O}^{17} + e^+ + \nu_e$$

Like the CNO-III, the CNO-IV is also only significant in massive stars. The reactions are started when one of the reactions in CNO-III results in a ${}_9\mathrm{F}^{19}$ and a gamma radiation instead of ${}_7\mathrm{N}^{15}$ and an alpha particle, and the sequence of reactions is as the following.

$${}_9\mathrm{F}^{19} + {}_1\mathrm{H}^1 \rightarrow {}_8\mathrm{O}^{16} + {}_2\mathrm{He}^4 \qquad {}_8\mathrm{O}^{16} + {}_1\mathrm{H}^1 \rightarrow {}_9\mathrm{F}^{17} + \gamma$$

$${}_9\mathrm{F}^{17} \rightarrow {}_8\mathrm{O}^{17} + e^+ + \nu_e \qquad {}_8\mathrm{O}^{17} + {}_1\mathrm{H}^1 \rightarrow {}_9\mathrm{F}^{18} + \gamma$$

$${}_9\mathrm{F}^{18} \rightarrow {}_8\mathrm{O}^{18} + e^+ + \nu_e \qquad {}_8\mathrm{O}^{18} + {}_1\mathrm{H}^1 \rightarrow {}_9\mathrm{F}^{19} + \gamma$$

2.4 Hot CNO cycle

Under conditions of higher temperature and pressure, such as those found in novae and x-ray bursts, the catalytic cycles are typically referred to as the hot CNO (HCNO) cycles. The difference between the HCNO-I cycle and the CNO-I cycle is that ${}_7\mathrm{N}^{13}$ captures a proton instead of decaying, leading to the sequence of reactions as the following.

$${}_6\mathrm{C}^{12} + {}_1\mathrm{H}^1 \rightarrow {}_7\mathrm{N}^{13} + \gamma \qquad {}_7\mathrm{N}^{13} + {}_1\mathrm{H}^1 \rightarrow {}_8\mathrm{O}^{14} + \gamma$$

$${}_8\mathrm{O}^{14} \rightarrow {}_7\mathrm{N}^{14} + e^+ + \nu_e \qquad {}_7\mathrm{N}^{14} + {}_1\mathrm{H}^1 \rightarrow {}_8\mathrm{O}^{15} + \gamma$$

$${}_8\mathrm{O}^{15} \rightarrow {}_7\mathrm{N}^{15} + e^+ + \nu_e \qquad {}_7\mathrm{N}^{15} + {}_1\mathrm{H}^1 \rightarrow {}_6\mathrm{C}^{12} + {}_2\mathrm{He}^4$$

The notable difference between the CNO-II cycle and the HCNO-II cycle is that ${}_9\mathrm{F}^{17}$ captures a proton instead of decaying, and neon is produced in a subsequent reaction on ${}_9\mathrm{F}^{18}$, leading to the sequence of reactions as the following.

$${}_7\mathrm{N}^{15} + {}_1\mathrm{H}^1 \rightarrow {}_8\mathrm{O}^{16} + \gamma \qquad {}_8\mathrm{O}^{16} + {}_1\mathrm{H}^1 \rightarrow {}_9\mathrm{F}^{17} + \gamma$$

$${}_9\mathrm{F}^{17} + {}_1\mathrm{H}^1 \rightarrow {}_{10}\mathrm{Ne}^{18} + \gamma \qquad {}_{10}\mathrm{Ne}^{18} \rightarrow {}_9\mathrm{F}^{18} + e^+ + \nu_e$$

$${}_9\mathrm{F}^{18} + {}_1\mathrm{H}^1 \rightarrow {}_8\mathrm{O}^{15} + {}_2\mathrm{He}^4 \qquad {}_8\mathrm{O}^{15} \rightarrow {}_7\mathrm{N}^{15} + e^+ + \nu_e$$

The HCNO-III cycle is an alternative to the HCNO-II cycle where a ${}_9\mathrm{F}^{18}$ captures a proton moving towards higher mass and using the same helium production mechanism as the CNO-IV cycle as the following.

$${}_9\mathrm{F}^{18} + {}_1\mathrm{H}^1 \rightarrow {}_{10}\mathrm{Ne}^{19} + \gamma \qquad {}_{10}\mathrm{Ne}^{19} \rightarrow {}_9\mathrm{F}^{19} + e^+ + \nu_e$$

$${}_9\mathrm{F}^{19} + {}_1\mathrm{H}^1 \rightarrow {}_8\mathrm{O}^{16} + {}_2\mathrm{He}^4 \qquad {}_8\mathrm{O}^{16} + {}_1\mathrm{H}^1 \rightarrow {}_9\mathrm{F}^{17} + \gamma$$

$${}_9\mathrm{F}^{17} + {}_1\mathrm{H}^1 \rightarrow {}_{10}\mathrm{Ne}^{18} + \gamma \qquad {}_{10}\mathrm{Ne}^{18} \rightarrow {}_9\mathrm{F}^{18} + e^+ + \nu_e$$

3. Helium burning

Older stars start to accumulate helium produced by the proton-proton chain reaction and the carbon-nitrogen-oxygen cycle in their cores. The products of further nuclear fusion reactions of helium with hydrogen or another helium nucleus produce lithium-5 and beryllium-8 respectively, both of which are highly unstable and decay almost instantly back into smaller nuclei. When the star starts to run out of hydrogen to fuse, the core of the star begins to collapse until the central temperature rises to 10^8 K. At this point the helium nuclei are fusing together faster than their product, beryllium-8, decays back into two helium nuclei.

3.1 Triple-alpha process

The triple-alpha process is a set of nuclear fusion reactions by which three $^4_2$He nuclei (alpha particles) are transformed into carbon.

Once beryllium-8 is produced a little faster than it decays, the number of beryllium-8 nuclei in the stellar core increases to a large number. Then in its core there are many beryllium-8 nuclei that can fuse with another helium nucleus to form carbon-12, which is stable:

$$_2\mathrm{He}^4 + \ _2\mathrm{He}^4 \rightarrow \ _4\mathrm{Be}^8 \qquad\qquad _4\mathrm{Be}^8 + \ _2\mathrm{He}^4 \rightarrow \ _6\mathrm{C}^{12}$$

Because the triple-alpha process is unlikely, it needs a long time to produce much carbon. One consequence of this is that no significant amount of carbon was produced in the Big Bang, because within minutes after the Big Bang, the temperature fell below the critical point for nuclear fusion.

Ordinarily, the probability of the triple alpha process is extremely small. However, the beryllium-8 ground state has almost exactly the energy of two alpha particles. In the second step, $_4\mathrm{Be}^8 + \ _2\mathrm{He}^4$ has almost exactly the energy of an excited state of $_6\mathrm{C}^{12}$. These resonances greatly increase the probability that an incoming alpha particle combine with beryllium-8 to form carbon.

As a side effect of the process, some carbon nuclei can fuse with additional helium to produce a stable isotope of oxygen.

$$_6\mathrm{C}^{12} + \ _2\mathrm{He}^4 \rightarrow \ _8\mathrm{O}^{16}$$

3.2 Alpha process

The alpha process, also known as the alpha ladder, is one of two classes of nuclear fusion reactions by which stars convert helium into heavier elements, the other being the triple-alpha process. While the triple-alpha process only requires helium, once some carbon is present, other reactions that consume helium are possible:

$$_6\mathrm{C}^{12} + {}_2\mathrm{He}^4 \rightarrow \ _8\mathrm{O}^{16} + \gamma \qquad\qquad _8\mathrm{O}^{16} + {}_2\mathrm{He}^4 \rightarrow \ _{10}\mathrm{Ne}^{20} + \gamma$$

$$_{10}\mathrm{Ne}^{20} + {}_2\mathrm{He}^4 \rightarrow \ _{12}\mathrm{Mg}^{24} + \gamma \qquad\qquad _{12}\mathrm{Mg}^{24} + {}_2\mathrm{He}^4 \rightarrow \ _{14}\mathrm{Si}^{28} + \gamma$$

$$_{14}Si^{28} + {_2}He^4 \rightarrow {_{16}}S^{32} + \gamma \qquad {_{16}}S^{32} + {_2}He^4 \rightarrow {_{18}}Ar^{36} + \gamma$$

$$_{18}Ar^{36} + {_2}He^4 \rightarrow {_{20}}Ca^{40} + \gamma \qquad {_{20}}Ca^{40} + {_2}He^4 \rightarrow {_{22}}Ti^{44} + \gamma$$

$$_{22}Ti^{44} + {_2}He^4 \rightarrow {_{24}}Cr^{48} + \gamma \qquad {_{24}}Cr^{48} + {_2}He^4 \rightarrow {_{26}}Fe^{52} + \gamma$$

$$_{26}Fe^{52} + {_2}He^4 \rightarrow {_{28}}Ni^{56} + \gamma \qquad {_{28}}Ni^{56} + {_2}He^4 + \gamma \rightarrow {_{30}}Zn^{60}$$

All these reactions have a very low rate at the temperatures and densities found in the stars and therefore do not contribute significantly to a star's energy production; with the elements heavier than neon (atomic number > 10), they occur even less easily due to the increasing Coulomb barrier.

Alpha process elements (or alpha elements) are so-called since their most abundant isotopes are integer multiples of four, the mass of the helium nucleus (the alpha particle). Stable alpha elements are: C, O, Ne, Mg, Si, S, Ar, Ca.

4. Burning of heavy elements

4.1 Lithium burning

Lithium is generally present in brown dwarfs and not in low-mass stars. Stars, which by definition must achieve the high temperature (2.5×10^6 K) necessary for fusing hydrogen, rapidly deplete their lithium. This occurs by a collision of lithium-7 and a proton producing two helium-4 nuclei. The temperature necessary for this reaction is just below the temperature necessary for hydrogen fusion.

From a study of lithium abundances in 53 T Tauri stars, it has been found that lithium depletion varies strongly with size, suggesting that lithium burning by the pp-chain. The pp-chain for lithium burning is as the following.

$$p + {_3}Li^6 \rightarrow {_4}Be^7 \text{ (unstable)} \qquad {_4}Be^7 + e^- \rightarrow {_3}Li^7 + \nu$$

$$p + {_3}Li^7 \rightarrow {_4}Be^8 \text{ (unstable)} \qquad {_4}Be^8 \rightarrow 2{_2}He^4 + \text{energy}$$

It does not occur in stars less than sixty times the mass of Jupiter.

4.2 Carbon-burning

The carbon-burning process or carbon fusion is a set of nuclear fusion reactions that take place in massive stars (at least 8 $M_\odot$ at birth) that have used up the lighter elements in their cores. It requires high temperatures ($> 5 \times 10^8$ K) and densities ($> 3 \times 10^9$ kg/m^3). The principal reactions are:

$$_6C^{12} + {_6}C^{12} \rightarrow {_{10}}Ne^{20} + {_2}He^4 \qquad {_6}C^{12} + {_6}C^{12} \rightarrow {_{11}}Na^{23} + {_1}H^1$$

$$_6C^{12} + {_6}C^{12} \rightarrow {_{12}}Mg^{24} + n \qquad {_6}C^{12} + {_6}C^{12} \rightarrow {_8}O^{16} + 2\,{_2}He^4$$

4.3 Neon-burning

The neon-burning process is a set of nuclear fusion reactions that take place in massive stars (at least 8 solar masses). Neon burning requires high temperatures and densities (around 1.2×10^9 K and 4×10^9 kg/m^3). At such high temperatures photo-disintegration becomes a significant effect, so some neon nuclei decompose, releasing alpha particles:

$${}_{10}\mathrm{Ne}^{20} + \gamma \rightarrow {}_{8}\mathrm{O}^{16} + {}_{2}\mathrm{He}^{4} \qquad {}_{10}\mathrm{Ne}^{20} + {}_{2}\mathrm{He}^{4} \rightarrow {}_{12}\mathrm{Mg}^{24} + \gamma$$

$${}_{10}\mathrm{Ne}^{20} + \mathrm{n} \rightarrow {}_{10}\mathrm{Ne}^{21} + \gamma \qquad {}_{10}\mathrm{Ne}^{20} + {}_{2}\mathrm{He}^{4} \rightarrow {}_{12}\mathrm{Mg}^{24} + \mathrm{n}$$

where the neutron consumed in the first step is regenerated in the second.

4.4 Oxygen-burning

The oxygen-burning process is a set of nuclear fusion reactions that take place in massive stars that have used up the lighter elements in their cores. It occurs at temperatures around 1.5×10^9 K and densities of 10^{10} kg/m^3. The principal reactions are:

$${}_{8}\mathrm{O}^{16} + {}_{8}\mathrm{O}^{16} \rightarrow {}_{14}\mathrm{Si}^{26} + {}_{2}\mathrm{He}^{4} \qquad {}_{8}\mathrm{O}^{16} + {}_{8}\mathrm{O}^{16} \rightarrow {}_{15}\mathrm{P}^{31} + {}_{1}\mathrm{H}^{1}$$

$${}_{8}\mathrm{O}^{16} + {}_{8}\mathrm{O}^{16} \rightarrow {}_{16}\mathrm{S}^{31} + \mathrm{n} \qquad {}_{8}\mathrm{O}^{16} + {}_{8}\mathrm{O}^{16} \rightarrow {}_{14}\mathrm{Si}^{30} + 2\,{}_{1}\mathrm{H}^{1}$$

$${}_{8}\mathrm{O}^{16} + {}_{8}\mathrm{O}^{16} \rightarrow {}_{15}\mathrm{P}^{30} + {}_{1}\mathrm{D}^{2} \qquad {}_{8}\mathrm{O}^{16} + {}_{8}\mathrm{O}^{16} \rightarrow {}_{16}\mathrm{S}^{32} + \gamma$$

$${}_{8}\mathrm{O}^{16} + {}_{8}\mathrm{O}^{16} \rightarrow {}_{12}\mathrm{Mg}^{24} + 2\,{}_{2}\mathrm{He}^{4}$$

The oxygen-burning process is the last nuclear reaction in the star's core which does not proceed via the alpha process.

4.5 Silicon-burning

In astrophysics, silicon burning is a very brief sequence of nuclear fusion reactions that occur in massive stars with a minimum of about 8 - 11 solar masses. Silicon burning is the final stage of fusion for massive stars that have run out of the fuels that power them for their long lives in the main sequence on the Hertzsprung-Russell diagram. It follows the previous stages of hydrogen, helium, carbon, neon and oxygen burning processes. Silicon burning begins when gravitational contraction raises the star's core temperature to (2.7 3.5) $\times 10^9$ K. The exact temperature depends on the mass. When a star has completed the silicon-burning phase, no further fusion is possible.

After a star completes the oxygen burning process, its core is composed primarily of silicon and sulfur. If it has sufficiently high mass, it further contracts until its core reaches temperatures in the range of (2.7 3.5) $\times 10^9$ K. At these temperatures, silicon and other elements can photo-disintegrate, emitting a proton or alpha particle. Silicon burning entails the alpha process, which creates new elements by adding one of these alpha particles per step in the following sequence:

$$_{14}Si^{28} + {}_2He^4 \rightarrow {}_{16}S^{32} \qquad {}_{16}S^{32} + {}_2He^4 \rightarrow {}_{18}Ar^{36}$$

$$_{18}Ar^{36} + {}_2He^4 \rightarrow {}_{20}Ca^{40} \qquad {}_{20}Ca^{40} + {}_2He^4 \rightarrow {}_{22}Ti^{44}$$

$$_{22}Ti^{44} + {}_2He^4 \rightarrow {}_{24}Cr^{48} \qquad {}_{24}Cr^{48} + {}_2He^4 \rightarrow {}_{26}Fe^{52}$$

$$_{26}Fe^{52} + {}_2He^4 \rightarrow {}_{28}Ni^{56} \qquad {}_{28}Ni^{56} + {}_2He^4 \rightarrow {}_{30}Zn^{60}$$

The entire silicon-burning sequence lasts about one day and stops when nickel-56 has been produced. The star can no longer release energy via nuclear fusion because a nucleus with 56 nucleons has the lowest mass per nucleon of all the elements in the alpha process sequence. Although iron-58 and nickel-62 have slightly higher binding energies per nucleon than iron-56, the next step up in the alpha process would be zinc-60, which has slightly more mass per nucleon and thus, is less thermodynamically favorable. Nickel-56 (which has 28 protons) has a half-life of 6.02 days and decays via $\beta+$ decay to cobalt-56 (27 protons), which in turn has a half-life of 77.3 days as it decays to iron-56 (26 protons).

During this phase of the contraction, the potential energy of gravitational contraction heats the interior to 5×10^9 K and this opposes and delays the contraction. However, since no additional heat energy can be generated via new fusion reactions, the final unopposed contraction rapidly accelerates into a collapse lasting only a few seconds. The central portion of the star is now crushed into either a neutron star or, if the star is massive enough, a black hole. The outer layers of the star are blown off in an explosion known as a Type II supernova that lasts days to months.

5. Production of elements heavier than iron

It is sill possible to produce the elements heavier than iron. It can be done through the neutron capture and the proton capture.

5.1 r-process

The r-process or rapid-process is a nucleosynthesis process that occurs in the core-collapse supernovae and is responsible for the creation of approximately half of the neutron-rich atomic nuclei heavier than iron. The process entails a succession of rapid neutron captures (hence the name r-process) by heavy seed nuclei, typically ^{56}Fe or other more neutron-rich heavy isotopes.

The most probable candidate sites for the r-process has long been suggested to be core-collapse supernovae, which may provide the necessary physical conditions for the r-process.

5.2 s-process

The s-process or slow-process is another nucleosynthesis process that occurs by means of slow captures of neutrons, primarily occurring in AGB stars. It occurs at relatively low neutron density and intermediate temperature conditions in stars. Under these conditions heavier nuclei are created by neutron capture, increasing the atomic weight of the nucleus by one. The s-process is secondary, meaning that it requires preexisting heavy isotopes as seed nuclei to be converted into other heavy nuclei.

The s-process is believed to occur mostly in asymptotic giant branch stars. In contrast to the r-process which is believed to occur over time scales of seconds in the explosive environments, the s-process is believed to occur over time scales of thousands of years, passing decades between neutron captures.

The main neutron source reactions are:

$$_{6}\mathrm{C}^{13} + \,_{2}\mathrm{He}^{4} \rightarrow \,_{8}\mathrm{O}^{16} + \mathrm{n} \qquad _{10}\mathrm{Ne}^{22} + \,_{2}\mathrm{He}^{4} \rightarrow \,_{12}\mathrm{Mg}^{25} + \mathrm{n}$$

The cycle that terminates the s-process is:

$$_{83}\mathrm{Bi}^{209} + \mathrm{n} \rightarrow \,_{83}\mathrm{Bi}^{210} + \gamma \qquad _{83}\mathrm{Bi}^{210} \rightarrow \,_{84}\mathrm{Po}^{210} + e^{-} + \overline{\nu_e}$$

$$_{84}\mathrm{Po}^{210} \rightarrow \,_{82}\mathrm{Pb}^{206} + \,_{2}\mathrm{He}^{4} \qquad _{82}\mathrm{Pb}^{206} + 3\mathrm{n} \rightarrow \,_{82}\mathrm{Pb}^{209}$$

$$_{82}\mathrm{Pb}^{209} \rightarrow \,_{83}\mathrm{Bi}^{209} + e^{-} + \overline{\nu_e}$$

The net result of this cycle therefore is that 4 neutrons are converted into one alpha particle, two electrons, two anti-electron neutrinos and gamma radiation:

$$4\mathrm{n} \rightarrow \,_{2}\mathrm{He}^{4} + 2\,e^{-} + 2\,\overline{\nu_e}$$

The process thus terminates in bismuth, the heaviest "stable" element, and polonium, the first non-primordial element after bismuth.

These r- and s-processes account for a majority of galactic chemical evolution of elements heavier than iron.

5.3 rp-process

The rp-process (rapid proton capture process) consists of consecutive proton captures onto seed nuclei to produce heavier elements. It is a nucleosynthesis process and, along with the s process and the r process, may be responsible for the generation of many of the heavy elements present in the universe.

The process has to occur in very high temperature environments (above 10^9 K) so that the protons can overcome the large Coulomb barrier for charged particle reactions. A hydrogen-rich environment is also a prerequisite due to the large proton flux needed.

Sites suggested for the rp-process are accreting binary systems where one star is a neutron star. In these systems the donor star is accreting material onto its compact partner star. The accreted material is usually rich in hydrogen and helium because of its origin from the surface layers of the donor star.

5.4 p-process

The p-process is a proton capture process which is the source of certain, naturally occurring, proton-rich isotopes of the elements from selenium to mercury. These nuclides are called p-nuclei. Proton-rich nuclides can be produced by sequentially adding one or more protons to an atomic nucleus. Such a nuclear reaction of type (p, γ) is called proton capture reaction. By adding a proton to a nucleus, the element is changed because the chemical element is defined by the proton number of a nucleus. At the same time the ratio of protons to neutrons is changed, resulting in a proton-richer isotope of the next element.

Such proton captures on stable nuclides (or nearly stable), however, are not very efficient in producing p-nuclei, especially the heavier ones, because the electric charge increases with each added proton, leading to an increased repulsion of the next proton to be added, according to Coulomb's law.

6. Review Questions

1. Discuss about the stellar nucleosynthesis.

2. Write notes on the following:

   (i) Hydrogen fusion

   (ii) Deuterium burning

   (iii) Proton-proton chain reaction

   (iv) CNO cycle

   (v) Helium burning

   (vi) Triple-alpha process

   (vii) Lithium burning

   (viii) Carbon-burning

   (ix) Neon-burning

   (x) Oxygen-burning

   (xi) Silicon-burning

   (xii) Production of elements heavier than iron

   (xiii) r-process

   (xiv) s-process

   (xv) rp-process

   (xvi) p-process

7. Multiple Choice Questions

1. Which is the catalytic process for the production of energy in the core of a star?

   A. pp-chain B. both pp-chain and CNO cycle

   C. CNO cycle D. neither pp-chain nor CNO cycle

   Ans. C

2. In a CNO cycle, the isotopes of which element act a catalyst?

   A. C only B. N only C. C and N only D. C, N, O all

   Ans. D

3. In the nucleosynthesis process in the stars, the heaviest element produced is

   A. He B. C C. Ca D. Fe

   Ans. D

10. Galaxies and Universe

We believe that the universe consists of several hundred billions of galaxies, though their estimation is a difficult task. These galaxies have various shapes, sizes and ways of their evolution. Our galaxy, Milky Way is one of them. As we live inside our galaxy, it is not possible to have direct information about its shape. In a night we see a dense stream of stars, which has very large number of stars in the winter nights as compared to those in the summer nights. We are in the northern hemisphere on the Earth and are situated in the disc at a distance of about 30,000 ly from the center of Milky Way. Hence, in the winter nights, we look toward the center whereas in the summer night we see away from the center. Though the galaxies are very large in number, but on the basis of their finite number of shapes, they are classified. Some galaxies are found active. Based on the gravitational effects, it is concluded that there is a large amount in the form of dark matter as compared to the normal matter. The recession of galaxies obtained from the red-shift provides information about the expanding universe. In this chapter, we have discussed about various galaxies.

1. Milky Way galaxy

Our galaxy, named Milky Way, is one of the several hundreds billions galaxies in the universe. It contains over 100 billion stars. In a dark night, one can even casually notice in the sky, an almost continuous band of light, extending from one end of horizon to other, inclined to some 62.6° with respect to the celestial equator. This band of light is a part of our Milky Way. Galileo was the first who realized this band having a large collection of stars.

1.1 Shape and size

Since we cannot go outside the Milky Way to see it from there, it is only possible to prepare a schematic drawing of the Milky Way, based on the observations made from the earth in various directions. A schematic drawing of the Milky Way is shown in Figure 1. The Milky Way may be divided into three major components: (i) a thin disc of thickness about 2000 ly and containing stars, gas, and dust, (ii) a central bulge and (iii) a halo which is roughly spherical and contains roughly 200 globular clusters. Our Sun is situated in the disc of the Milky Way at a distance of about 30,000 ly from the center.

As discussed earlier, because of the difference between the sidereal day and the mean solar day by 3 min and 56 sec, stars in the summer nights observed at a place are different from those in the winter nights. On the Earth, we are situated in the northern hemisphere on the Earth and in the summer nights, we see towards the galactic center (center of the Milky Way) whereas in the winter nights, we see opposite to the galactic center. Thus, to an observer in the northern hemisphere, our Milky Way appears more prominent during the summer nights than that during the winter nights. In the southern hemisphere, the case is the reverse and the Milky Way appears more prominent during the winter nights than that during the summer nights.

1.2 Stellar populations

The stars found in the disc are mostly Population I (with high abundance of heavy-elements) stars. The central bulge has a mixture of Population I and Population II (with low abundance of heavy-elements) stars. The halo of the galaxy virtually contains Population II stars.

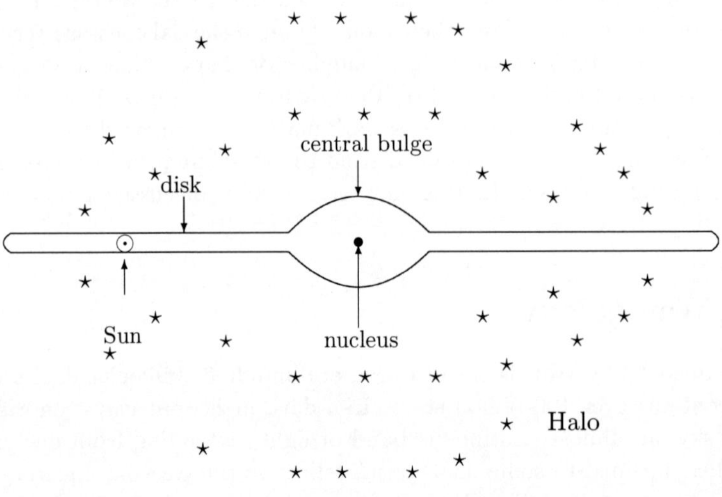

Figure 1: A schematic drawing of the Milky Way viewed wedge-on by an observer outside the galaxy. Our Sun is at a distance of about 30,000 ly from the center of the Milky Way.

1.3 Stellar motions

Our Milky Way is a bound stellar system, where in order to balance the gravitational force by the centrifugal force, the stars must go in a rotational motion. In the disc, the rotational speeds of the stars are larger as compared to the small random motions. This high value of rotational speed makes the disc highly flattened. In the central bulge, the rotational speeds of stars are not very large and therefore they move in a random manner. Thus, these stars form a roughly spherical distribution. In the galactic halo, the stars have very high random velocities and they are less tightly bound to the galaxy as compared to those in the central bulge. It is not known if there is a smooth transition from the stars in the central bulge to the stars in the halo.

1.4 Missing mass

From the measured gravitational field, it is possible to calculate the mass of the material required locally to provide the observed gravitational force. On the other side, the mass of the visible stars, gas, and dust can be estimated, and is found about one-half of the mass required for the observed gravitational field. This is known as the missing mass problem of the Milky Way. This is one of the major unresolved problems in the modern astronomy. The missing mass problem exists in all the galaxies and thus there is a question:

- Does there exist in the universe a large amount of non-luminous matter which is undetected (and perhaps is undetectable) except for its gravitational influence?

- Scientists have an opinion that there is dark matter in our Milky Way whose amount is more than that of the visible matter.

1.5 Local mass-to-light ratio

From the star counts, in the solar neighbourhood, we can estimate the mean surface brightness of galactic disc. If we divide the surface mass density (mass per unit area, denoted by μ) by the surface brightness (luminosity per unit area, denoted by L), we can get the local mass-to-light ratio μ/L. In the solar units, the local mass-to-light ratio is found to be 5. That is,

$$\frac{\mu}{L} = 5\,\frac{M_\odot}{L_\odot}$$

This value is typical for the visible part of disc galaxies in general. The mass-to-light ratio for our Sun is 1 in solar units. The fact that the ratio is 5 in the immediate vicinity of the Sun has some interesting implications.

(i) It can be interpreted that the mean gravitating mass in the galaxy has a lower efficiency for producing light per unit mass than in the Sun. This conclusion is consistent with the fact that most of the stars in the galaxy are faint, K or M dwarfs. Such stars have low mass and further low luminosities.

(ii) The second interpretation is concerned with the production of helium in the stars. The Sun with the mass-to-light ratio 1 could convert only $\sim$ 10% of its hydrogen mass into the helium mass through the nuclear processes in $\sim 10^{10}$ years. Thus, the matter with the mass-to-light ratio 5 could manage to convert only 2% of its hydrogen mass into the helium mass through the nuclear processes in $\sim 10^{10}$ years, which is roughly the age of the Milky Way.

1.6 Mass of halo

The observed mass of all the luminous stars in the halo of Milky Way is estimated to be less than a few percent of the mass of the disc. It is believed that the halo may contain a very large and very massive invisible component. Astronomers believe that the total mass of the halo is about 10 times the mass of the disc. This again leads to the situation of missing mass in the Milky Way, which is still an unresolved problem.

1.7 Differential rotation

Out of the three components of the Milky Way, the disc is the most studied one. A body gets flattened when it spins about some axis. Thus, the flattened shape of the disc provides a clue that the stars as well as the interstellar gas clouds in the disc rotate about the center of the galaxy. Random motions of the stars as well as of the interstellar gas clouds are superimposed on the respective circular motions. The flattening of the disc however does not provide any clue whether the rotation is uniform or non-uniform. It was found that the rotation of the disc is not uniform but differential, *i.e.*, the disc does not rotate like a solid body. The inner parts of the disc rotate faster than the outer ones.

Our Sun being part of the disc also rotates around the center of the Milky Way. The mean motion of the disc's material in the neighbourhood of the Sun is defined as the local standard of rest (LSR). Though the Sun has a slight motion relative to the LSR, but the non-circular component of the Sun's motion is negligibly small and we can take the motion of the Sun to be the LSR. Stars outside the solar circle appear as in a slower lane and the stars inside the solar circle appear as in a faster lane than the stars in the solar circle. Individual disc stars also possess a small excesses or deficits of velocities as compared with the mean of all stars in their neighbourhood.

In the solar neighbourhood, there are few stars which have large velocity relative to the LSR. Since these high-velocity stars move in a random manner, they exhibit an asymmetry.

1.8 Spiral structure

In the universe, many galaxies having disc are found to have spiral structure. Since the Milky Way is considered to have a disc, it is quite reasonable to assume that it also has a spiral structure. The galaxy has a beautiful spiral pattern having two trailing spiral arms. The spiral arms have brilliant young stars looking like beads in a string. Hence, the spiral arms show a beautiful scenario. The beads are actually the giant H II regions where the embedded OB stars cause them to shine with a large power output.

We are aware of the fact that the main sequence, the lifetime of the OB stars is a few million years whereas the age of the galaxy is about 10 billion years. Since we find OB stars in the spiral arms, it is only possible when the process of the formation of such stars is going on regularly. The newly formed stars become visible whereas the old ones disappear and produce the dark matter.

The main question now arises: Why new stars are born in a grand spiral pattern? The spiral arms delineated by the strongest H II regions are often narrow and extend to a size of more than 100 ly from one end to another. What could be the mechanism responsible for the simultaneous formation of stars throughout such a narrow and long front?

The OB stars form a spiral pattern as the gas and dust, which is raw material for the star formation, are always found in the spiral arms. In order to answer about the spiral structure, differential rotation of the disc was suggested to be responsible. This concept cannot be acceptable because of the reason which can be understood with the help of Figure 2.

For understanding the proposed winding process in a galaxy, consider the gas clouds lined up in the form of a long string stretched radially in the galaxy at time t_1. We are aware of the differential rotation of the disc of the galaxy. The inner layers rotate faster than the outer ones. Suppose, at time t_2, when the innermost layers completes one-half circle, the line of the clouds forms the trailing spiral arc. By the time t_3, when the inner most layers completes one full circle, the outermost layers complete only a fraction of a circle.

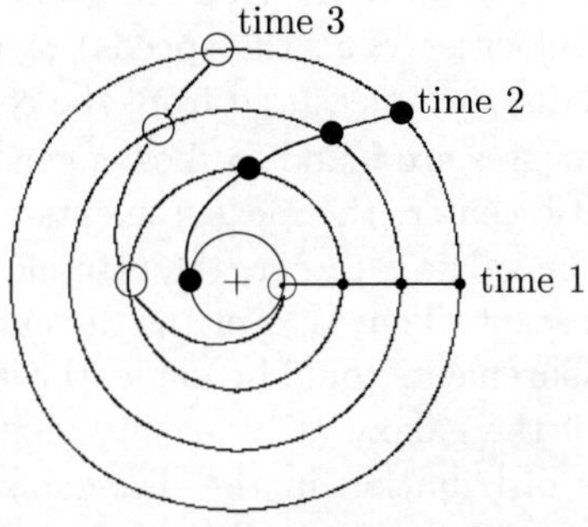

Figure 2: Concept of winding process in a galaxy.

It gives an impression of the formation of spiral structure. The time-period of rotation of the inner most layers is generally less than 10^8 years. Typical age of a galaxy is about 10^{10} years. It shows that in the lifetime of a galaxy, the inner most layers would have completed about 100 rounds. But, in the most of the galaxies, we do not have such large number of turns in the spiral arms. It is a rare event when more than one or two turns are available in a spiral galaxy. Hence, the differential rotation cannot be a mechanism for the spiral structure. A way out for this dilemma was suggested in the form of the Density-Wave theory.

2. Oort constants

By analyzing the local rotational properties of Milky Way, Jan Oort derived empirically two parameters, generally denoted as A and B, and known as the Oort constants, expressed as

$$A = \frac{1}{2}\left(\frac{V_0}{R_0} - \frac{dv}{dr}\bigg|_{R_0}\right) \quad \text{and} \quad B = -\frac{1}{2}\left(\frac{V_0}{R_0} + \frac{dv}{dr}\bigg|_{R_0}\right)$$

where V_0 and R_0 are the rotational velocity and distance from the Galactic center, respectively, measured at the position of the Sun. These constants depend only on the motions and positions of stars in the solar neighborhood. As of 1997, the derived most accurate values of these constants are $A = 14.82 \pm 0.84$ km/s/kpc and $B = -12.37 \pm 0.64$ km/s/kpc. From the Oort constants, it is possible to determine the orbital properties of the Sun, such as the orbital velocity and period, and to infer the local properties of the Galactic disk, such as the mass-density and how the rotational velocity changes as a function of radius measured from the Galactic center.

3. Galaxy rotation curve

For a disc-galaxy, the rotation curve (also known as the velocity curve) is a plot of the magnitude of the orbital velocities (*i.e.*, the speeds) of visible stars or gas in that galaxy versus their radial distance measured from the center of that galaxy.

So far, the galaxy rotation curves are found to show a general behaviour that up to the distance form the galactic center, the measurements can be made, with the increase of the distance from the galactic center, the orbital speed of stars and gas increases or remains almost constant. That is, over the distances measured from the center of a disc-galaxy, the measurements could be made, the stars and gas are found to revolve around the center of the galaxy at a velocity that increases or remains almost constant. If the mass distribution in the disc-galaxies is similar to that observed from the distribution of stars and gas, then on increasing the distance, the orbital speed would always decrease in the same manner as in other systems where the most of their mass is concentrated in the center, such as the Solar System or the moons of the Jupiter.

For the spiral galaxies, the rotation curves are known to be asymmetric. Then the observational data from each side of a galaxy are averaged. The asymmetry in the rotation curve is a normal feature rather than exceptional.

The theoretical calculations based on the assumption of centrally dominated mass associated with the observed luminous material for the galaxy rotation curve differs much from the observations.

The calculations of mass profiles of galaxies from the luminosity profiles and mass-to-light ratios in the stellar disks do not match with the masses derived from the observations of the rotation curves and the law of gravity. This discrepancy can be considered by postulating a large amount of dark matter that permeates the galaxy and extends into the galaxy's halo.

Though the assumption of dark matter is the most accepted explanation for the resolution for discrepancy for the galaxy rotation curve, other proposals are offered with varying degrees of success.

3.1 Halo density profiles

To accommodate a flat rotation curve, a density profile for a galactic environ must be different from that for a centrally concentrated. According to the Newton's version, for the Kepler's third law, the radial density profile $\rho(r)$ should be as the following.

$$\rho(r) = \frac{3v(r)^2}{4\pi G r^2}\left(1 + 2\,\frac{\mathrm{d}\log_{10} v(r)}{\mathrm{d}\log_{10} r}\right)$$

where $v(r)$ is the radial orbital velocity at a distance r and G is the gravitational constant. Observations did not match with such a simple profile. Several scientists remarked that a "gently changing logarithmic slope" for a density profile could also accommodate approximately flat rotation curves over the large scales. Then, the famous Navarro-Frenk-White profile, which is consistent with N-body simulations and observations both, was derived as the following.

$$\rho(r) = \frac{\rho_0\; R_s}{r\,(1\; +\; r/R_s)^2}$$

where the central density, ρ_0, and the scale radius, R_s, are parameters which vary from halo to halo. Here, the slope of the density profile diverges at the center, and therefore, other alternative profiles were proposed.

For the spiral galaxies, the rotation curves $v(r)$, for obtaining their mass structure, is as the following.

$$v(r) = \left(r\,\frac{\mathrm{d}\Phi}{\mathrm{d}r}\right)^{1/2}$$

where Φ denotes the gravitational potential of galaxy. As the rotation curves do not show a Keplerian fall-off at large distances, they do not match the distribution of luminous matter. This implies that the spiral galaxies contain large amounts of dark matter or, in alternative, at the galactic scales, the exotic physics exists. The additional invisible component becomes progressively more conspicuous in each galaxy at the outer radii and among galaxies in the less luminous ones.

Cosmology predicts that about 84% of the mass of the Universe is in the form of the dark matter, a massive component which cannot emit radiation, but it dominates the gravitational potential of the galaxies and the cluster of galaxies. Galaxies are baryonic condensations made by stars and gas (namely H and He) lying at the centers of much larger dark halos made by unknown matter (dark matter), affected by a gravitational instability caused by the primordial density fluctuations.

In order to avoid the presence of dark matter, some attempts were made for solving the problem of galaxy rotation curves by modifying gravity. One of the most discussed model is the Modified Newtonian Dynamics (MOND), where the Newtonian force law is modified at the low accelerations, in order to enhance the effective gravitational attraction. MOND has been found to show considerable success in the prediction of the rotation curves of low-surface-brightness galaxies.

4. Dark matter

The dark matter is a hypothetical form of matter, which cannot be seen with a telescope but is considered to account for most of the matter in the universe. It is one of the greatest mysteries in modern astrophysics. The existence and properties of dark matter are derived mainly from its gravitational effects observed on the visible matter, radiation, and the large-scale structure of the universe.

The dark matter neither emits nor absorbs any kind of electromagnetic radiation at any significant level. The observations of the Planck mission, for the standard model of cosmology, have been concluded that the total mass-energy of the known universe contains 4.9% ordinary matter, 26.8% dark matter and 68.3% dark energy. Hence, the invisible matter is estimated to constitute 95.1% of the total mass-energy content of the universe.

Although the existence of dark matter is generally accepted by the mainstream scientific community, some alternative theories of gravity have been proposed, such as MOND. However, these theories cannot account for the properties of galaxy clusters.

The first person to interpret the evidence and infer the presence of dark matter was Dutch astronomer Jan Oort, a pioneer in radio astronomy, in 1932. Later on. much of the evidence for dark matter comes from the study of the motions of galaxies. Many of them appear to be fairly uniform, and according to the Virial theorem, the total kinetic energy should be half the total gravitational binding energy of the galaxies. From the observations, however, the total kinetic energy is found to be much larger, when assumed that the gravitational mass is due to only the visible matter of the galaxy. The stars far from the center of galaxies have much higher velocities than predicted by the Virial theorem.

Gravitational lensing observations of galaxy clusters allow direct estimates of the gravitational mass based on its effect on the light from the background galaxies, since large collections of matter (dark or otherwise) deflects the light due to gravitational effect.

5. Classification of galaxies

Our universe is made up of several hundreds billions of galaxies. Hubble studied a large number of galaxies and noticed that though the number of galaxies is very large, but their geometrical shapes, which are projections on the sky-plane, can be classified into a limited number of categories. The first point he noticed that a majority of galaxies had a regular shape whereas a few percent of them had an irregular shape. Based on this criterion, he classified the galaxies into two categories: (i) those having regular shape, called the regular galaxies, and (ii) those having irregular shape, called the irregular galaxies.

On the basis of their shapes, the regular galaxies were further classified into two categories: (a) elliptical galaxies, also called the ellipticals, having a shape of an ellipse, and (b) spiral galaxies, also called the discs or spirals, having spiral arms. The spiral galaxies were again divided into two kinds: ordinary spirals also called the normal spirals, and the barred spirals. Classification of galaxies is shown in Figure 3. Hubble's morphological classification of galaxies is shown in Figure 4, which is of the form of a tuning fork. Hence, this diagram is known as the Hubble's tuning fork diagram.

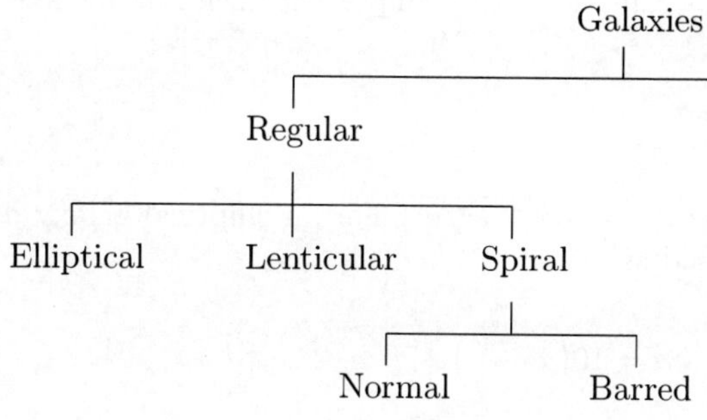

Figure 3: Classification of galaxies.

A large majority of galaxies, about 70%, are spirals (ordinary plus barred); irregular galaxies are a few percent; and the rest are elliptical galaxies. Thus, a bulk of the mass of the visible universe is contained in the regular galaxies.

5.1 Irregular galaxies

These are the galaxies in which the stars are distributed in an irregular manner without showing any regularity in the distribution. They are known as the irregular galaxies.

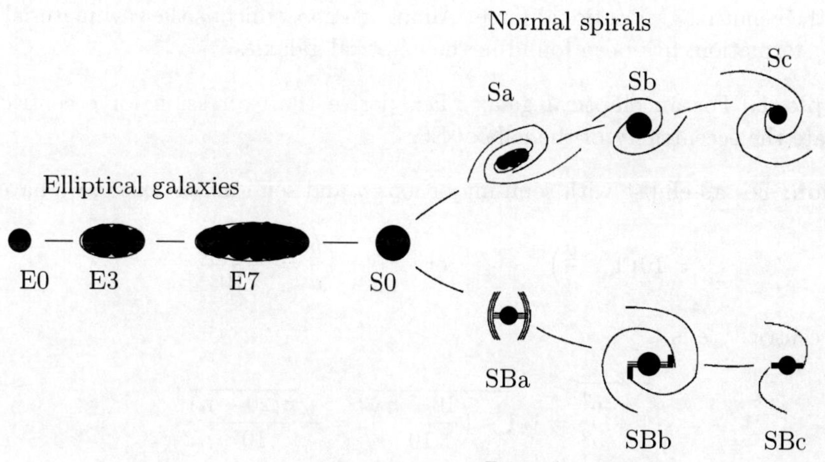

Figure 4: Hubble's tuning fork diagram for classification of regular galaxies.

5.2 Elliptical galaxies

These are galaxies which look like an ellipse when projected on the sky-plane. Depending on the flatness of the ellipse of the galaxy, they are classified into eight groups denoted by from E0 to E7, where E0 denotes a circular form (zero value of eccentricity) and E7 denotes the available most flattened form (the largest available value of the eccentricity). Equation of an ellipse with semi-major axis a and semi-minor axis b is

$$\frac{x^2}{a^2} + \frac{y^2}{b^2} = 1$$

and eccentricity e of the ellipse is $e = \sqrt{1 - b^2/a^2}$. For an elliptical galaxy denoted by En, the integer n is expressed as

$$n = 10\left(1 - \frac{b}{a}\right)$$

Earlier, the reason for flatness of an elliptical galaxy was assumed to be its rotation about its own axis. The speed of rotation was assumed to increase when we go from E0 to E7.

However, later on the observations found that the reason for flattening was not the rotation, but it was due to that the large random velocities of constituent stars are anisotropic (*i.e.*, the velocities of constituent stars have different values in different directions). From the model calculations of the stellar dynamics, this effect of flattening of a galaxy due to anisotropy in velocities was known for a long time, but it was not accepted earlier.

When we neglect the disc of a spiral galaxy, the bulge of the galaxy shows a resemblance with an elliptical galaxy. About the dynamics of stars in an elliptical galaxy, it can be said that the stars are moving in a more or less random manner under their mutual gravitational force. Almost no gas, which is the raw material for the star formation, has been found in the elliptical galaxies.

Example 1: For an elliptical galaxy En, derive the expression for eccentricity. Calculate the eccentricity of the galaxy E6.

Solution: For an ellipse with semi-major axis a and semi-minor axis b, we have

$$n = 10\left(1 - \frac{b}{a}\right) \qquad \text{or} \qquad \frac{b}{a} = \frac{10 - n}{10}$$

The eccentricity e is

$$e = \sqrt{1 - \frac{b^2}{a^2}} = \sqrt{1 - \left(\frac{10 - n}{10}\right)^2} = \frac{\sqrt{n(20 - n)}}{10}$$

For the galaxy E6, we have $n = 6$, and thus

$$6 = 10\left(1 - \frac{b}{a}\right) \qquad \text{or} \qquad \frac{b}{a} = 0.4$$

Thus, eccentricity e of the ellipse is

$$e = \sqrt{1 - b^2/a^2} = \sqrt{1 - 0.16} = 0.917$$

5.3 Spiral galaxies

These are the galaxies which have spiral arms. They are known as the spiral galaxies. Our galaxy Milky Way also is assumed to be a spiral galaxy. The spiral galaxies can be classified into two categories: (i) normal spirals, denoted by S, and (ii) barred spirals, denoted by SB. The normal spirals, besides the spiral arms, have a disc whereas the barred spirals, besides the spiral arms, have a bar. Each of these types is further divided into three sub-categories: for normal spirals, we have Sa, Sb and Sc; for barred spirals, we have SBa, SBb and SBc.

In each of the sub-category, the spiral arms are increasingly more loosely wound. The tightness of spiral winding correlates closely with the relative size of the central part: the bigger the central part, the more tightly bound are the spiral arms. Thus, Sa and SBa galaxies have larger central part than the Sc and SBc galaxies. However, some observations show that this correlation may be more complicated.

It has been suggested that the spirals may be losing their spiral arms and turning to S0 and SB0 type galaxies. This suggestion is however not accepted by most of the scientists. Although not all present S0 and SB0 galaxies might have emerged from spirals, all spirals which avoid merging with other galaxies apparently one day would become S0 and SB0 type galaxies.

It is universally accepted that the amount of gas in the spiral galaxies is gradually decreasing in the formation of new stars. Though during the death of stars, some of the gas may return to the interstellar medium, but in each cycle a net amount of gas is locked in the end stages (white dwarfs, neutron stars and black holes) of the stars. Confinement of birth of stars to the spiral arms suggests that most of the spiral galaxies have become so diffuse that only in the regions where the gas is compressed, the formation of new stars takes place. This fact has two important implications:

1. A long ago when the galaxies were nearly gas rather than nearly stars, as in the present days, the star formation rate would have been very high.
2. Present estimations show that in the few times 10^{10} years, the star formation process would exhaust all the remaining gas and in about 10^{11} years the gas in the spiral arms would become so rarefied that no more star formation would take place. Thus, all the dying spiral galaxies eventually be transformed to S0 and SB0 type galaxies. This scenario may explain the existence of present S0 and SB0 type galaxies. It is also found that on average S0 and SB0 type galaxies tend to have large bulges than the spiral galaxies.

Though astronomers have an outline of theory for evolution of galaxies, but they are still very far from making definite calculations explaining the scenario of the galaxies.

Tully-Fisher relation

In astronomy, the TullyFisher relation (TFR) is an empirical relationship between the mass or intrinsic luminosity of a spiral galaxy and its angular velocity or emission line width. The luminosity is calculated by multiplying the galaxy's apparent brightness by $4\pi d^2$, where d is its distance from us, and the spectral-line width is measured using long-slit spectroscopy.

Several different forms of the TFR exist, depending on which precise measures of mass, luminosity or rotation velocity one takes it to relate. Its modified form, known as the Baryonic TullyFisher relation (BTFR) states that baryonic mass is proportional to velocity to the power of roughly 3.5-4.

The TFR can be used to estimate the distance to spiral galaxies by allowing the luminosity of a galaxy to be derived from its directly measurable line width.

5.4 Lenticular galaxies

The galaxies at the transition between the elliptical and spiral galaxies come under the class of lenticular galaxies, denoted by S0 as well as SB0.

5.5 Major differences between the spirals and ellipticals

The spirals and ellipticals have the following major differences:

1. In the elliptical galaxies, the ratio of the random velocities to the rotational velocities is much larger than that in the spiral galaxies.

2. Elliptical galaxies have almost no gas and no dust. On the other side, most of the spirals have gas as well as dust.

3. Since gas and dust, which are raw materials for the formation of new stars, are absent in the ellipticals, there are neither young stars nor the formation of stars is going on. On the other side, young stars are being regularly formed in the spirals as the required raw material is present there in a large amount.

4. There are no spiral arms in the elliptical galaxies. On the other side, almost all the spirals have spiral arms.

5.6 Various distributions in regular galaxies

(i) Distribution of light

Although much information about the galaxies can be obtained by looking at the optical photographs of the galaxies, but this simple approach has some short comings. For example, a photographic film is usually most sensitive to blue light than to the red one. In order to avoid such problem, astronomers make quantitative measurements of the distribution of surface brightness in the galaxies. This technique is known as the surface photometry of galaxies. The resolution in the microelectronics and surface photometry in several colours, it is possible to make rapid automated measurements. Following the observation, the surface brightness L (luminosity per unit area) of an elliptical galaxy as a function of distance r from the center along the major-axis could be fitted through the expression

$$L(r) = L(0) \exp[-(r/r_0)^{1/4}] \tag{10.1}$$

This equation (10.1) is also applicable for the light-distribution of the central bulge of a spiral galaxy. For the disc of a spiral galaxy, a sinusoidal variation of light was found in the azimuthal direction. However, when the light is averaged over the circles, the light distribution in the disc of a spiral galaxy is found to follow an exponential law

$$L(r) = L(0)\ \exp[-(r/r_0)] \tag{10.2}$$

Equations (10.1) and (10.2) give the distribution of light in the regular galaxies. Numerical values of the central surface luminosity $L(0)$ and the scale length r_0 are adjusted to obtain the best fit for the observed surface photometry of a galaxy (or various components of a galaxy).

When such fits were performed for various galaxies, the scale length r_0 was found to vary greatly from one galaxy to another, as expected because the galaxies are found to have different sizes. On the other side, the central surface luminosity $L(0)$ was found to have smaller variation in its value. From the study of about two dozen elliptical galaxies, the value of $L(0)$ was found to be about 2×10^4 $L_\odot$/(light-year)^2, whereas for about three dozen disc galaxies, $L(0)$ was found to be about 15 $L_\odot$/(light-year)^2. It is also found that the constancy of $L(0)$ does not hold for dwarf systems, and that even for the giant galaxies. Nevertheless, the near constancy of $L(0)$ may be important for studies of the large scale structure of the universe.

(ii) Distribution of mass

Let the measured velocities in the elliptical galaxies pertain to the speed in random velocities of stars whereas in the spiral galaxies to the rotational velocities of either the stars or the gas about the center of the galaxy. For a galaxy if v denotes either kind of velocity, then the mass $M(r)$ contained in a sphere of radius r can be roughly expressed through the relation

$$M(r) = r\ v^2/G \tag{10.3}$$

where G is the universal gravitational constant. Equation (10.3) gives the distribution of mass in a regular galaxy. In application of the relation (10.3), there are some difficulties. First, this relation requires numerical correlation, a coefficient of order one, whose exact value depends on the details of the kinematics and the mass distribution within the system. Second, we directly measure the angular displacement θ from the galactic center and not the linear distance r. Hence, to get r from θ, the distance of the galaxy is needed. Third, the astronomical measurement of v itself is a nontrivial matter.

(iii) The $L \propto v^4$ law for ellipticals and spirals

Because of very large distances of galaxies, angular size of most of them is too small and the detailed kinematics of the structure of a galaxy cannot be investigated. From line profile of radiation from the visible part of a galaxy, it is possible to measure the total spread v of velocity. In the ellipticals, the velocity v results from the random velocities of the stars whereas in the spirals, it is from the rotational velocities of the stars and gas. It was interesting to find that in the optical spectrum, the velocity spread v of an elliptical galaxy or a spiral galaxy is related to its luminosity L through the relation

$$L \propto v^4$$

6. Active galaxies

So far we have discussed about such galaxies which have very little activity. These galaxies may be described as the quiet galaxies. A small percentage of the observed galaxies show violent activity much beyond the norms shown by the quiet galaxies. These galaxies are known as the active galaxies. The active galaxies may be classified into various categories and some of them are as the following.

1. Seyfert galaxies or Seyferts
2. Radio galaxies
3. Quasars
4. QSOs

6.1 Seyfert galaxies

Carl Seyfert was the first person who discovered anomalies in a class of galaxies with very bright centers. These galaxies are now known as the Seyfert galaxies. In many of such galaxies, the bright nucleus outshines the rest part of the galaxy put together. In these galaxies not only the nucleus is very bright, but the energy output from the nucleus can often vary by more than twice in the span of less than one year. Since coherent changes in an object on a time scale t can take place when its size is smaller than ct - the distance traveled by light in time t. This observation puts a limit on the size of the central light source to be less than one light-year. It is interesting to find that the size of the nucleus of a Seyfert galaxy is less than the average separation between two stars in a typical galaxy, and its energy output is equal to that of a system having 100 billion stars.

Further, the light coming from the nucleus of a Seyfert galaxy is quite different in nature from that emitted by a typical galaxy. For a typical galaxy, the optical light is mostly star-like which generally consists of a thermal continuum plus absorption lines, and the same is true for the light combined from a large number of stars in a typical galaxy. On the other side, a Seyfert nucleus exhibits strong emission lines like those from H II regions and planetary nebulae (recombination spectra) and a continuum which looks non-thermal (in that $\lambda^2 f_\lambda$ increases with increasing λ). It gives an impression that the basic energy source of nucleus of a Seyfert galaxy probably does not derive from the thermonuclear power of ordinary stars. This point got a support from the observations that Seyfert galaxies emit also strongly in radio, infrared and x-rays.

It is well known fact that the infrared radiation are emitted by the dust which absorbs high energy photons. From the observations, a conclusion has been derived that the nuclei of Seyfert galaxies are surrounded by a large amount of dust and gas of mass several times the solar mass might being ejected out per year at the speeds of several hundred kilometers per second. It is obvious that the presence of gas and dust surrounding a nucleus would obscure a direct view of the nucleus of the Seyfert galaxy. Thus, it is believed that the real size of the central power house in the Seyfert galaxies may be considerably less than one light-year.

6.2 Radio galaxies

The interest in radioactive galaxies was stimulated by the fact that a large number of such galaxies emit strongly in the radio continuum. Up to the development of radio interferometry techniques, poor angular resolution had been a problem for radio astronomy. In the beginning, scientists were mainly engaged in the mapping of the radio continuum background of the Milky Way galaxy.

Besides our Sun, Cygnus A was the first source found to have radio emission. Distance of this galaxy was then estimated from its red shift. From its distance and observed radio power, the radio luminosity of the source was then estimated. Radio luminosity of Cygnus A was found to be million times that of the radio luminosity of a typical galaxy.

Deep photographs show Seyfert galaxies to have extended optical images, but their radio appearance is essentially point-like. On the other side, some of the strongest extragalactic sources have extended radio structures. This structure of radio galaxies typically consists of two radio lobes which are the extended parts of the central galaxy whose nucleus may or may not be abnormally bright in the radio and the visible. The radio lobes typically lie at a distance of millions of light years on the diametrically opposite sides of the central galaxy, with the sizes of lobes being about a third to fifth of the spacing. The central part of the galaxy is generally a giant elliptical or a super giant elliptical, but never a spiral.

The first observed radio source Cygnus A is a radio double. The double lobed radio sources contain intrinsically bright radio emitting objects in the universe. The mechanism of radio emission in the lobes is the synchrotron radiation. The amount of energy in the cosmic ray particles and magnetic fields required to produce a given amount of synchrotron radiation depends on the magnetic field strength B in the region. The total energy is the minimum when the magnetic field and the relativistic particles have comparable energies. When this is applied to radio galaxies, this estimate of energy resulted in tremendous amount of total energy in the lobes, as much as 10^{61} erg, which is equivalent to the amount of energy released by ten billion supernova explosions. Whatever the source behind this amazing phenomenon is, it probably involves more than the power of mere supernova explosions.

With the improved resolution, very long baseline interferometry (VLBI) further indicated some kind of extraordinary activity in the nucleus of the central galaxy when there is a central radio source between the two radio lobes.

6.3 Quasars

The word quasar stands for quasi-stellar radio source. Owing to poor resolution in radio astronomy, quasars were first detected as unresolved radio sources. After development of the very long baseline interferometry (VLBI) technique, radio images of a few quasars were resolved. The quasars were occasionally found to show a classic double-lobed structure, similar to that of strong radio galaxies. However, the major apparent difference between the quasars and radio galaxies is that the central optical image of a quasar is not obviously a galaxy. It may be a galaxy, but it is not obviously a galaxy. Optical spectrum of the central source of a quasar contains a continuum with a non-thermal slope superimposed with broad emission lines and narrow absorption lines. The spectrum is different from that produced by a collection of stars (thermal continuum plus absorption lines); it is similar to that produced by a Seyfert galaxy.

Astronomers have found about 20 radio-quiet star-like objects whose optical properties are similar to those of quasars. These objects are known QSO's (quasi-stellar objects). Both QSO's and quasars are believed to be similar kinds of systems except that QSO's do not generate appreciable non-thermal radio emission to give large amount of optical, infrared and x-ray radiation. The resemblance of the quasars to radio galaxies and of QSO's to the nuclei of Seyfert galaxies, is so strong that most (but not all) astronomers believe that the quasars and QSO's arise from nuclear activity in the galaxies.

Puzzling pattern of emission lines is quasars was first realized in 1963 by Schmidt. He found that the observed pattern of 3C273 could be accounted for when he postulated a red shift $z = 0.158$. Such a large amount of red shift was very surprising for an object, which was earlier thought to be a star. If the observed red-shifts were due to Doppler effect, then the velocity of recession of quasars comes out to be about $0.9c$. This is an enormous speed, the largest ever encountered for any macroscopic object.

In this cosmological interpretation, the quasars would then be the most distant massive objects yet seen in the universe located at about 10^9 or 10^{10} light-year distance from the Earth. Even at such enormous distances, the quasars and QSO's appear as bright as a faint star. The intrinsic brightness of quasars must be truly tremendous if they are the cosmological distances - and they are. The optical luminosity of some quasars is estimated to be about 10^{46} erg/sec if we assume them to be at the distances shown by their redshifts.

This amount of energy is more than a trillions times the output of the Sun. Recent observations show that the x-ray luminosity of quasars may be even more impressive, as much as 10^{47} erg/sec with large variations in as little as three hours. The cosmological interpretation of quasars tells about their central workings to be objects that are about the size of the solar system and which produces energy at a rate greater than ten trillion suns. It is not surprising that some astronomers were reluctant to accept such a conclusion.

6.4 BL Lac objects

A BL Lacertae object or BL Lac object is a type of active galaxy with an active galactic nucleus (AGN). In contrast to other types of the AGNs, the BL Lacs are characterized by the rapid and large-amplitude flux variability and significant optical polarization. Due to these properties, the prototype of the class (BL Lac) was originally believed to be a variable star. When compared to the more luminous active nuclei (quasars) with strong emission lines, the BL Lac objects have spectra dominated by a featureless non-thermal continuum.

Some examples of BL Lac objects are BL Lacertae itself, OJ 287, AP Librae, PKS 2155-304, PKS 0521-365, Markarian 421, 3C 371, W Com, ON 325 and Markarian 501.

6.5 Optically violent variables

An optically violent variable quasar (often abbreviated as OVV quasar) is a type of highly variable quasar. It is a subtype of blazar that consists of a few rare, bright radio galaxies, whose visible light output can change by 50% in a day.

They are similar in appearance to BL Lacs but generally have a stronger broad emission line, and tend to have higher red shift components.

6.6 Unified model for various active galaxies

For a long time it has been argued that an AGN must be powered by accretion of mass onto massive black holes (10^6 to 10^{10} times the solar mass). AGN are both compact and persistently extremely luminous. Accretion can potentially give very efficient conversion of potential and kinetic energy to radiation, and a massive black hole has a high Eddington luminosity, and as a result, it can provide the observed high persistent luminosity. Supermassive black holes are now believed to exist in the centers of most if not all massive galaxies since the mass of the black hole correlates well with the velocity dispersion of the galactic bulge (the M-sigma relation) or with bulge luminosity.

In the standard model of AGN, cold material close to a black hole forms an accretion disc. Dissipative processes in the accretion disc transport matter inwards and angular momentum outwards, while causing the accretion disc to heat up. The radiation from the accretion disc excites cold atomic material close to the black hole and this in turn radiates at particular emission lines. A large fraction of the AGN's radiation may be obscured by interstellar gas and dust close to the accretion disc, but (in a steady-state situation) this is re-radiated at some other waveband, most likely the infrared.

Image taken by the Hubble Space Telescope of a 5000 ly long jet ejected from the active galaxy M87. The blue synchrotron radiation contrasts with the yellow starlight from the host galaxy. Some accretion discs produce jets of twin, highly collimated, and fast outflows that emerge in opposite directions from close to the disc. The direction of the jet ejection is determined either by the angular momentum axis of the accretion disc or the spin axis of the black hole.

7. Interaction of galaxies

So far, we have discussed an isolated galaxy which had no interaction between them. In fact, in the universe, various galaxies interact with one another, as was the case of interaction among various stars in a star cluster. Thus, like stars clusters, there are clusters of galaxies. Let us look into the interacting galaxies.

7.1 Interacting binary galaxies

Though a large percentage of galaxies behave like isolated galaxies, but a small percentage of galaxies are found having strongly interacting pairs of galaxies. These interacting pairs of galaxies produce intriguing structures which are very different from those of single galaxies. With the advent of computers, numerical simulations for galactic systems can be carried out and it has advanced greatly our theoretical understanding about such systems. Computer simulations of closely gravitating disc galaxies are especially interesting. The greatest commotion results from orbits which cause the spinning near edge of a participating disc to travel in the same direction as the passing galaxy.

If masses of two interacting galaxies are very different, the smaller galaxy can pull out material from the near side of the bigger one, forming a bridge between the two galaxies. When the two interacting galaxies have nearly equal masses, each of the galaxies develops a long tail that generally extends away from the system. Such structure arises because the tidal interaction between the galaxies is much stronger than that of the known tidal force for the Earth-Moon system. The tails are due to the mass loss in the direction of outer Lagrangian points. When the orbits of two interacting galaxies are bound, they can produce long bridges and tails. It is because the two systems are not flying past one another too fast when they reach closest approach.

It is very rare case when two interacting galaxies interpenetrate and exotic looking ring-system can be formed.

7.2 Mergers

When two galaxies encounter very closely, internal motions of each of them are modified. Following the law of conservation of energy, the energy required to make such changes obviously must come from the orbital motions in the galaxies. In a system of collection of gravitating stars, the kinetic energy of motion cannot be dissipated in the form of radiation. The energy would only be shared among the stars. During the encounter between the two galaxies, individual stars suffer gravitational deflections by the entire collection of stars.

The interstellar gas clouds may collide together inelastically, but the collection of stars would tend to conserve total energy of the system. The stars may however transform orbital energy into the energy of random motions and vice versa. On the statistical grounds, it can be argued that the repeated collisions between two bound galaxies would tend to bring the two galaxies close together following the laws of conservation of total energy and total angular momentum. From the second law of thermodynamics, it can be expected that for the system of trillions of stars the 'order' tends to be replaced by the 'disorder'.

A close encounter of two bound spiral galaxies, having a lot of energy in the form of orbital and angular motions, must tend eventually to produce a merger into a single pile of stars, having much more energy in the disordered form of random motions. Such merger process involves a form of 'violent relaxation', in which violently changing gravitational fields help to produce a relaxed (final) smooth distribution of stars. Except in having relatively great amounts of interstellar gas and dust, the collection of stars strongly look like an elliptical galaxy. The NGC 7252 has been found an excellent example for such a merger process. When the gas and dust is completely used up to from more stars, the final form of the system would probably look like an ordinary elliptical galaxy.

It was also speculated that perhaps all elliptical galaxies were formed in this manner. Following this speculation, mergers of spiral galaxies would cause 10% of all galaxies to become elliptical galaxies. The observed fraction of elliptical galaxies is somewhat 20% or 30%, and many of these elliptical galaxies are found in rich clusters. It is discussed that mergers have been more common in the past than now, or is more common in the clusters than in the general field.

One would definitely expect the merged product to have fewer central stars than the constituent spirals as the excess orbital energy must be absorbed into the product. But, the present central surface brightness of the giant ellipticals is much higher than those of giant spirals. It is assumed that perhaps the process of dynamical friction brings the dense core of the merging galaxies into a common bright center.

7.3 Clusters of galaxies

Likewise the stars, in addition to doublets, galaxies are also found in small groups and rich clusters. A small group of galaxies may have about one dozen to two dozen galaxies extending over a radius of about a million light-years. An example of such small group is the Local Group whose list of members includes our galaxy Milky Way as well as Andromeda galaxy. On the other side, rich clusters of galaxies may contain several thousand galaxies extending over a radius of about ten million light-years.

On even at a larger scale, say, extending over a radius of about hundred million light-years, another level of this hierarchical clustering, called a super cluster, has been discovered. A super cluster contains about one hundred clusters. It is not yet known whether any cluster of super clusters exists or not. Detailed statistical analysis of galaxy counts have found clustering of all sizes up to 60 million light-years, beyond which the amount of clustering drops dramatically.

7.4 Expansion of universe

Cosmological interpretation for red-shifts of spectral lines from galaxies is that the universe is expanding and the velocity of recession of a galaxy is proportional to its distance from an observer. Thus, the larger the red-shift of a galaxy, the large the distance of it from the observer. The red-shift provides the velocity of galaxy.

8. Hubble's law

It is remarkable to have a note that observed wavelengths of all radiations coming from the space are longer than the wavelengths of corresponding transitions measured in the laboratories. If λ_0 is the wavelength of a radiation (emission line) measured in a laboratory and if the observed wavelength of the corresponding line coming from a cosmic object is λ, then the red-shift z of the object is defined as

$$z = \frac{\lambda - \lambda_0}{\lambda_0}$$

According to Doppler effect, the wavelength of a radiation emitted by a cosmic object receding away from an observer with a velocity v is found to increase as compared to its wavelength measured in a laboratory. The relation between the red-shift z of a cosmic object and its receding velocity v is

$$z + 1 = \frac{\sqrt{1 + v/c}}{\sqrt{1 - v/c}}$$

For non-relativistic case ($v << c$), we have $z = v/c$.

Since all the cosmic objects show red-shift (no object shows blue-shift) and if the Doppler effect is valid, it shows that all the objects are moving away from us, *i.e.*, the universe is expanding.

For some galaxies, it is possible to estimate their distances. Hubble realized that there was a relation between the distance of a galaxy and its red-shift. He found that the velocity of recession v of a galaxy was proportional to its distance r from the Earth. Thus, the Hubble's law can be stated as

$$v = H_0\ r$$

where the proportionality constant H_0 is known as the Hubble's constant. When it was calculated first time by Hubble, its value was found to be 536 (km/s)/Mpc. Various scientists from time to time obtained different values of the Hubble's constant. The latest value measured is 70.1 ± 1.3 (km/s)/Mpc.

Example 2: A galaxy has a red-shift $z = 0.02$. What is its distance if we adopt the Hubble's constant H_0 to be 20 km s^{-1}/Mly.

Solution: We have

$$v = H_0 r \qquad 0.02\ c = 20\ r$$

$$0.02 \times 3 \times 10^5 = 20\ r \qquad r = 300 \text{ Mly}$$

Example 3: Two galaxies A and B, moving with velocities v_A and v_B, respectively, relative to the observer O, are at distances d_A and d_B, respectively, from the observer as shown below. The galaxies A and B follow the Hubble law when observed from O. Show that the galaxy B follows the Hubble law when observed from A.

Solution: The distance of B from A is $d' = d_B - d_A$ and velocity of B relative to A is $v' = v_B - v_A$. As the galaxies A and B follow the Hubble law when observed from O, we have

$$v_A = H_0 d_A \qquad v_B = H_0 d_B$$

On subtracting the first equation from the second, we have

$$(v_B - v_A) = H_0(d_B - d_A) \qquad \text{or} \qquad v' = H_0 d'$$

It shows that the galaxy B follows the Hubble law when observed from A.

Example 4: When a quasar has red shift of 4.5 how fast it is moving away and how far away it is. Hubble constant $H = 6.5 \times 10^4$ m/s.

Solution: A red shift of 4.5 implies that $\Delta\lambda/\lambda$ is 4.5. This is the case of relativistic Doppler effect. We have

$$\frac{\Delta\lambda}{\lambda} = 4.5 = \left(\frac{1 + v/c}{1 - v/c}\right)^{1/2} - 1$$

$$5.5^2 = \frac{1 + v/c}{1 - v/c} \qquad \frac{v}{c} = 0.94$$

The quasar is receding at a speed of 94% of the speed of light or 2.8 $\times 10^8$ m/s. For this value, we have

$$d = \frac{v_r}{H} = \frac{2.8 \times 10^8}{6.5 \times 10^4} == 4300 \text{ Mpc}$$

The uncertainty in Hubble's constant leads to an actual distance for the quasar to be between 3 billion parsecs and 6 billion parsecs.

9. Cosmic microwave background

In the scenario of expanding universe, since we are not at the outermost surface of the universe, there should be a background around us. Around 1950, George Gamow proposed that the present-day cosmos must be filled with the black-body radiations at a kinetic temperature of 10 K. Hence, there should be cosmic microwave background (CMB) around us. At that time, radio-astronomical technology was not efficient enough to record these radiations, corresponding to a low temperature of the order of 10 K.

Scientists, however, got interested in recording these black-body radiations (now known as the CMB radiations) corresponding to a low temperature of the order of 10 K. Efforts were put to improve the technology for radio astronomical observations. Improved calculations for the temperature of the CMB were performed, and it was found to be around 4 K. The CMB radiation was discovered coincidentally in 1965 by A.A. Penzias and R.W. Wilson. They detected weak radio frequency radiation at $\lambda = 7.35$ cm (4.08 GHz) and were honoured by the Noble Prize in Physics in

1978. They found isotropic unpolarised radiations coming from all directions which were independent of the time of day or of the season. It was later on confirmed that these radiations were in fact the remnant of the big-bang. It was interesting to notice that, in the entire accessible wavelength region, all the radiations follow Planck's law at $T = 2.73 \pm 0.05$ K.

10. Formation and evolution of galaxies and clusters

The matter is distributed throughout the universe in an inhomogeneous manner, and a large part of the universe is empty. But, embedded in these empty regions, there are concentrations of gases, concentrations of stars, and concentrations of galaxies. There are neutral and ionized clouds of gas, groups of stars, clusters of stars, small and big galaxies, small groupings of galaxies, and clusters of galaxies. Evidently, a number of different physical processes compete in the determination of the characteristics and makeup of these different types of objects. The physical properties which lead to the substantial differences that distinguish globular clusters from galaxies, and individual galaxies from clusters of galaxies, are completely unknown. It is not known why there should be distinctions when we switch from a scale of 10^{20} cm to 10^{23} cm, and finally to 10^{25} cm. These are the dimensions of globular clusters, galaxies and clusters of galaxies, respectively. Larger yet, by a factor of $\sim 10^3$, is the scale of the entire universe whose horizon is roughly 10^{28} cm distant.

Before talking about the differences due to the sizes, let us make sure that they really exist. For example, we can think of very small galaxies whose characteristics are similar to those of the largest globular clusters. However, since galaxies would be very faint, we would not see them at large distances, and cannot tell about their existence.

In order to resolve this problem, we can undertake an exhaustive study of all discoverable clusters and groupings in the immediate neighbourhood of our own galaxy, where we may be able to determine the properties of galaxies and clusters in some details. In particular, we would like to have answers to such questions as:

(a) If there are small galaxies whose size are intermediate between those of globular clusters and of large galaxies such as the Andromeda Nebula, our nearest spiral galactic neighbour in space.

(b) If yes, whether such galaxies show the same kind of structure as a globular cluster, or whether there are more like galaxies in that they actually contain a number of subordinate globular clusters as part of their makeup. Further, if they contain interstellar gas like the larger galaxies have.

(c) If the globular clusters sometimes behave like independent galaxies, in the sense that they are not gravitationally bound to any galaxy, but exist in space by themselves. Whether such clusters would have their own quota of interstellar gas.

Such types of queries can be fortunately answered. For the Galaxy and the Andromeda nebula do not form a pair of galaxies completely isolated from all others.

A careful search has shown that there is a Local Group of objects containing some 21 identified members. There are some such objects which are obscured by the absorbing matter within the Galaxy, and probably many more that simply have been too faint to be observed by the present day techniques.

It has been discussed that the Local Group is beginning to throw light on some of the questions posed at the beginning. Further, there is no absolutely sharp cut-off distinguishing galaxies from globular clusters. We sense that there are differences; but the criteria we choose do not always provide a clear distinction in borderline cases.

Let us ask how galaxy formation may take place. We know that galaxies exist in all parts of the universe. At the largest distances accessible with modern techniques, galaxies can still be found in apparently unchanged form and undiminished numbers.

The normal approach of astrophysicists has been to postulate that galaxies are condensations formed out of previously tenuous material. Parallel of this, there are other approaches taken in the theories of star formation. In both cases, there are difficulties.

For galaxy formation, the prime difficulty is the cosmic expansion. For understanding galaxy formation, we would like to think of gravitation, or some other force, acting to concentrate matter into a small volume. This action, however, is always countered by the continuing recession of any two or more elements of matter, as the expansion of the universe forces them further apart. There results a tug-of-war between the cosmic expansion and the gravitation. Gravitation evidently wins but we do not observe the process anywhere and so we cannot be certain about it.

Let us now re-examine our main concept: How sure we are about the condensation process through which the galaxies are formed. Not very sure. There is an alternate approach given by Ambartsumian. He noticed that there are many pairs of galaxies observed in the sky and that the differences in radial recession velocity sometimes is so great that the galaxies are escaping from one another. Their mutual gravitational attraction is too weak to hold them together, unless there is some large unobserved mass that makes the total aggregate much more massive than we find from the brightness of individual galaxies alone.

Ambartsumian argued that such galaxies might be very young, perhaps recently formed in a violent explosion. There is no doubt that peculiar galaxies, quite irregular in shape, exist in abundance. About 3% of all galaxies exhibit large scale irregularities that could become smoothed out on a time scale τ of one galactic rotation, or several hundred million years. If $\tau = 3 \times 10^8$ y, and the apparent age of the universe is 10^{10} y, then the 3% occurrence of disruptions is consistent with the idea that each galaxy goes through such a violent stage once in its life. Ambartsumian suggested that this stage represents the birth of galaxies.

The discovery of quasars also was interpreted by some astrophysicists as possible evidence in favour of present day galaxy formation. The quasars are very bright and they are probably consuming their energy resources at a very fast rate. It shows that they cannot continue to shine for longer than a few million years with their present luminosity.

A prevailing view proposes that much of the mass in the universe is in the form of dark matter, so called because it makes itself apparent only through its gravitational attraction and emits no electromagnetic radiation. The evidence for the dark matter comes from rotation curves in spiral galaxies for which centrifugal forces are excessive when the only mass present were that of stars and gas clouds. Similarly, some elliptical galaxies should be unable to gravitationally bind their observed halos of hot, x-ray emitting plasma, when they were not far more massive than their stellar component suggests.

As there is some evidence for low rates of galaxy formation at the present epoch, we believe that most of the galaxies were formed at a much earlier stage, and that contemporary galaxy formation could only represent the closing stages of a much more vigorous period.

For evolving cosmological models, galaxy formation from an intergalactic medium could be most easily understood in a Lemaitre universe. This is a cosmological model in which a rapid expansion following the birth of the cosmos is followed by a slowing down, and a quiescent stage of zero expansion. Thereafter, the expansion picks up again. During the quiescent period, the universe is roughly described by a static state in which galaxies might well be able to form, as there is no cosmic expansion to interfere with the condensation process. When instabilities leading to clumping of the extragalactic gas occur at all, this is the kind of universe in which the instabilities would be most likely to set in.

It is not clear if it is a sufficient reason to support Lemaitre models. Most modern theories are concerned not just with the formation of galaxies, but also of galaxy clusters and large intergalactic voids. The formation process in all such models therefore have to be theoretically analyzed not only with a view toward understanding how condensation can arise at all, but also with the idea of determining whether condensations of the right-size range are formed. So far this problem has not been successfully solved. While galaxy formation then constitutes a serious problem in astrophysics, it does not have the great promise of leading to greater insights into processes on a very large scale in cosmology. If we can understand how galaxies are formed, we can also understand a great deal more about the origin - provided there ever was one - of the universe and about the dynamic laws that hold on a scale some 10^{10} times larger than the scale of the solar system on which our current laws of dynamics are tested.

11. Review Questions

1. Discuss various parts of Milky Way, stellar populations and motions of stars in the Milky Way.

2. Discuss the differential rotation and spiral structure of Milky Way.

3. Discuss the problem of dark matter on our Milky Way.

4. Describe in brief the classification of galaxies.

5. Discuss the spiral galaxies and elliptical galaxies. Describe the major differences between them.

6. Discuss the distribution of light and distribution of mass in a regular galaxy.

7. Discuss the active galaxies.

8. Discuss the interaction of galaxies.

9. For the expanding universe, describe about the Hubble law.

10. Write notes on the following:

    (i) Milky Way galaxy
    (ii) Dark matter
    (iii) Differential rotation of Milky Way
    (iv) Oort constants
    (v) Galaxy rotation curve
    (vi) Classification of galaxies
    (vii) Hubble's tuning fork diagram
    (viii) Irregular galaxies
    (ix) Elliptical galaxies
    (x) Spiral galaxies
    (xi) Tully-Fisher relation
    (xii) Lenticular galaxies
    (xiii) Active galaxies
    (xiv) Seyfert galaxies
    (xv) Radio galaxies
    (xvi) Quasars
    (xvii) BL Lac objects
    (xviii) Interaction of galaxies
    (xix) Clusters of galaxies
    (xx) Expansion of the universe
    (xxi) Hubble's law
    (xxii) Cosmic microwave background

12. Multiple Choice Questions

1. A person in the north hemisphere on the Earth sees towards the center of the Milky Way in the night of

   A. Summer B. Autumn C. Winter D. Spring

   Ans. A

2. A person in the north hemisphere on the Earth sees opposite to the center of the Milky Way in the night of

   A. Summer B. Autumn C. Winter D. Spring

   Ans. C

3. A person in the south hemisphere on the Earth sees towards the center of the Milky Way in the night of

   A. Summer B. Autumn C. Winter D. Spring

   Ans. C

4. A person in the south hemisphere on the Earth sees opposite to the center of the Milky Way in the night of

   A. Summer B. Autumn C. Winter D. Spring

   Ans. A

5. Typical size of a globular cluster is

   A. 10^{15} cm B. 10^{20} cm C. 10^{25} cm D. 10^{28} cm

   Ans. B

6. Typical size of a galaxy is

   A. 10^{20} cm B. 10^{23} cm C. 10^{25} cm D. 10^{28} cm

   Ans. B

7. Size of the entire universe is

   A. 10^{20} cm B. 10^{23} cm C. 10^{25} cm D. 10^{28} cm

   Ans. D

8. Dark matter

   A. emit radiations in the visible range.

   B. is experienced by the gravitational attraction.

   C. emits radio waves having long wavelength.

   D. is experienced by Coulomb attraction.

   Ans. B

9. The ratio of the local mass-to-light ratio for the Milky Way to that of the Sun is 5. If in the period of 10^{10} years, in the Sun only 10% of its hydrogen mass is converted into the helium mass through the nuclear processes. In the Milky Way, the percent of hydrogen mass converted into the helium mass through the nuclear processes in the same period is

   A. 10 % B. 5 % C. 2 % D. 50 %

   Ans. C

10. The class of the stars, the disc of the Milky Way contains mainly is

    A. Population I B. Population I & II both

    C. Population II D. neither Population I nor II

    Ans. A

11. The class of the stars, the central bulge of the Milky Way contains is

    A. Population I B. Population I & II both

    C. Population II D. neither Population I nor II

    Ans. B

12. The class of the stars, the halo of the Milky Way contains mainly is

    A. Population I B. Population I & II both

    C. Population II D. neither Population I nor II

    Ans. C

13. In the disc of Milky Way, the rotational speed and random speed, respectively, are

    A. large, large B. large, small C. small, large D. small, small

    Ans. B

14. In the central bulge of Milky Way, the rotational speed and random speed, respectively, are

    A. large, large B. large, small C. small, large D. small, small

    Ans. D

15. In the halo of Milky Way, the rotational speed and random speed, respectively, are

    A. large, large B. large, small C. small, large D. small, small

    Ans. C

11. Cosmic Distance Ladder

By looking the sky, we can measure the angular separation between two objects situated on the celestial sphere, but cannot say which one of them is close and which one is far away. Measurement of distance of an cosmic object, in general, is a difficult task, particularly when we go to the large distances. For nearby objects, parallax method can be used for determination of distance of object. For distant objects, various methods are used. In the present chapter, we have discussed various methods used for the measurement of distance of cosmic objects.

1. Size of solar system

In order to establish of a cosmic distance scale, the first requirement is to measure the distances within our solar system with accuracy. The basic step in this direction is to measure the distance of Venus from the Earth. Being very close to the Sun, the detection of Mercury from the Earth is a bit difficult task. The most reliable and precise way to obtain the distance of Venus is with the help of radar techniques. In this method, a radar pulse is communicated towards the Venus and is received back after its reflection from the Venus surface. The time interval between the communication of pulse and its reception back is measured. As the time measurements can be made to large accuracy, the distance of Venus and the dimensions of its orbit can be calculated within an accuracy of one kilometer.

Example 1: A radar pulse sent to the Venus was received back after 26 min 40 sec. Calculate the distance of Venus from the Earth. Speed of light is 3×10^8 m/s.

Solution: Time spent by the pulse for one way is 13 min 20 sec = 800 sec.

Distance = Speed of light × Time = $(3 \times 10^8) \times 800 = 2.4 \times 10^{11}$ m.

The distance of Venus from the Earth at that time is 2.4×10^8 km.

1.1 Parallax method

The distance of Venus from the Earth may be calculated by measuring the angle on the star background, when the measurements are taken from two different positions on the Earth separated by a known distance. Two examples are as the following.

(i) Two observers separated by a distance of 104 km along a line perpendicular to the direction of Venus from the Earth, find the position of Venus on the star background to differ by $49''$. Calculate the distance of Venus from Earth.

Let us consider two observers A and B on the surface of Earth. The separation between the A and B is 10^4 km. The angle subtended by the two observers at the Venus V is $\angle$ AVB. The distance of Venus from the Earth is much larger than the separation between the A and B. Thus, the $\angle$ AVB is very small. It is 49″. On expressing the angle in radians, we have

$$\text{Angle} = \frac{\text{Separation}}{\text{Distance}} \qquad \text{or} \qquad \frac{49 \times \pi}{60 \times 60 \times 180} = \frac{10^4}{d}$$

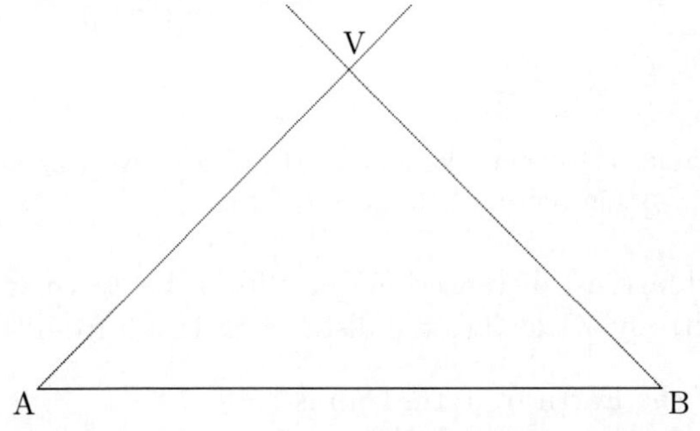

On solving the value of d is 4.21×10^7 km. Thus, the distance d of Venus from the Earth is 4.21×10^7 km.

(ii) Consider that both the Venus and Earth are revolving in their circular orbits around the Sun. At the closest approach of Venus to the Earth, its angular diameter is 64″, and at the greatest separation of Venus from the Earth, its angular diameter is 10″. Calculate the radii of the orbits of Venus and Earth. Diameter of Venus is 1.21×10^4 km.

The distance of Venus from the Earth is much larger than its diameter. Thus,

$$\text{Angle} = \frac{\text{Diameter}}{\text{Distance}}$$

The closest distance a and the farthest distance b are

$$\frac{64 \times \pi}{60 \times 60 \times 180} = \frac{1.21 \times 10^4}{a} \qquad \frac{10 \times \pi}{60 \times 60 \times 180} = \frac{1.21 \times 10^4}{b}$$

On solving, we get $a = 3.9 \times 10^7$ km and $b = 2.5 \times 10^8$ km. The radius of the orbit of Earth is

$$R_E = \frac{b + a}{2} = 1.44 \times 10^8 \text{ km}$$

The radius of the orbit of Venus is

$$R_V = \frac{b - a}{2} = 1.05 \times 10^8 \text{ km}$$

The distance of Venus at the closest approach a and at the most distant separation b can be thus measured (Figure 1). These measurements are repeated over a number of times for several years. These data can be used for determination of the diameter and eccentricity of both the Earth and Venus. The mean distance between the Earth and Sun is $(a+b)/2$.

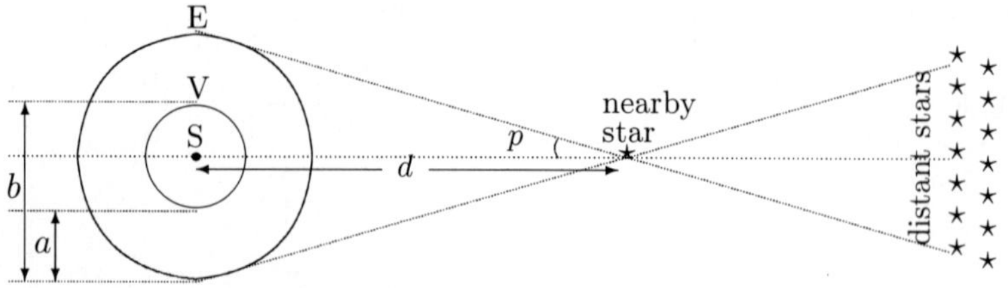

Figure 1: Here, S denotes the Sun. The Venus (V) and Earth (E) are revolving around the Sun in the circular orbits with common center at the center of Sun.

Example 2: The closest and farthest distances of Venus from the Earth are 4×10^7 km and 2.6×10^8 km, respectively. Calculate the distance of Earth from the Sun.

Solution: The distance d of the Earth from the Sun is

$$d = \frac{4 \times 10^7 + 2.6 \times 10^8}{2} = 1.5 \times 10^8 \text{ km}$$

2. Trigonometric parallax

This is same as the parallax, but the observations are made from opposite extremes in the Earth's orbit about the Sun. It is already discussed in Chapter 4. The nearby stars appear to be displaced relative to the more distant stars in the same part of the sky. The parallax, p is defined as half the apparent angular displacement measured in this manner. The distance d of the star from the Earth is

$$d = \frac{\text{astronomical unit}}{\tan p} \qquad \text{or} \qquad d = \frac{1.5 \times 10^{13}}{\tan p} \text{ cm}$$

2.1 Limitation

The trigonometric parallax can be reliably for determination of distance up to about 50 pc, where the parallax is $0.002''$. The parallax less than $0.002''$ cannot be measured with accuracy.

3. Spectroscopic parallax

After measuring the distance to nearby stars, the absolute brightness with the spectral type can be correlated. The bright stars of recognizable spectral type are then used as distance indicators across large distances where only the brightest stars are individually recognized and the trigonometric parallax cannot be used.

4. Moving cluster method

In the Hyades star cluster, the members are close together and move through space as a group. The moving cluster method can be used for this cluster by making the following measurements:

(a) Using the spectral displacement (Doppler shift) method, the radial velocity v of the cluster stars along the line of sight is measured.

(b) By inter-comparing the positions of individual stars in the cluster on the photographic plates taken many decades apart, the proper motions (apparent motions perpendicular to the line of sight) of individual stars in the cluster are measured. This motion has the form of an apparent contraction of the cluster and provides the fractional rate of increase of the cluster's distance from the Sun, as shown in Figure 2.

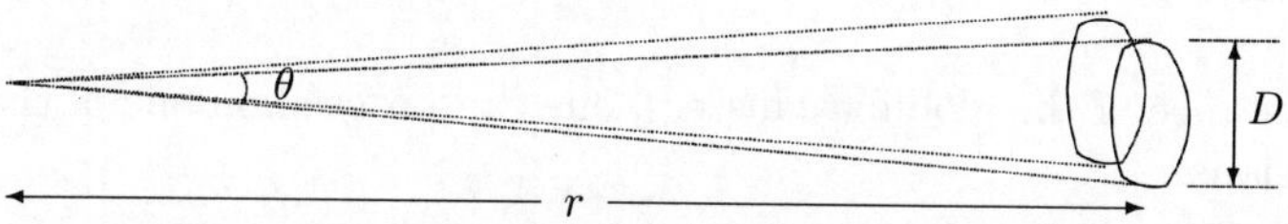

Figure 2: The Hyades cluster at a distance r recedes away from us with a radial velocity component v. The cluster diameter D remains constant, and therefore the angle θ decreases.

Here, the cluster's distance r can be expressed in terms of the angular size θ, the rate of change of angular size $\dot{\theta}$ and the receding velocity v of the cluster as the following. When the angle θ is small and is expressed in radians, we have

$$\theta = \frac{D}{r}$$

where D denotes the diameter of the cluster. Considering the D as constant, the differentiation of this relation with respect to the time gives

$$\dot{\theta} = -\frac{D\,\dot{r}}{r^2}$$

The differentiation is denoted by putting a dot over the quantity. Using $\dot{r} = v$ and $D/r = \theta$, we have

$$\dot{\theta} = -\frac{Dv}{r^2} = -\frac{\theta v}{r} \qquad \text{or} \qquad r = -\frac{\theta v}{\dot{\theta}}$$

The distance r of the Hyades cluster can be measured in terms of the three quantities, which can be measured directly. This simple method is restricted to its use for the Hyades stars, because no other cluster is near enough to obtain accurate proper motions. The use of this method obviously depends on the dynamical stability of the Hyades group, as stars are gravitationally bound.

5. Superposition of main sequences

In this method, one assumes that in all the galactic clusters, the main sequence stars have identical properties. That is, for all such clusters, the slop of the main sequence is the same and the main sequence stars of a given spectral type or colour have the same absolute brightness in all the clusters. On the basis of this assumption, one can compare the brightness of the main sequence of the Hyades and any other galactic cluster. The vertical shift necessary to bring the two main sequences into their superposition provides the relative distances of the clusters.

Suppose, the shift in the apparent magnitudes of galactic cluster and of Hyades is $\Delta m = m_{GC} - m_{Hya}$, the relative distances r_{GC} of galactic cluster and r_{Hya} of Hyades are related as

$$\Delta m = 5 \log_{10}\left(\frac{r_{GC}}{r_{Hya}}\right) + A' \tag{11.1}$$

where A' is a correction for the difference in interstellar reddening of the galactic cluster, and the Hyades. It can be determined with the help of the stellar line spectra.

For finding distance of the globular clusters, one can proceed on one of the three different assumptions:

(a) The Hertzsprung-Russell diagram of the globular cluster has a segment that runs essentially parallel to the galactic cluster main sequence. One can assume that this segment coincides with the main sequence of the Hyades cluster. Then one can use equation (11.1) to calculate the distance of the globular cluster.

(b) One can assume that the segment of the H-R diagram of the globular cluster coincides with the main sequence, defined by a group of dwarf stars in the Sun's immediate neighbourhood. The distance to these dwarf stars can be determined with the help of the trigonometric parallax.

(c) One can assume that the mean absolute magnitude for the short-period variables (RR Lyrae variables) is the same as the globular clusters and in the solar neighbourhood.

None of these three choices is safe in itself. However, when applied to the globular cluster M3 (the third object in the messier catalog), all the three methods are found to give distances in fair agreement with each other. This verifies that the main sequences of different groupings of stars coincide reasonably well and can be used as a distance indicator.

6. Brightness of RR Lyrae variables

It has been observed that the apparent brightness of all the RR Lyrae variables in a given globular cluster is the same, irrespective of the variable period. As these stars are intrinsically bright, and their short period make them stand out, they can serve as the ideal distance indicators. One can assume that the absolute brightness of these stars is the same not only within a given cluster, but in all the clusters. Thus, equation (11.1) can be used to calculate the relative distance of two clusters.

7. Brightness of Cepheid variables

The Cepheid variables in the Magellanic clouds were found to have periods that are a function of brightness. The Magellanic clouds are dwarf companions to the Galaxy. There are small galaxies in their own right and are so compact that all their stars can be considered to be at essentially the same distance from the Sun. By comparing the brightness of Cepheid in the Magellanic clouds to that in the globular clusters, one can get relative distances to these objects.

There has been however a problem in this comparison. In the clouds, the Cepheid are population I stars, normally found in the disk of the galaxy. On the other hand, the globular clusters, belong to the halo component that are more or less spherically distributed about the center of a galaxy. Generally, population I consists of bright early stars, and dwarf late stars that lie on the main sequence. Population II is characterized by the dwarf late stars and late-type giants.

8. Brightness of novae and H II regions

Novae have an absolute brightness which is related to the decay rate of the brightness after an outburst. The great intrinsic brightness of a nova may be used as an useful indicator for distance of nearby galaxies. The diameters of bright H II (ionized hydrogen) regions also form good yardsticks though which the distances of such galaxies can be estimated.

9. Seeliger's theorem

Regarding the estimation of distance of stars, the Seeliger's theorem may be expressed in the following manner. Consider a set of emitting objects is distributed homogeneously in space, then the ratio $N_m/N_{m\text{-}1}$ of the number of objects having apparent magnitude less than m to the number of those having apparent magnitude is less than m-1 is 3.98.

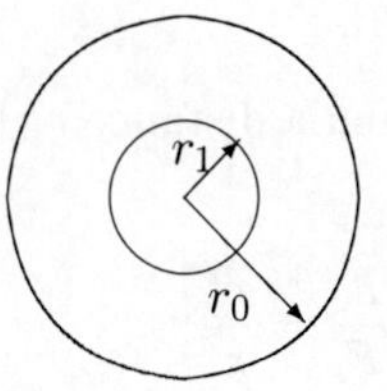

Figure 3: When a star moves away from a distance r_1 to r_0, its apparent magnitude changes from m-1 to m.

The proof of the theorem can be given in the following manner. Consider a star at a distance r_1 from an observer having apparent magnitude m-1, as shown in Figure 3. Suppose, on moving the star to the distance r_0, its apparent magnitude becomes m. The radiant fluxes F_1 and F_0 received from the distances r_1 and r_0, respectively are given as (the star is the same)

$$\frac{F_0}{F_1} = \frac{r_1^2}{r_0^2}$$

The fluxes are related to the magnitudes as

$$\frac{F_0}{F_1} = 10^{0.4(m_1 - m_0)}$$

Using $m_1 = m$ and $m_0 = m$-1, these two equations can be rearranged as

$$\frac{r_1^2}{r_0^2} = 10^{-0.4} \qquad \text{or} \qquad r_0 = (2.512)^{1/2}\ r_1$$

If the stars have a fixed brightness and are distributed uniformly in the space, they appear brighter than the apparent magnitude m out to a distance r_0, but brighter than m-1 only out to a distance r_1. The ratio of the number of stars brighter than a certain magnitude, $N_m/N_{m\text{-}1}$, is proportional to the volume occupied, than a certain magnitude. The ratio of the number of stars, $N_m/N_{m\text{-}1}$, is proportional to the volume occupied, and we have

$$\frac{N_m}{N_{m\text{-}1}} = \frac{4\pi r_0^3/3}{4\pi r_1^3/3} = \frac{r_0^3}{r_1^3}$$

Using the value of r_0/r_1, we have

$$\frac{N_m}{N_{m\text{-}1}} = (2.512)^{3/2} = 3.98$$

This is the proof of the theorem.

Example 3: Consider a space where identical stars are distributed homogeneously throughout. Suppose, a star has magnitude m_1 when it is at a distance r_1 from the observer. On moving the star to a distance r_2, its magnitude becomes m_2. Calculate the ratio of the number of stars having magnitude less than m_1 to in the number of those having magnitude less than m_2.

Solution: When a star is moved from a distance r_1 to r_2, its corresponding radiant fluxes F_1 and F_2 are expressed as

$$\frac{F_2}{F_1} = \frac{r_1^2}{r_2^2}$$

In terms of the magnitudes m_1 and m_2, the fluxes are related as

$$\frac{F_2}{F_1} = 10^{0.4(m_1 - m_2)}$$

These two equations can be rearranged as

$$\frac{r_1^2}{r_2^2} = 10^{0.4(m_1 - m_2)} \qquad \text{or} \qquad \frac{r_1}{r_2} = 10^{0.2(m_1 - m_2)}$$

The stars within the sphere of radius r_1 have magnitude m_1 or less and the stars within the sphere of radius r_2 have magnitude m_2 or less. The ratio of the number of stars brighter than a certain magnitude, N_1/N_2, is proportional to the volume occupied, than a certain magnitude, and we have

$$\frac{N_1}{N_2} = \frac{4\pi r_1^3/3}{4\pi r_2^3/3} = \frac{r_1^3}{r_2^3}$$

Using the value of r_1/r_2, we have

$$\frac{N_1}{N_2} = 10^{0.6(m_1 - m_2)}$$

10. Review Questions

1. A radar pulse sent from the Earth to the Venus was received back after 6 min 40 sec. Calculate the distance of Venus from the Earth. $c = 3 \times 10^8$ m/s.

2. Discuss about the parallax method for determination of distance of Venus from the Earth.

3. Discuss about the trigonometric parallax for determination of distance of nearby stars and its limitation.

4. Describe the moving cluster method used for determination of distance of Hyades star cluster.

5. Describe the method for determination of distance with the help of the superposition of main sequences.

6. Describe the method for determination of distance of RR Lyrae variables.

7. State the Seeliger's theorem and prove it.

8. Write notes on the following:

   (i) Trigonometric parallax method
   (ii) Spectroscopic parallax
   (iii) Superposition of main sequences method
   (iv) Moving cluster method
   (v) Distance of RR Lyrae variables
   (vi) Distance of Cepheid variables
   (vii) Distance of novae and H II regions
   (viii) Seeliger's theorem

11. Multiple Choice Question

1. The limitation of trigonometric parallax is about

   A. 10″ B. 1″ C. 0.01″ D. 0.002″

   Ans. D

12. Cosmology

The word *cosmos* derives from the Greek word *kosmos*, which means the *world*. The cosmology is a branch of science which deals with the study of the origin of universe, its evolution, and its eventual fate. So, under the cosmology, scientists study about the origin and evolution of the universe, large scale structures and their dynamics, and the ultimate fate of the universe. It deals with the scientific laws also which govern these realities. There is a difference between *Cosmology* and *Astronomy*. Cosmology deals with the universe as a whole whereas Astronomy deals with individual celestial objects in the universe. The modern cosmology is based on the Big Bang theory, where the universe is considered as emerged out of the Big Bang, which occurred about 13.8 billion years ago. The cosmology assumes the homogeneous and isotropic universe, which is justified on the scales of larger than 100 Mpc. Thus, apart from the local irregularities, the universe looks the same from all positions in space. For the evolution of universe, various models are given by scientists from time to time. This chapter deals with the cosmology.

1. Olbers paradox

The Olbers' paradox, also known as the *dark night-sky paradox*, is an argument that the darkness of night-sky conflicts with the assumption of an infinite and eternal static universe. Thus, the darkness of the night sky is an evidence for a dynamic universe, such as the Big Bang model.

The paradox may be stated that a static, infinitely large universe having infinite number of stars distributed uniformly would be bright rather than dark. In order to show this, let us consider the universe which is static, homogeneous at a large scale, and is populated by an infinite number of stars. Now, we divide the universe into a series of concentric spherical shells, each of thickness of 1 ly. The volume of a spherical shell of radius r and thickness dr is $4\pi r^2\mathrm{d}r$. Suppose, ρ denotes the uniform density of stars in the universe. The number of stars in the shell lying between n ly and $(n+1)$ ly (n is very large number, say 10^9) is

$$N_1 = 4\pi r^2 \mathrm{d}r\rho = 4\pi n^2\rho$$

where the distance is measured in ly and dr = 1 ly. The number of stars in the shell lying between $2n$ ly and $(2n+1)$ ly is

$$N_2 = 4\pi r^2\ \mathrm{d}r\rho = 4\pi(2n)^2\rho$$

Thus, the ratio of the number of stars in the two shells is

$$\frac{N_2}{N_1} = 4$$

Suppose, the stars are identical and I is the light-energy emitted by each star. The amount of energy received by the observer situated at the common center of shells per unit area from each star in the first shell is

$$E_1 = \frac{I}{4\pi n^2}$$

where n is the radius of the first shell. The amount of energy received per unit area from each star in the second shell is

$$E_2 = \frac{I}{4\pi (2n)^2}$$

where $(2n)$ is the radius of the second shell. Thus, the ratio of light received from a star in two shells is

$$\frac{E_2}{E_1} = \frac{1}{4}$$

It shows that $E_1 N_1 = E_2 N_2$, *i.e.*, the total energy received by the observer from each shell is the same. Thus each shell of a given thickness provides the same amount of light regardless of how far away it is from the observer. That is, the light of each shell adds to the total amount received by the observer. Thus the more shells, the more light; and with infinitely many shells, there would be a bright night sky.

While dark clouds could obstruct the light, these clouds would heat up, until they were as hot as the stars, and then radiate the same amount of light. We have dark nights and it contradicts the existence of static, infinitely large universe having infinite number of stars distributed uniformly. It may be concluded that the universe is dynamic and not the static.

2. Age of the universe

The estimates about the ages of Earth, Sun and star clusters are important cosmological observations and they do not depend on the specific cosmological models. From the decay of radioactive isotopes, the age of the Earth is estimated as 4.6 billion years. The age of the Sun is thought to be slightly more than the age of the Earth. The ages of the oldest star clusters in the Milky Way are estimated to be from 10 billion years to the age of the universe.

Example 1: The speed of expansion of particles in a planetary nebula is measured spectroscopically to be 20 km/s. The planetary nebula has a diameter of 0.1 pc. Calculate when it started to form.

Solution: For solving the problem, we assume that the particles are moving with the same speed throughout their journey. It is then simple matter to calculate how long ago they set off. As the diameter of nebula is 0.1 pc then particles have traveled 0.05 pc (the distance from the center). Their speed is 20 km/s. The time elapsed is

$$\text{time} = \frac{\text{distance}}{\text{speed}} = \frac{0.05 \text{ pc}}{20 \text{ km/s}}$$

$$= \frac{0.05 \times 9.5 \times 10^{15}}{20 \times 10^3} = 2.4 \times 10^{10} \approx 800 \text{ years}$$

3. Relative helium abundance

The Big Bang theory is treated as a natural way to explain the present abundance of the elements. After about 2 to 3 minutes of the Big Bang, the expanding universe had cooled to below about 109 K so that protons and neutrons could fuse to make stable deuterium nuclei that would not be broken by the energetic photons. Remember that the deuterium was produced in the fusion chain process in the core of stars for the fusion of hydrogen nuclei to make a helium nucleus. But, because of the abundant free neutrons in the early universe, the fusion chain process in the early universe was slightly different as compared to what occurs in the stars.

However, the general process is the same as the following. Two protons combine to from a deuteron.

$${}_1\text{H}^1 + {}_1\text{H}^1 \rightarrow {}_1\text{H}^2$$

The deuteron combines with a proton to form helium isotope.

$${}_1\text{H}^2 + {}_1\text{H}^1 \rightarrow {}_2\text{He}^3$$

Two helium isotopes obtained from two sets of reactions combine to form a helium nucleus.

$${}_2\text{He}^3 + {}_2\text{He}^3 \rightarrow {}_2\text{He}^4 + {}_1\text{H}^1 + {}_1\text{H}^1$$

These fusion chain reactions could not take place until the universe had cooled down enough. The exact value of temperature depends sensitively on the density of protons and neutrons at that time.

Beside the ${}_2\text{He}^4$, extremely small amounts of ${}_3\text{Li}^7$ were also produced during the early universe nucleosynthesis process and no other elements heavier than helium were produced. The elements heavier than helium were produced in the cores of stars.

After about 15 minutes from the Big Bang, the universe had expanded and cooled so much that there was no more fusion process. The composition of the universe was 10% helium and 90% hydrogen in terms of numbers or 25% helium and 75% hydrogen in terms of mass. Stars also produce some of the helium visible today, but not most of it.

In this context, the deuterium nucleus ${}_1\text{H}^2$ is of great importance because its production is sensitive to the density of protons and neutrons, and the temperature in the early universe. The number of deuterium nuclei which later on do not participate in the fusion reaction to make ${}_2\text{He}^3$ nuclei also depends sensitively on the temperature, and density of protons and neutrons. A denser universe would have had more deuterium fused to form helium. A less dense universe would have had more deuterium remaining. The amount of the final $_2He^4$ product is not as sensitive to the ordinary matter density of the early universe, and therefore the amount of remaining deuterium seen today is used as a probe of the early density.

4. Cosmological principle

Before development of a theory about the universe, we have to have some postulates granted about it. These postulates must be such that they are not only internally consistent but their consequences must explain the observed facts. One of the postulates about the universe is the *Cosmological principle*, which can be expressed in a number of forms. The cosmology deals at large scales and the universe is assumed to be homogeneous (no change during linear motion) and isotropic (no change during angular motion) at any instant of time. This assumption is a form of the Cosmological principle. It can be justified on the scales of larger than 100 Mpc. Thus, following the Cosmological Principle, apart from the local irregularities, the universe looks the same from all positions in space.

The *Perfect Cosmological Principle* may be described that for an observer situated at an arbitrary position in space, at an arbitrary time in the history of the universe, the cosmos gives exactly the same results as those obtained by some other observer situated at some other position in space, at the same or other time. This principle has far reaching consequences and it directly leads to the steady state model of the universe. In this scenario, the average properties of the universe become well defined. The average density of the universe is constant.

5. Homogeneous and isotropic universe

Results of observations made so far of the outer space tell that there are no preferred directions or unusually dense regions in the universe. The data are nearly same as for a homogeneous and isotropic universe. Hence, in the universe, there are no preferred locations or directions. An observer situated at a location in the universe would find the distant galaxies red-shifted, showing recession away from the observer, irrespective of the direction of observation. That is, the galaxies situated in all directions of an observer appear to be receding away from the observer.

For having a model of such universe, we assume that the red shift indicates a genuine expansion indeed. Let us attempt to understand a model of the universe in which an observer situated at a point sees all other observers in distant galaxies receding away from him/her.

A simple model in two dimensions may be a rubber sheet of length l (along x-axis) and breadth b (along y-axis) as shown in Figure 1 (a). Let the spots be painted on this sheet in a random manner. The two spots A and B at the positions (0, 0) and (x, y), respectively, are separated by the distance r, expressed as

$$r = \sqrt{x^2 + y^2}$$

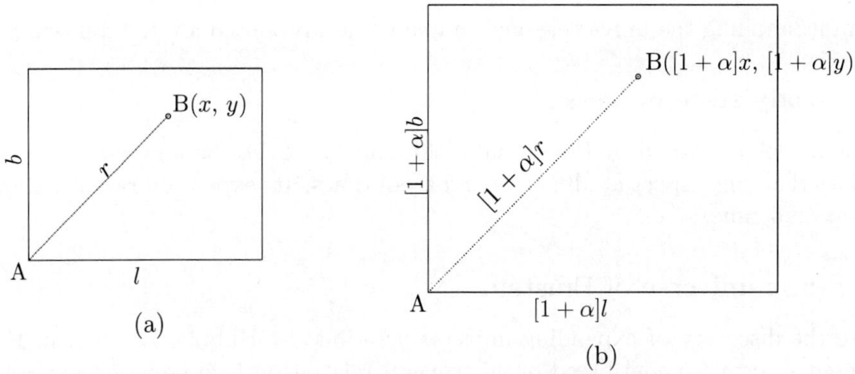

Figure 1: Two-dimensional model of universe.

Now, the sheet is stretched in length and breadth by a fractional amount α, so that the sides of the sheet are $[1+\alpha]l$ and $[1+\alpha]b$, respectively, as shown in Figure 1(b). Moreover, the coordinates of the spots A and B are (0, 0) and ($[1+\alpha]x$, $[1+\alpha]y$), respectively, and the separation between them is $[1+\alpha]r$, such that

$$([1+\alpha]r)^2 = ([1+\alpha]x)^2 + ([1+\alpha]y)^2$$

which is equivalent to

$$r = \sqrt{x^2 + y^2}$$

showing each part is stretched by the same fractional amount.

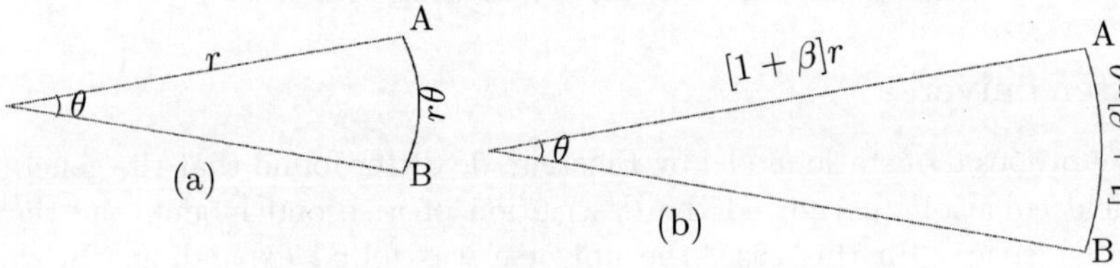

Figure 2: Three-dimensional model of universe.

A flat rubber sheet is not the only two-dimensional model for an expanding homogeneous isotropic universe. It can be a three-dimensional also. Let us have a rubber balloon and paint spots on it in a random manner to represent the galaxies. Suppose, the radius of balloon is r at a given instant, as shown in Figure 2(a). Let the angle subtended by two galaxies A and B on the surface of balloon be θ at the center of the balloon. The distance between the galaxies A and B measured along the surface is the arc length $r\theta$.

Now, the balloon is blown so that the radius of the balloon increases to some new value say $[1+\beta]r$, where β is the fractional increase in the radius of the balloon. For the constant value of angle θ, the distance between the galaxies is now $[1+\beta]r\theta$, as shown in Figure 2(b). The fractional increase is independent of the angle θ. This means that if the universe is homogeneous and isotropic at a given instant, and isotropic expansion keeps it that way.

6. Some simple models of the universe

For understanding the universe, some simple models proposed are the following.

6.1 Steady state universe

In this model, described by Bondi and Gold, and by Hoyle, the universe is flat and provides the same aspect at all places and at all times. Its expansion rate is uniform in space and time.

6.2 Static universe of Einstein

Before the discovery of expanding universe described by Hubble, Einstein in 1917 proposed a cosmic model based on his general relativistic field equation expressed as:

$$R_{\mu\nu} - R\, g_{\mu\nu} + \Lambda g_{\mu\nu} = \frac{8\pi G}{c^4}\, T_{\mu\nu}$$

where $R_{\mu\nu}$ is the Ricci curvature tensor, R the scalar curvature, $g_{\mu\nu}$ the metric tensor, Λ the cosmological constant, G the gravitational constant, c the speed of light in vacuum and $T_{\mu\nu}$ the stress-energy tensor.

This model was static (not expanding). The Einstein universe is spherical and has constant radius of curvature a. It was later found that if Einstein had not taken Λ, he would have found the expanding universe.

In 1930, Lemaitre and Eddington discovered that the Einstein universe is unstable. A small deviation from the perfect conditions taken by Einstein, the universe results in either a continuing expansion or an accelerating collapse.

6.3 de Sitter universe

Soon after the proposal of static model by Einstein, de Sitter found that the general relativistic field equations permitted the description of a second model, one that existed in a flat space. In this case, the universe was found expanding. In the beginning, it had no more than an academic interest. But, in the late 1920s, Hubble's discovery of the cosmic expansion revived interest in this model. The main drawback found in this model is that the density must be zero.

6.4 Eddington universe

Starting of this universe is from the Einstein state. Then an initially uniform distribution of gas becomes perturbed by the processes involving the formation of galaxies and then goes over into a uniform expansion. One difficulty in this model is the uncertainty whether the formation of galaxies should not lead to an instability that would give rise to contraction rather than expansion. This model is found interesting as it focused attention on the fact that the cosmology is not just a matter of geometry. A model in principle must also be able to account for the physical state of matter available in the universe. Galaxies are considered to have condensed out of an initially uniform gas. In case this gas was in rapid expansion, how it could be possible to counteract the expansion in order to force the gas to contract into galaxies? The answer is not known. But, Eddington and Lemaitre tried to make a plausible guess.

6.5 Lemaitre universe

Following Lemaitre, the universe starts out in a highly contracted state which initially expands at a rapid rate. Later on, the expansion is slowed down and brought to a halt in a state. Galaxies are formed at this stage and give rise to a new expanding phase that continues indefinitely.

A noticeable feature of Lemaitre's model is the high initial density of the universe, very similar to the Big Bang. We can compute the temperature and pressure that must have existed at the initial time, and thus establish the nuclear reactions that should have occurred. We obtain information about the chemical composition of matter at early stages of the universe before the formation of galaxies.

This gives the importance of the chemical characteristics of the universe. A cosmological model must be able to simulate not only the overall density and pressure of the universe, or the existence of galaxies, it must also be able to predict in some detail the chemical composition of matter for the formation of earliest stars. The chemical composition of evolved stars need not be predicted in this way, because nuclear reactions take place in the interior of stars so that early generation of stars can produce heavier chemical elements that later are distributed in the interstellar space through ejection or explosion. Subsequent generations of the stars can incorporate this newly formed matter and their atmospheres contain these chemical elements.

6.6 Friedmann universe

From the Einstein field equations, Friedmann derived two equations for homogeneous and isotropic universe as the following.

$$\frac{\dot{a}\ +kc^2}{a^2} = \frac{8\pi G\rho + \Lambda c^2}{3} \qquad \text{and} \qquad \frac{\ddot{a}}{a} = -\frac{4\pi G}{3}\left(\rho + \frac{3p}{c^2}\right) + \frac{\Lambda c^2}{3}$$

where $a(t)$, a function of time t, is the scale factor, $\dot{a}$ and $\ddot{a}$ are the first and second derivatives, respectively, of a with respect to the time t. The Hubble parameter (constant) H is expressed as $H = \dot{a}\ /a$. The G, Λ and c are, respectively, the universal gravitational constants, the cosmological constant, and the speed of light in vacuum. For a particular case, the intrinsic curvature k remains constant, but may vary from one case to another. Here, the a, H, ρ and p are functions of time t. The ratio k/a^2 is the spatial curvature in any time slice of the universe. The solution of equations is independent of coordinate system chosen for the spatial slices. There are two commonly used choices for a and k which describe the universe.

- The spatial curvature k is considered to have three integer values: $+1$, 0, -1.

  For $k = +1$, the universe is a closed three-dimensional sphere. Here, a denotes the radius of curvature of the universe (sphere).

For $k = 0$, the universe is flat (Euclidean space). Here, a may be given any arbitrary positive number at one particular time.

For $k = -1$, the universe is an open three-dimensional hyperboloid. Here, ia denotes the radius of curvature of the universe.

- The parameter a is the scale factor which is taken at the present time. If R_t is the radius of curvature of hypothetical universe at a time t (R_0 in the present day radius of curvature), the $a = R_t/R_0$.

These equations may be simplified by replacing

$$\rho \to \rho - \frac{\Lambda c^2}{8\pi G} \qquad \text{and} \qquad p \to p + \frac{\Lambda c^4}{8\pi G}$$

to give

$$H^2 = \left(\frac{\dot{a}}{a}\right)^2 = \frac{8\pi G}{3}\rho - \frac{kc^2}{a^2} \qquad \text{and} \qquad \dot{H} + H^2 = \frac{\ddot{a}}{a} = -\frac{4\pi G}{3}\left(\rho + \frac{3p}{c^2}\right)$$

The Hubble parameter can change over time when other parts of the equation are time dependent (in particular the mass density, vacuum energy, or the spatial curvature). Evaluation of Hubble parameter at the present time give the Hubble's constant, which is the proportionality constant of Hubble's law. Applied to a fluid with a given equation of state, the Friedmann equations yield the time evolution and geometry of the universe as a function of the fluid density.

Some cosmologists call the second of these two equations as the Friedmann acceleration equation and the first equation as the Friedmann equation.

6.7 Density parameter

In the Friedmann universe, the density parameter Ω is expressed as the ratio of actual (or observed) density ρ to the critical density ρ_c. This parameter determines the overall geometry of the universe. When $\Omega = 1$, the universe is flat (Euclidean).

To date, the critical density is estimated to be approximately five atoms (of mono-atomic hydrogen) per cubic meter, whereas the average density of ordinary matter in the universe is believed to be 0.2 atoms/m^3. A large contribution to density comes from the unidentified dark matter. Both the ordinary matter and dark matter lead to the contraction of the universe. The largest contribution to the density comes from the so-called the dark energy, which accounts for the cosmological constant term. Although the total density is equal to the critical density (exactly up to the measurement error), the dark energy does not lead to the contraction of the universe, but rather accelerates its expansion. Therefore, the universe expands forever.

An expression for the critical density is found by assuming Λ to be zero and setting the normalized spatial curvature k equal to zero. Substitution of these values in the Friedmann's first equation gives

$$\rho_c = \frac{3H^2}{8\pi G}$$

The density parameter is thus defined as

$$\Omega = \frac{8\pi G\rho}{3H^2}$$

Assuming a zero vacuum energy density, when the Ω is larger than unity, the space sections of the universe are closed; the universe eventually stops expanding, then collapses. When the Ω is less than unity, they are open, and the universe expands forever.

6.8 Mathematical forms of models

Interesting to note that all the models have similar mathematical forms: Lemaitre, Robertson and Walker, independently, found that in an isotropic homogeneous space the field equations of Einstein reduce to two simple differential equations in the radius of curvature or scale factor a. The first of these gives the mass density, including all radiation energy, in terms of the mass density of the vacuum, the expansion or contraction kinetic energy, and a curvature energy:

$$\kappa\rho = -\Lambda + 3\Big(\frac{\dot{a}\ +kc^2}{c^2a^2}\Big)$$

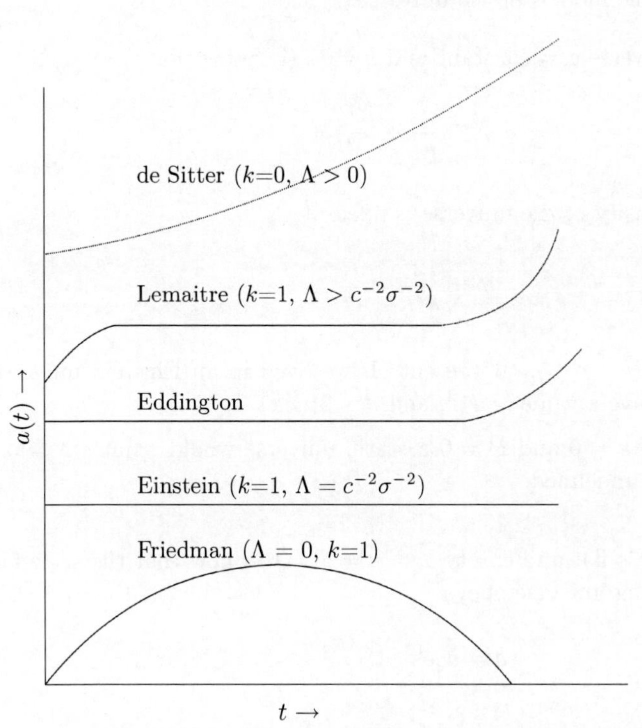

Figure 3: Some cosmological models. The scales of $a(t)$ and of t are not the same for different curves. These are qualitative natures of variation of $a(t)$ versus t in different models.

The second gives the change in this density on expansion or contraction, including work done against the pressure of the fluid. This is the energy conservation equation.

$$\frac{d\rho}{dt} + 3\left(\rho + \frac{P}{c^2}\right) = 0 \qquad \text{or} \qquad \frac{\kappa P}{c^2} = \Lambda - \left(\frac{2a\,\ddot{a} + \dot{a}^2 + kc^2}{c^2 a^2}\right)$$

Here, κ is $8\pi G/c^2 = 1.86 \times 10^{-27}$ cm/g. κ is sometimes called the Einstein gravitational constant. Λ is the cosmological constant, and ρ and P are the density and pressure of the universe. Dots represent differentiation with respect to world time. A qualitative nature of the variation of $a(t)$ versus time t for various models is shown in Figure 3.

7. Review Questions

1. State and describe the Olbers paradox.

2. Describe the cosmological principle.

3. Describe the homogeneous and isotropic universe.

4. Discuss some simple models of the universe.

5. For Einstein's universe $a =$ constant and $k = 1$. (i) Show that

$$\Lambda = \frac{1}{a^2} + \frac{\kappa P}{c^2}$$

   and that the density of the universe is fixed at

$$\rho = \frac{2}{a^2 \kappa} - \frac{P}{c^2}$$

   We know that $P/c^2 << \rho$, at present. If we lived in an Einstein universe, Λ would have to have a value $\sim a^{-2}$, and $\rho \sim 2(a^2\kappa)^{-1}$.

   (ii) Show that if $k = 0$ and $P = 0$ a static universe would require $\Lambda = 0$ and $\rho = 0$, leaving a undefined.

6. De Sitter universe is flat and empty, $k = \rho = P = 0$. Show that the scale factor a of the expending universe obeys

$$a = a_0 e^{(\Lambda c^2/3)^{1/3} t}$$

   and the age of the universe judged from the Hubble constant H is

$$\frac{1}{H} = \sqrt{\frac{3}{\Lambda c^2}}$$

7. From the field equations with cosmological constant $\Lambda = 0$ and with zero pressure, obtain the relations

$$\frac{\ddot{a}}{a} = -\frac{4\pi G\rho}{3} \qquad \text{and} \qquad H^2 q_0 = \frac{4\pi G\rho}{3}$$

8. Prove the instability of the Einstein universe for $P = 0$.

9. Prove for Friedman universe ($\Lambda = 0$) that

   Note that an infinitesimal expansion makes $\rho < 2(\kappa c^2 a^2)^{-1} - P/c^2$, so that even with $d = 0$, we have $\ddot{a} > 0$ and the expansion must continue. The proof for an initial contraction is similar.

   (a) with $k = 1$ and an initially dense universe for which $P/c^2 = \rho/3$, the solution has the parametric form

   $$a = b_0 c \sin x, \qquad t = b_0(1 - \cos x)$$

   where x is a parameter.

   (b) For late stages of the universe, with $k = 1$, $P = 0$

   $$a = a_0 c(1 - \cos x), \qquad t = a_o(x - \sin x)$$

   Note that x grows monotonically with t, so that $(1 - \cos x)$ eventually must approach zero. The universe first grows, but later on collapses.

   (c) For the dense stage of a hyperbolic universe ($k = -1$)

   $$a = b_0 c \sinh x, \qquad t = b_o(\cosh x - 1)$$

   (d) For late stage of a hyperbolic universe

   $$a = a_0 c(\cosh x - 1), \qquad t = a_o(\sinh x - x)$$

10. Write notes on the following:

    (i) Olbers paradox
    (ii) Age of the universe
    (iii) Relative helium abundance in the universe
    (iv) Cosmological principle
    (v) Homogeneous and isotropic universe
    (vi) Steady state universe
    (vii) Static universe of Einstein
    (viii) de Sitter model for the universe
    (ix) Eddington model for the universe
    (x) Lemaitre model for the universe
    (xi) Friedmann models for the universe

8. Multiple Choice Questions

1. Which aspect is not dealt with under Cosmology.

   A. origin of universe
   B. evolution of universe
   C. evolution of Sun
   D. ultimate fate of universe

   Ans. C

2. The age of the Earth is about

   A. 4.6×10^3 years
   B. 4.6×10^6 years
   C. 4.6×10^9 years
   D. 4.6×10^{12} years

   Ans. C

13. Life in the Universe

Our Earth is the only place in the universe where the existence of life is known so far. Scientists have always been interested in the search of life in the universe, other than that we know on the surface of Earth. This motto of scientists may be rephrased as: "Are we alone in the universe?". An Earth-based mission to explore, understand and explain the origin, nature and prevalence of life outside the Earth in the universe is the SETI (Search for ExtraTerrestrial Intelligence). In the SETI mission, researchers study various aspects related to the existence of life, its origin, the environment in which a life develops, and its ultimate fate. Because of the interest, this topic has been widely studied by the scientists worldwide.

The very first question in this context may arise about the existence of life on the surface of Earth. In our solar system, the Earth is situated at such an advantageous position that it has an atmosphere having the constituents, which are necessary for the formation of life and for its protection from the intense and hazardous radiations coming from the outer space. Keeping in view our Earth, the scientists have opinion that some other stars may have planetary system and some of the planets may be at an advantageous position, similar to our Earth. Consequently, the search of the exoplanets (planets in other solar systems) is going on extensively worldwide. Some scientists have the opinion that the life may be found in some other parts in our own solar system. Thus, the mission for the SETI may be classified into two parts:

(i) Search of life in our own solar system

(ii) Search of life on the exoplanets

1. Search of life in our solar system

Our solar system is the only place where life has been found so far in the universe. Some scientists believe that, besides on the surface of Earth, the life may be found in some parts of our solar system. Keeping in view the physical conditions prevailing there, the places where life is being searched are: (i) Mars, and (ii) Titan.

1.1 Missions for Mars

The Viking mission was launched by NASA in 1975 to scan the Martian surface for its structure and also the atmospheric composition in order to search some evidence for the existence of life there. This mission was composed of two spacecraft, namely, Viking 1 and Viking 2. The Viking 1 was launched on August 20, 1975 and arrived at Mars on June 19, 1976, whereas the Viking 2 was launched on September 9, 1975 and entered the Mars orbit on August 7, 1976. The mission revealed many features of Mars which have been studied in detail by the Spirit (Mars exploration Rover-A) and Opportunity (Mars exploration Rover-B) launched in 2003. Later on, other missions were launched, but so far no evidence for the existence of life on the Mars has been witnessed.

1.2 Mission for Titan

Titan (the largest moon of the planet Saturn) being the only natural satellite in our solar system with atmosphere and adequate density, has been studied thoroughly by the NASA with the help of Cassini-Huygens mission since 2004. The duration of the mission has been from 15 October 1997 to 15 September 2017. Being rich in organic compounds, Titan has been studied for the traces of life. Miller experiment in 2010 confirmed the possibility of nucleotides and amino acids with similar conditions as on the Earth in a laboratory. In April 2013, NASA reported that complex organic molecule could arise on the Titan, on the basis of the simulation of the atmosphere of Titan. In June 2013, polycyclic aromatic hydrocarbons (PAHs) were detected in the upper atmosphere of Titan.

Similar study has been done for the Europa (the satellite of Jupiter) where a vast reservoir of water is predicted by the scientists.

1.3 Voyager mission

The Voyager mission was launched in 1979. The Voyager 1 and Voyager 2 spacecraft crossed our solar system in July 2013. They were launched with the sole motto of searching of extraterrestrial life in the deep space.

2. Search of life in exoplanets

In 1584, Catholic Monk Giordano Bruno asserted that there are countless suns and countless earths, and all these earths are revolving around their respective suns. When he said this in public, he was removed from the Church by charging that he was against the religion. But, this fact is now found true, as we have information about a large number of exoplanets revolving around their respective suns.

The extrasolar planets (also known as the exoplanets) are the planets which are orbiting around the respective stars outside our solar system. A total of 3758 such planets (in 2808 planetary systems, including 627 multiple planetary systems) have been identified as of April 1, 2018. This number is increasing day by day and the latest information can be obtained on the websites. The discovery of first two exoplanets was announced on 9 January 1992 as orbiting the pulsar PSR 1257+12. Direct detection of exoplanets is very difficult because they are so much dimmer than the host star they are orbiting. For example, the Sun is billion (10^9) times brighter than the Jupiter and 10 billion (10^{10}) times brighter than the Earth. The nearest known exoplanet is at a distance of 12 ly from the Earth, in the habitable zone of Proxima Centauri.

There has been Kepler mission with the aim to discover Earth-size extrasolar planets orbiting around other stars in or near the **habitable zone** and to estimate how many of the billions of stars in our galaxy have such planets. The Kepler spacecraft was launched on March 7, 2009. On February, 2011, The Kepler space observatory mission team released a list of 1235 extrasolar planets, including 54 that may be in the *habitable zone.* Some of these candidates were Earth - sized and super Earth - sized (defined as less than or equal to twice of Earth radii. Six of these candidates were in the **habitable zone**.

For the detection of exoplanets, the following methods are generally used.

2.1 Radial velocity method

This method uses the fact that if a star has a planet (or planets) revolving around it, it is not strictly correct to say that the planet orbits around the star. Instead, the planet and the star orbit together about their common center-of-mass. When the star is so much more massive than the planet, the center-of-mass is within the star and the star appears to wobble slightly on the traveling of the planet around it. Astronomers can measure this wobble with the help of the spectroscopic methods. If a star is traveling towards us, its light appears as blue shifted, and if it is traveling away from us, the light appears as red shifted. Spectroscopy can show this change in colour from a star as it moves towards and away from us while orbiting around the center-of-mass of the star-planet system. Many of the planets have been discovered by this method and they are very large and very close to their stars. These planets, called the super-Jupiters, are much larger than the Jupiter, and have their period of revolution of a few days. Such a large planet results a large wobble which is one of the features along with the short orbit time, to detect them rather than a smaller planet or planets which is farther from its star.

2.2 Transit method

This method only works for the star-planet systems which have orbits aligned in such a way that, as seen from Earth, the planet travels between us and the star, and consequently blocks some of the light from the star. If a planet crosses (or transits) in front of its parent star disk, then the observed brightness of the star drops by a small amount. The amount by which the star dims depends on the size of the planet. This has been the second most productive method of detection of exoplanets.

The transit method tells us about the radius of a planet, and it has the benefit that it sometimes allows a planet's atmosphere to be investigated through the spectroscopy. A planet does not block much light from a star, but this is detectable. Like the radial velocity method, this method has a bias towards the discovery of a large number of planets orbiting close to their stars, because the larger planets block more light and transit more frequently so that they are easier to detect. This method does not work for all the systems because only 10 percent of the super-Jupiters are aligned in such a way, so that we see their transit.

Smaller planets in larger orbits are even less likely to be aligned in such a way that we can observe transits.

2.2.1 Transit timing variation (TTV)

When more than one planet (multiple planets) are revolving around a common star, each one of them slightly perturbs the orbits of other planets. Small variations in the times of transit for one planet can thus indicate the presence of other planet, which itself may or may not transit.

2.2.2 Transit duration variation (TDV)

When a planet orbits multiple stars or the planet has moons, its transit time can significantly vary from transit to transit. Although no new planets or moons are discovered with this method, it has been used for successful confirmation of many transiting circumbinary planets.

2.3 Gravitational microlensing

Microlensig occurs when the gravitational field of a star acts like a lens, magnifying the light of a distant background star. Planets orbiting the lensing star can cause detectable anomalies in the magnification as it varies over time. Gravitational microlensing relies on chance events where one star passes in front of the another. When the farther star is a bright one and the nearer one is a normal one, we could not ordinarily see from the Earth. When the nearer star passes in front of the farther star, its gravity causes the light from the farther star to bend and the star is magnified, as seen from the Earth. If, during the event, the background star appears to be magnified even more for a short time that means a planet orbiting the smaller star is increasing the effect of magnification. This method has resulted in the detection of some exoplanets but it has the advantage of being sensitive to planets at large separations from their parent stars specifically.

2.4 Astrometry

This method gives accurate measurements of the position of a star in the sky and the changes in its position over a period of time. The movement of a star due to the gravitational effect of a planet may be noticeable. But, as the motion is so small, this method is not yet so effective. It has predicted only a few disputed detections. However, it has been successfully used for the determination of the properties of planets found with the help of other methods. Astrometry is found to be more sensitive to the planets having large orbits. This is an advantage over other methods. No extrasolar planet discoveries by this method have been confirmed by other methods, but the field is quite promising.

2.5 Direct imaging

Direct imaging method uses infrared wavelengths for observing the planets. This method works because at the infrared wavelengths, a Sun-like star is only 100 times brighter than the Jupiter. This method works for the planets which are very far from their stars, and therefore, a period of hundreds or thousands of years may be required for a planet to be discovered with the help of this method. Astronomers have to ascertain that the planet and the star move simultaneously across the space to show that the planet revolving the star since to observe the complete orbit takes much long. This method does not permit astronomers to compute the mass of a planet instantly, but they can use the spectrum and luminosity to obtain data regarding its surface temperature and diameter.

2.6 Pulsar timing

A pulsar emits the radio waves very frequently as it rotates. A planet orbiting a pulsar causes a bit abnormality in the timing of its observed radio pulses.

3. Drake equation

Formation of stars is a regular process in the galaxies. Some stars may have planets orbiting around them and out of them some planets may have possibility of the existence of life there. Drake in 1961 suggested an equation, known as the Drake equation, which is a probabilistic argument used to estimate the number of active, communicative extraterrestrial civilizations in the Milky Way galaxy. The number of such civilizations N may be expressed in the from of the product of various parameters:

$$N = R_* \cdot f_p \cdot n_{e \cdot f_l \cdot f_i \cdot f_c} \cdot L$$

where

R_* denotes the average rate of star-formation in our galaxy.

f_p is the fraction of formed stars that have planets.

n_e is the average number of planets (for stars that have planets) that can potentially support the life.

f_l is the fraction of those planets that actually develop the life.

f_i is the fraction of planets bearing life on which intelligent, civilized life, has developed.

f_c is the fraction of these civilizations that have developed communications.

L is the length of time over which such civilizations release detectable signals.

The equation was written in 1961 by Frank Drake, not for the purposes of quantifying the number of civilizations, but as a way to stimulate scientific dialogue at the first scientific meeting on the SETI.

Criticism made on the Drake equation focused not on the equation itself, but on the fact that the estimated values for several of its factors are highly conjectural. The combined effect being that the uncertainty associated with any derived value is so large that the equation cannot be used to draw any firm conclusions.

There is considerable disagreement on the values of above parameters used in the Drake equation. However, the *educated guesses* used by Drake and his colleagues in 1961 were as the following.

$R_* = 1$ yr^{-1} (1 star formed per year, on the average over the life of the galaxy)

$f_p = 0.2$ to 0.5 (one fifth to one half of all stars formed have planets)

$n_e = 1$ to 5 (stars with planets have between 1 and 5 planets capable of developing life)

$f_l = 1$ (100% of these planets develop life)

$f_i = 1$ (100% of which develop intelligent life)

$f_c = 0.1$ to 0.2 (10 - 20% of which is able to communicate)

$L = 10^3$ to 10^8 years

On inserting the minimum numbers in the Drake equation, we get

$$N = 1\ (0.2)\ 1\ (1)\ 1\ (0.1)\ 10^3 = 20$$

On inserting the maximum numbers in the Drake equation, we get

$$N = 1\ (0.5)\ 1\ (1)\ 1\ (0.2)\ 10^8 = 5 \times 10^7$$

Drake stated that, given the uncertainties, the original meeting concluded that $N = L$, and there were probably between 10^3 and 10^8 civilizations in the Milky Way galaxy.

4. Earth similarity index

Similarity index is a mathematical tool that can be applied for a set of data. This method is a measure of deviation from a reference system, usually on a scale lying between zero and one. The Earth similarity index (ESI) is a quantitative measure of Earth-likeness. One way to understand the ESI may be as the following. Consider a physical quantity having value x_0 on the surface of Earth. Suppose, the value of this quantity varies from x_a to x_b, such that $x_a < x_0 < x_b$, on other Earth. Then, the percentage variations p and q are expressed as

$$p = \frac{x_0 - x_a}{x_0} \times 100 \qquad \text{and} \qquad q = \frac{x_b - x_0}{x_0} \times 100$$

Now, we define a threshold value T as

$$T = \left[1 - \left(\frac{x_0 - x_a}{x_0 + x_a}\right)^m\right]^{w_a} \tag{13.1}$$

and

$$T = \left[1 - \left(\frac{x_b - x_0}{x_b + x_0}\right)^m\right]^{w_b} \tag{13.2}$$

where w_a and w_b denote the weight exponents and m is a free parameter. The threshold is the limiting value of ESI above which a planet is considered similar to the Earth. Further, the threshold value T can be expressed as

$$T = \left(1 - \frac{t}{100}\right)$$

where t can assume positive values between zero and 100, so that T is positive, having the value between zero and 1. Thus, we have taken T also as another free parameter. The parameter m may be considered to assume the values from 0.6 to 1.2.

From equation (13.1) and (13.2), we get

$$w_a = \frac{\ln T}{\ln[1 - (\frac{p}{200-p})^m]} \qquad \text{and} \qquad w_b = \frac{\ln T}{\ln[1 - (\frac{q}{200+q})^m]}$$

The weight exponent w_x may be taken as the geometrical mean of w_a and w_b. That is

$$w_x = \sqrt{w_a \times w_b}$$

Here, the subscript x denotes a physical quantity. The basic ESI_x for a physical quantity x is expressed as

$$ESI_x = \left(1 - \left|\frac{x - x_0}{x + x_0}\right|^m\right)^{w_x}$$

For the planetary property x, the terrestrial reference is x_0. The ESI_x can assume a value between zero (no similarity with the Earth) and one (identical to the Earth). One can calculate ESI_r, ESI_ρ, ESI_e and ESI_T corresponding to radius, density, escape velocity and temperature, respectively.

The ESI_r and ESI_ρ, for the mean radius r and the bulk density ρ, respectively, are used to define the interior Earth similarity index ESI_I, expressed as

$$ESI_I = (ESI_r \times ESI_\rho)^{1/2} \tag{13.3}$$

Thus, the ESI_I is the geometrical mean of ESI_r and ESI_ρ. The ESI_e and ESI_T for the escape velocity v and the mean surface temperature T, respectively, are used to define the surface Earth similarity index ESI_S, expressed as

$$ESI_S = (ESI_e \times ESI_T)^{1/2} \tag{13.4}$$

Thus, the ESI_S is the geometrical mean of ESI_e and ESI_T. The ESI (sometimes called the global ESI) is expressed as

$$ESI = (ESI_I \times ESI_S)^{1/2} \tag{13.5}$$

Using equations (13.3) and (13.4) in (13.5), we have

$$ESI = (ESI_r \times ESI_\rho \times ESI_e \times ESI_T)^{1/4}$$

On considering the Earth as the most superior planet, the value of T can never be greater than 1. The variations of various parameters, relative to those of the mean values on the Earth, may be taken as the following. The definitional limits for the radius are from 0.5 to 1.5 times the Earth's radius. The range for the mass of an existing planet is from 0.1 to 10 times that of the Earth. For the density, the definitional range is from 0.7 to 1.5 times the Earth's density. The temperature variation is taken from 273 K to 232 K. The range for the escape velocity is considered from 0.4 to 1.4 times that of the Earth. For these limits, one can calculate the Earth similarity indexes corresponding to the radius, density, escape velocity and temperature, and finally the Earth similarity index.

Table 1. Values of various ESI for some solar planets and satellites and exoplanets with $m = 0.8$ and $T = 0.8$.

S. No.	Body	Radius (EU)	Density (EU)	Esc. Vel. (EU)	Temp K	ESI_I	ESI_S	ESI
1	Earth	1.000	1.000	1.000	288	1.000	1.000	1.00
2	Mars	0.532	0.713	0.449	227	0.864	0.739	0.80
3	Mercury	0.382	0.984	0.379	440	0.877	0.614	0.73
4	Moon	0.272	0.606	0.212	220	0.751	0.624	0.68
5	Venus	0.949	0.950	0.925	730	0.985	0.44	0.66
6	Io	0.285	0.639	0.228	130	0.766	0.394	0.55
7	Callisto	0.378	0.352	0.218	134	0.692	0.402	0.53
8	Jupiter	10.973	0.240	5.379	152	0.480	0.439	0.46
9	Ganymede	0.412	0.352	0.244	110	0.701	0.338	0.49
10	Ceres	0.074	0.376	0.045	167	0.523	0.345	0.42
11	Europa	0.244	0.546	0.180	102	0.719	0.293	0.46
12	Saturn	9.140	0.124	3.225	134	0.405	0.432	0.42
13	Titan	0.404	0.341	0.235	94	0.693	0.284	0.44
14	Uranus	3.980	0.230	1.910	76	0.573	0.260	0.39
15	Neptune	3.864	0.297	2.105	72	0.620	0.241	0.39
16	Titania	0.123	0.301	0.069	60	0.543	0.130	0.27
17	Enceladus	0.039	0.291	0.021	75	0.427	0.126	0.23
18	Pluto	0.180	0.371	0.109	40	0.618	0.088	0.23
19	Triton	0.212	0.379	0.130	38	0.640	0.086	0.23
20	HD69830 d	4.190	0.250	2.100	312	0.582	0.844	0.70
21	55 Cnc c	5.680	0.250	2.840	310	0.551	0.803	0.67
22	55 Cnc f	4.910	0.390	3.060	310	0.641	0.792	0.71
23	61 Vir d	3.680	0.460	2.500	375	0.703	0.709	0.71
24	HIP 57050 b	6.640	0.320	3.780	250	0.574	0.722	0.64
25	Mu Ara d	8.380	0.280	4.450	327	0.529	0.706	0.61
26	HD 142 b	11.120	0.240	5.430	286	0.479	0.740	0.60
27	HD 96167 b	9.360	0.260	4.810	334	0.507	0.683	0.59
28	HD 108874 b	12.490	0.220	5.890	294	0.456	0.720	0.57
29	HD 210277 b	11.980	0.230	5.720	275	0.466	0.711	0.58
30	HD 147513 b	10.990	0.240	5.380	263	0.480	0.697	0.58
31	HD 69830 c	2.820	0.540	2.070	549	0.764	0.525	0.63

EU stands for the Earth Unit.

For the given values of parameters for the planets in our solar system, one can calculate ESI, where m and T are considered the free parameters. Table 1 has the values of physical parameters, interior-ESI, surface-ESI and ESI for 31 objects. The abbreviation EU stands for the Earth Unit. The ESI for some objects is within 10% limit from the threshold value, supporting the opinion for the existence of life in the universe.

5. Prebiotic molecules

The amino acids are constituents of life. A number of molecules pertaining to life (responsible for production of amino acids) have been identified in the interstellar medium. Some of the prebiotic molecules are:

1. Formic acid ($HCOOH$)
2. Glycolaldehyde
3. Glycine (NH_2CH_2COOH) - the simplest amino acid.
4. Ethylene oxide (c-C_2H_4O)
5. 2-Aminopyridine ($H_2NC_5H_4N$)
6. 2-Aminoethanol ($NH_2CH_2CH_2OH$)
7. Alanine ($NH_2CH_2CH_2COOH$) - amino acid.
8. Amino acetonitrile (NH_2CH_2CN)
9. Methanimine (CH_2NH)

Some reactions pertaining to the molecules in the interstellar medium are as the following.

The methanimine (CH_2NH) is a precursor of aminoacetonitrile (NH_2CH_2CN)

$$CH_2NH + HCN \rightarrow NH_2CH_2CN$$

The aminoacetonitrile can be hydrolysed to give the glycine (NH_2CH_2COOH).

$$NH_2CH_2CN + H_2O \rightarrow NH_2CH_2OH + HCN$$

$$NH_2CH_2OH + HCN + H_2O \rightarrow NH_2CH_2COOH + NH_3$$

The formic acid has great importance, as it is a precursor for the amino acids (Glycine and Alanine, for example) :

$$\left.\begin{array}{l}\text{Aminomethanol } (NH_2CH_2OH) \\ \text{Aminoethanol } (NH_2CH_2CH_2OH)\end{array}\right\} + \begin{array}{l}\text{Formic Acid} \\ (HCOOH)\end{array} \longrightarrow \left\{\begin{array}{l}\text{Glycine} \\ \text{Alanine}\end{array}\right.$$

6. Review Questions

1. Describe the places where the life, other than the Earth, is being searched.
2. Describe the Earth similarity index.

3. Write notes on the following:
   (i) Gravitational microlensing
   (ii) Exoplanets
   (iii) Drake equation
   (iv) Earth similarity index
   (v) Prebiotic molecules

7. Multiple Choice Questions

1. The place in our solar system where the life is being searched is

   A. Moon B. Jupiter C. Mars D. Saturn

   Ans. C

2. The place in our solar system where the life is being searched is

   A. Moon B. Mars C. Jupiter D. Saturn

   Ans. B

APPENDICES

A. Radial and transverse components for a motion along a curved path

Let us consider a particle moving along a curved path (Figure 1). Suppose at time t, the particle is at a position P (with polar coordinates r, θ). Let it reach the position Q (with polar coordinates $r + \mathrm{d}r$, $\theta + \mathrm{d}\theta$) after a time interval dt. The velocity $\vec{v}$ of the particle can be resolved into two components: (i) radial component, v_r and (ii) transverse component, v_θ as

$$\vec{v} = v_r \hat{r} + v_\theta \hat{\theta}$$

where the radial and transverse components, respectively, of velocity are

$$v_r = \frac{\mathrm{d}r}{\mathrm{d}t} \qquad \text{and} \qquad v_\theta = r\,\frac{\mathrm{d}\theta}{\mathrm{d}t} \tag{A1}$$

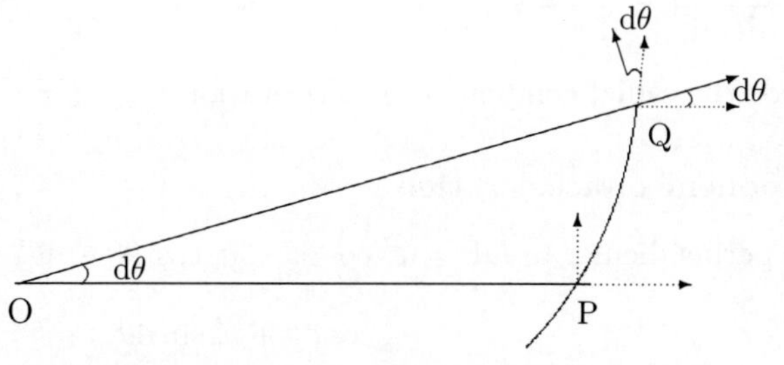

Figure 1: A particle moving along a curved path is at the position P at a time t. After a time-interval dt, the particle reaches at the position Q.

While the particle is moving along a curved path, its direction of motion is changing continuously. Hence, the particle undergoes an acceleration, which can be resolved into two components: (i) radial component and (ii) transverse component.

(i) Radial component of acceleration

Let the particle move from a position P to another position Q in the time interval dt and the line joining the particle and the focus of the ellipse describes an angle dθ. Let at P, the radial and transverse components of velocity are v_r and v_θ, and the radial and transverse components of velocity at Q are v'_r and v'_θ, respectively.

At the Q, the velocity parallel to OP $= v'_r \cos \mathrm{d}\theta + v'_\theta \cos(90 + \mathrm{d}\theta)$

$$= v'_r \cos \mathrm{d}\theta - v'_\theta \sin \mathrm{d}\theta$$

The change in velocity parallel to OP $= v'_r \cos \mathrm{d}\theta - v'_\theta \sin \mathrm{d}\theta - v_r$

$$\text{Acceleration parallel to OP, } \alpha_r = \frac{v'_r \cos \mathrm{d}\theta - v'_\theta \sin \mathrm{d}\theta - v_r}{\mathrm{d}t} \tag{A2}$$

Let dv_r and dv_θ, respectively, denote the change in radial and transverse components of velocity during the motion from P to Q. Hence, we have

$$v'_r = v_r + dv_r \qquad \text{and} \qquad v'_\theta = v_\theta + dv_\theta \tag{A3}$$

Using equation (A3) in (A2), we get

$$\alpha_r = \frac{(v_r + dv_r)\cos d\theta - (v_\theta + dv_\theta)\sin d\theta - v_r}{dt}$$

Since $d\theta$ is small, we have $\cos d\theta = 1$ and $\sin d\theta = d\theta$ and thus we get

$$\alpha_r = \frac{(v_r + dv_r) - (v_\theta + dv_\theta)\, d\theta - v_r}{dt} = \frac{dv_r}{dt} - v_\theta\,\frac{d\theta}{dt} - \frac{dv_\theta}{dt}\,d\theta \tag{A4}$$

The last term on the right side is very small and can be neglected. Using equation (A1) in (A4), we get

$$\alpha_r = \frac{d}{dt}\left(\frac{dr}{dt}\right) - r\,\frac{d\theta}{dt}\,\frac{d\theta}{dt} = \frac{d^2r}{dt^2} - r\left(\frac{d\theta}{dt}\right)^2 \tag{A5}$$

This is an expression for the radial component of acceleration.

(ii) Transverse component of acceleration

At the Q, the velocity perpendicular to OP $= v'_\theta \cos d\theta + v'_r \cos(90 - d\theta)$

$$= v'_\theta \cos d\theta + v'_r \sin d\theta$$

The change in velocity perpendicular to OP $= v'_\theta \cos d\theta + v'_r \sin d\theta - v_\theta$

$$\text{Acceleration perpendicular to OP, } \alpha_\theta = \frac{v'_\theta \cos d\theta + v'_r \sin d\theta - v_\theta}{dt} \tag{A6}$$

Using equation (A3) in (A6), we get

$$\alpha_\theta = \frac{(v_\theta + dv_\theta)\cos d\theta + (v_r + dv_r)\sin d\theta - v_\theta}{dt}$$

Since $d\theta$ is small, we have $\cos d\theta = 1$ and $\sin d\theta = d\theta$ and thus we get

$$\alpha_\theta = \frac{(v_\theta + dv_\theta) + (v_r + dv_r)\, d\theta - v_\theta}{dt} = \frac{dv_\theta}{dt} + v_r\,\frac{d\theta}{dt} + \frac{dv_r}{dt}\,d\theta \tag{A7}$$

The last term on the right side is very small and can be neglected. Using equation (A1) in (A7), we get

$$\alpha_\theta = \frac{d}{dt}\left(r\,\frac{d\theta}{dt}\right) + \frac{dr}{dt}\,\frac{d\theta}{dt} = r\,\frac{d^2\theta}{dt^2} + 2\,\frac{dr}{dt}\,\frac{d\theta}{dt} = \frac{1}{r}\,\frac{d}{dt}\left(r^2\,\frac{d\theta}{dt}\right)$$

This is expression for the transverse component of acceleration.

B. Ellipse

An ellipse (Figure 2) is a locus of the points satisfying the equation

$$r + r' = 2a \tag{B1}$$

where a is a constant, called the semi-major axis of the ellipse, and r' and r are the distances of a point P on the ellipse from the focal points O′ and O, respectively. When O′ and O coincide, we have $r' = r$ and equation (B1) gives $r = a$, which is the equation of a circle. It shows that a circle is a special case of an ellipse. For $r = r'$, we have the point Q and the distance of Q from the center C is b, known as the semi-minor axis of the ellipse. The distance of either focal point from the center of the ellipse is ea, where e is known as the eccentricity of the ellipse ($0 \leq e \leq 1$). Obviously, for a circle, we have $e = 0$.

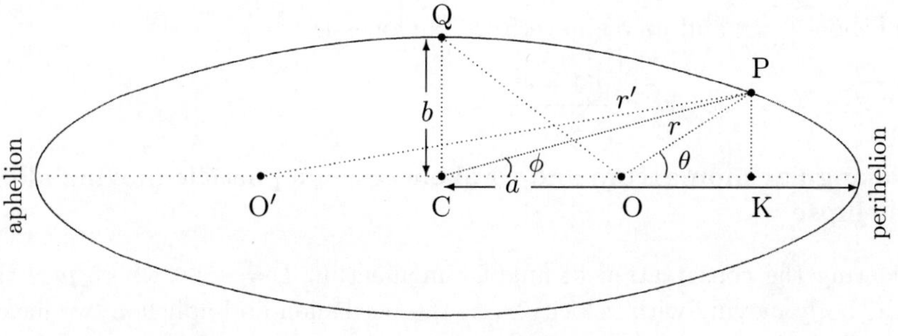

Figure 2: The distances a and b measured from the center C are the semi-major axis and semi-minor axis, respectively, of the ellipse. When the polar coordinates (r, θ) are measured from O, the closest point on the ellipse is known as perihelion whereas the farthest point is known as aphelion.

For getting relation between a, b and e, let us consider the point Q at the end of the minor axis. At Q, we have $r' = r$ and thus $r = a$. In the right-angle triangle OCQ, we have

$$\text{OQ}^2 = \text{CO}^2 + \text{CQ}^2 \qquad a^2 = a^2e^2 + b^2 \qquad b^2 = a^2(1 - e^2) \tag{B2}$$

Thus, the eccentricity of the ellipse is

$$e = \sqrt{1 - (b^2/a^2)}$$

and the equation of ellipse, in the Cartesian coordinates with the origin at the center C of the ellipse, is

$$\frac{x^2}{a^2} + \frac{y^2}{b^2} = 1$$

When ϕ is the angle PCO, Cartesian coordinates of a point on the ellipse are $(a\cos\phi,\ b\sin\phi)$.

It is often convenient to express orbit of an object, moving on an elliptical path, in terms of the polar coordinates $(r,\ \theta)$, where r is the distance measured from the focus O of the ellipse and the angle θ is measured anticlockwise from the major axis of the ellipse. Using the Pythagorean theorem ($O'P^2 = O'K^2 + PK^2$) in the triangle O′PK, we have

$$r'^2 = (2ae + r\cos\theta)^2 + r^2\sin^2\theta = r^2 + 4a^2e^2 + 4aer\cos\theta \qquad \text{(B3)}$$

Using equation (B1) in (B3), we get

$$(r - 2a)^2 = r^2 + 4a^2e^2 + 4aer\cos\theta \qquad \text{and therefore} \qquad r = \frac{a(1-e^2)}{1+e\cos\theta}$$

This is polar form of the equation of ellipse. At the perihelion $\theta = 0$ and its distance from the focus is

$$r_p = \frac{a(1-e^2)}{1+e} = a(1-e)$$

At the aphelion $\theta = \pi$ and its distance from the focus is

$$r_a = \frac{a(1-e^2)}{1-e} = a(1+e)$$

(i) Total angular momentum and total energy of a particle moving along the ellipse

Considering the conservation of angular momentum $L = \mu r v$, where μ is the mass of the body moving with velocity v, at the perihelion and aphelion, we have

$$\mu r_p v_p = \mu r_a v_a \qquad \text{and therefore} \qquad \frac{v_p}{v_a} = \frac{r_a}{r_p}$$

Using the expressions for r_a and r_p, we get

$$\frac{v_p}{v_a} = \frac{1+e}{1-e} \qquad \text{(B4)}$$

Considering the conservation of total energy (kinetic + potential) at the perihelion and aphelion, assuming that the body of mass μ is moving around the mass M, we have

$$\frac{1}{2}\mu v_p^2 - \frac{GM\mu}{r_p} = \frac{1}{2}\mu v_a^2 - \frac{GM\mu}{r_a}$$

or

$$v_p^2 = v_a^2 + 2GM\left[\frac{1}{r_p} - \frac{1}{r_a}\right]$$

Using here the expressions for r_a and r_p, we get

$$v_p^2 = v_a^2 + \frac{2GM}{a}\left[\frac{1}{1-e} - \frac{1}{1+e}\right] = v_a^2 + \frac{4GMe}{a(1-e^2)}$$

Using equation (B4) here, we get

$$v_p^2 = v_p^2\frac{(1+e)^2}{(1-e)^2} + \frac{4GMe}{a(1-e^2)} \qquad \text{and thus} \qquad v_p^2 = \frac{GM}{a}\frac{1+e}{1-e}$$

and from equation (B4), we have

$$v_a^2 = \frac{GM}{a}\,\frac{1-e}{1+e}$$

Thus, the angular momentum is

$$L = \mu r_p v_p = \mu a(1-e)\sqrt{\frac{GM}{a}\,\frac{1+e}{1-e}} = \mu\,\sqrt{GMa(1-e^2)}$$

It shows that for a circular orbit ($e = 0$), the angular momentum would be maximum. Total energy is

$$E = \frac{1}{2}\,\mu v_p^2 - \frac{GM\mu}{r_p} = \frac{\mu}{2}\,\frac{GM}{a}\,\frac{1+e}{1-e} - \frac{GM\mu}{a(1-e)} = -\frac{GM\mu}{2a}$$

(ii) Area of the ellipse

Area of the ellipse can be obtained by using the Cartesian equation

$$\frac{x^2}{a^2} + \frac{y^2}{b^2} = 1$$

The area A of the ellipse is

$$A = 4\int_0^a y\ \mathrm{d}x = 4b\int_0^a \sqrt{1 - x^2/a^2}\ \mathrm{d}x$$

Putting $x = a\sin\theta$, we get

$$A = 4b\int_0^{\pi/2} \sqrt{1-\sin^2\theta}\ a\cos\theta\ \mathrm{d}\theta = 4ab\int_0^{\pi/2} \cos^2\theta\ \mathrm{d}\theta = \pi ab$$

(iii) Auxiliary circle

A circle of radius a concentric to the ellipse is known as the auxiliary circle (Figure 3). Cartesian coordinates of a point P on the ellipse are $(a\cos\phi,\ b\sin\phi)$, where ϕ is the angle subtended by the point P at the center of the ellipse. The coordinates of a point Q on the auxiliary circle are $(a\cos\theta,\ a\sin\theta)$ where θ is the angle subtended by the point Q at the center of the circle. Draw a vertical line QPK meeting the semi-major axis at the point K.

$$\mathrm{OK} = \mathrm{OP}\ \cos\phi \qquad \mathrm{PK} = \mathrm{OK}\ \sin\phi \qquad \mathrm{OK} = \mathrm{OQ}\ \cos\theta \qquad \mathrm{QK} = \mathrm{OQ}\ \sin\theta$$

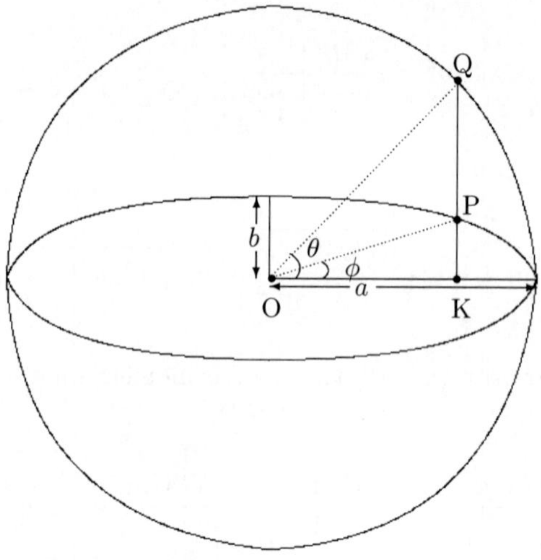

Figure 3: Ellipse and auxiliary circle.

C. Constants

Physical constants (c.g.s. units)

Speed of light	$c = 3.00 \times 10^{10}$ cm/s
Gravitational constant	$G = 6.67 \times 10^{-8}$ gm^{-1} cm^3 s^{-2}
Planck's constant	$h = 6.63 \times 10^{-27}$ erg s
Boltzmann constant	$k = 1.38 \times 10^{-16}$ erg/K
Stefan-Boltzmann constant	$\sigma = 5.67 \times 10^{-5}$ erg cm^{-2} s^{-1} K^{-4}
Charge of electron	$e = 4.80 \times 10^{-10}$ e.s.u.
Mass of electron	$m_e = 9.11 \times 10^{-28}$ gm
Mass of proton	$m_p = 1.67 \times 10^{-24}$ gm

Physical constants (SI units)

Speed of light	$c = 3.00 \times 10^{8}$ m/s
Gravitational constant	$G = 6.67 \times 10^{-11}$ kg^{-1} m^3 s^{-2}
Planck's constant	$h = 6.63 \times 10^{-34}$ J s
Boltzmann constant	$k = 1.38 \times 10^{-23}$ J/K
Stefan-Boltzmann constant	$\sigma = 5.67 \times 10^{-8}$ J m^{-2} s^{-1} K^{-4}
Charge of electron	$e = 1.6 \times 10^{-19}$ C
Mass of electron	$m_e = 9.11 \times 10^{-31}$ kg
Mass of proton	$m_p = 1.67 \times 10^{-27}$ kg

D. Molecules in interstellar space

Because of the assumed harshness (presence of γ-rays, x-rays) of interstellar environment, scientists were against the opinion of existence of molecules in the interstellar medium. Since the molecules do not emit radiations in the optical region, their presence in the interstellar space could not be noticed directly. In 1932, Karl Jansky, found that radiations of long wavelengths also reach to the Earth's surface from the interstellar medium. This discovery opened an entirely new astronomical world, known as the radio-window, having a range from few millimeter to a few decameters in wavelength. Hence, with the discovery of radio-window, detection of molecules in the interstellar medium could be possible, as the transitions between rotational levels in a molecule produce radiations lying in the radio-window. However, for three decades after the discovery of radio-window, no molecules were detected in the interstellar medium. It supported the opinion of scientists who were against the existence of molecules in the interstellar space.

Weinreb *et al.* (1963) detected the first molecule OH in the interstellar medium through its 18 cm radiation. With the detection of OH molecule in the interstellar medium, scientists were astonished as their assumption of harshness of high frequency radiations (γ-rays, x-rays) in the medium shattered down. However, they got interested in the search of more molecules in the interstellar medium. As of today, at least 254 molecules, reported in Table 1, have been identified in the interstellar medium.

Table 1. Molecules identified in the interstellar space*.

Molecules of two atoms:

AlCl, CO, CP, CS, H_2, HCl, N_2, NH, AlF, C_2, CH, CH^+, CN, KCl, NO, NaCl, NP, SO^+, SiN, NS, OH, SO, SiC, SiO, SiS, CO^+, HF, PO, FeO, SH, TiO, SH^+, CF^+, FeH, OH^+, CN^-, AlO, O_2, H_2^+, MgH, ND, NaH, $O_2{}^+$, HCl^+, ArH^+, CeO, TiH

Molecules of three atoms:

C_2H, C_2S, H_3^+, HCN, HNC, HCO, DCO^+, HNO, H_2O, N_2H^+, OCS, SO_2, HCS^+, c-SiC_2, NaCN, NaOH, HDO, C_3, CO_2, CH_2, N_2O, CNO, MgCN, HOC^+, MgNC, D_2H^+, HCP, CCD, DCN, DNC, AlNC, SiCN, SiNC, N_2D^+, H_2O^+, H_2Cl^+, FeCN, KCN, HO_2, HCO^+, D_2S, CCP, D_2O, OCN^-, NH_2, CCO, $CO_2{}^+$, TiO_2, AlOH, HNSi, CaNC

Molecules of four atoms:

l-C_3H, c-C_3H, C_3O, $HCNH^+$, H_2CO, C_3N^-, C_3S, H_2CS, HNCS, H_3O^+, NH_3, HNCO, D_2CO, HC_2N, H_2CN, NH_2D, C_2H_2, CH_3, C_4, HDCO, HDCS, ND_2H, NCCN, HSCN, C_2HD, HOOH, $HOCO^+$, HCNO, ND_3, PH_3, l-C_3H^+, D_2CS, HCO_2^+, H_2SiO, HMgNC, C_3N, HCCO

Molecules of five atoms:

C_4Si, CH_2CN, HCOOH, HC_3N, H_2CCO, H_2NCN, CH_4, *c*-C_3H_2, SiH_4, C_3HD, C_4D, C_5, H_2COH^+, HCCNC, HNCCC, DC_3N, CH_3D, HCCCN, CH_2NH, H_2CCC, *l*-C_3H_2, CH_3O, HNCNH, CNCHO, C_4H^-, GeH_4, NH_3D^+

Molecules of six atoms:

C_2H_4, C_5H, CH_3CN, CH_3OH, CH_3SH, HC_2CHO, H_2C_4, NH_2CHO, HC_3NH^+, CH_3OD, C_5N, H_2CCCC, HC_4N, E-HNCHCN, *c*-H_2C_3O, CH_2CNH, CD_3OH, CH_2DOH, CHD_2OH, CH_3NC, C_5N^-, C_5S, SiH_3CN

Molecules of seven atoms:

HC_5N, C_6H, CH_3NH_2, c-C_2H_4O, CH_3C_2H, CH_3CHO, CH_3CCD, HC_4OH, C_6H^-, HC_4CN, C_3H_4, HCO_2NH_2, CH_2DCCH, DC_5N, HCCCCCN, CH_2CHCN, H_2C_5, H_2CHOH, c-C_2H_5

Molecules of eight atoms:

CH_3C_3N, $HCOOCH_3$, CH_2CHCHO, CH_3COOH, $(NH_2)_2CO$, C_7H, C_3H_3NO, H_2C_6, CH_2OHCHO, $HCOCH_2OH$, CH_3CHNH, C_4H_4, C_2H_6, CH_2CCHCN, H_2NCH_2CN, H_2CCHCN, CH_3C_2NC, CH_3SiH_3

Molecules of nine atoms:

CH_3C_4H, $(CH_3)_2O$, HC_7N, C_2H_5OH, C_8H, CH_3CH_2CN, $H(C{\equiv}C)_3CN$, C_8H^-, CH_3CONH_2, CH_3CH_2OH, CH_3CHCH_2, CH_3CH_2SH

Molecules of ten and more atoms:

$HC_{11}N$, $(CH_3)_2CO$, CH_3C_5N, C_{60}, $C_{14}H_{10}$, $C_{16}H_{10}$, CH_3C_6H, CH_3COCH_2OH, $C_{60}{}^+$, C_{24}, $(CH_2OH)_2$, C_{70}, C_6H_6, HC_8CN, CH_3COOCH_3, C_3H_7CN, HC_9N, C_3H_8, CH_3CH_2CHO, $C_{14}H_{10}{}^+$, C_2H_5OCHO, *c*-C_4H_5N, NH_2CH_2COOH

---

*Here, we have not included isotopomers of molecules.

Weinreb, S., Barrett, A.H., Meeks, M.L. & Henry, J.C.: Radio observations of OH in interstellar medium, Nature, **200**, 829 -831 (1963).

E. Cosmic molecular laboratories

In the interstellar medium some regions are found to have molecules. These regions may be called the **cosmic molecular laboratories**. Names and positions of some of them are given in Table 2.

Table 2. Cosmic molecular laboratories.

Object	Right ascension	Declination
Orion Nebula	$05^h\ 35^m\ 17.3^s$	$-05°\ 23'\ 28''$
IRC+10216	$09^h\ 47^m\ 57.406^s$	$13°\ 16'\ 43.56''$
Epsilon Persei	$03^h\ 57^m\ 51.232^s$	$40°\ 00'\ 36.7752''$
HD 124314	$14^h\ 15^m\ 1.606^s$	$-16°\ 42'\ 24.38''$
Zeta Persei	$03^h\ 54^m\ 7.922^s$	$31°\ 53'\ 1.0812''$
Zeta Ophichuhi	$16^h\ 37^m\ 9.539^s$	$-10°\ 34'\ 1.52''$
Rho Ophiuchi	$16^h\ 28^m\ 06^s$	$-24°\ 32'\ 30''$
NGC 7822	$0^h\ 3^m\ 00^s$	$68°\ 37'\ 00''$
TMC 1	$4^h\ 38^m\ 38.6^s$	$25°\ 36'\ 18''$
B1-b Cloud	$3^h\ 33^m\ 20.8^s$	$31°\ 07'\ 34''$
L134 (Dark Cloud)	$15^h\ 51^m\ 26^s$	$-2°\ 43'\ 31''$
IC 443G Nebula	$06^h\ 13^m\ 42^s$	$22°\ 33'\ 30''$
Sgr B2	$17^h\ 44^m\ 9.5^s$	$-28°\ 21'\ 20''$
W51	$19^h\ 21^m\ 27^s$	$14°\ 24'\ 30''$
NGC 7027	$21^h\ 07^m\ 1.7^s$	$42°\ 14'\ 11''$
V Y Canis Majoris	$7^h\ 22^m\ 58.32^s$	$-25°\ 46'\ 03.23''$
16 Cygni B	$19^h\ 41^m\ 48.95^s$	$50°\ 31'\ 30.2''$
NGC 6210	$16^h\ 44^m\ 29.5^s$	$23°\ 47'\ 59.7''$
IRAS 16293-2422	$16^h\ 32^m\ 22.6^s$	$-24°\ 28'\ 33.0''$
W 49	$19^h\ 10^m\ 17^s$	$9°\ 06'\ 00''$
NGC 1952	$5^h\ 34^m\ 30^s$	$22°\ 01'\ 00''$
NGC 2024	$5^h\ 41^m\ 54^s$	$-01°\ 51'\ 00''$
NGC 2264	$06^h\ 41^m\ 06^s$	$09°\ 53'\ 00''$
Rosette Nebula	$06^h\ 33^m\ 45^s$	$04°\ 59'\ 54''$
SK 66-169	$05^h\ 36^m\ 54.50^s$	$-66°\ 38'\ 25''$
NGC 1333	$3^h\ 29^m\ 02^s$	$31°\ 20'\ 54''$
NGC 1333 IRAS 4A	$3^h\ 29^m\ 10.49^s$	$31°\ 13'\ 31''$
NGC 1333 IRAS 2	$3^h\ 28^m\ 49.7^s$	$31°\ 14'\ 47''$
NGC 4945	$13^h\ 05^m\ 27.5^s$	$-49°\ 28'\ 06''$
NGC 7538	$23^h\ 13^m\ 45.7^s$	$61°\ 28'\ 21''$
W 33A	$18^h\ 14^m\ 14.4^s$	$-17°\ 55'\ 50''$
CRL 618	$4^h\ 42^m\ 53.64^s$	$36°\ 00'\ 53.5''$
GL 2591	$20^h\ 27^m\ 35.8^s$	$40°\ 01'\ 14''$
CLD 2	$4^h\ 38^m\ 21^s$	$25°\ 41'\ 00''$
Horse head Nebula	$5^h\ 40^m\ 59^s$	$-02°\ 27'\ 30''$
Benard 1	$3^h\ 33^m\ 20.8^s$	$31°\ 7'\ 34''$
IRAS 04368+2557	$4^h\ 39^m\ 53.59^s$	$26°\ 03'\ 5.5''$
Cernis 52	$3^h\ 43^m\ 0.3^s$	$31°\ 58'\ 26''$
NGC 7023	$21^h\ 00^m\ 05^s$	$68°\ 10'\ 00''$
DR21(OH)	$20^h\ 39^m\ 02^s$	$42°\ 19'\ 38''$
Red rectangle nebula	$6^h\ 19^m\ 58.21^s$	$-10°\ 38'\ 14.69''$
HD 27778	$4^h\ 23^m\ 59.76^s$	$24°\ 18'\ 3.563''$
Cassiopeia A	$23^h\ 23^m\ 26^s$	$58°\ 48'\ 00''$
NGC 6334	$17^h\ 19^m\ 58^s$	$-35°\ 57'\ 47''$
55 Cygni	$20^h\ 48^m\ 56.29^s$	$46°\ 06'\ 50.88''$
W3 (OH)	$2^h\ 27^m\ 4.1^s$	$61°\ 52'\ 22''$
Cygni OB2-12	$20^h\ 32^m\ 40.95^s$	$41°\ 14'\ 29.28''$

F. Lane-Emden equation

The Lane-Emden equation is a dimensionless form of Poisson equation for the gravitational potential of a Newtonian self-gravitating, spherically symmetric, polytropic fluid. The equation can be expressed as the following.

$$\frac{1}{\xi^2}\frac{d}{d\xi}\left(\xi^2\frac{d\theta}{d\xi}\right)+\theta^n=0$$

where ξ is a dimensionless radius and θ is related to the density (and thus the pressure) through the relation $\rho=\rho_c\,\theta^n$, where ρ_c denotes the central density. The index n is the polytropic index that appears in the polytropic equation of state,

$$P=K\rho^{(n+)/n}$$

where P denotes the pressure and K is a constant of proportionality. For the θ, the standard boundary conditions are $\theta(0)=1$ and $\theta'(0)=0$. The solutions of this equation are known as the polytropes of index n.

Exact solutions

For some values of the polytropic index n, the Lane-Emden equation can be solved exactly.

(i) For n = 0, the equation becomes

$$\frac{1}{\xi^2}\frac{d}{d\xi}\left(\xi^2\frac{d\theta}{d\xi}\right)+1=0$$

On rearranging and integrating once, we get

$$\xi^2\frac{d\theta}{d\xi}=C_1-\frac{1}{3}\,\xi^3$$

Dividing both sides by ξ^2 and integrating again, we get

$$\theta(\xi)=C_0-\frac{C_1}{\xi}-\frac{1}{6}\,\xi^2$$

On applying the boundary conditions $\theta(0)=1$ and $\theta'(0)=0$, the constants of integration are obtained as $C_0=1$ and $C_1=0$. Thus, the solution is

$$\theta(\xi)=1-\frac{1}{6}\,\xi^2$$

(ii) For n = 1, the Lane-Emden equation can be expanded in the form

$$\frac{d^2\theta}{d\xi^2}+\frac{2}{\xi}\frac{d\theta}{d\xi}+\theta=0$$

For this, we assume a power series solution:

$$\theta(\xi) = \sum_{n=0}^{\infty} a_n \xi^n$$

where a_n are the expansion coefficients. This leads to a recursive relation between the expansion coefficients:

$$a_{n+2} = -\frac{a_n}{(n+3)(n+2)}$$

This relation can be solved leading to the general solution:

$$\theta(\xi) = a_0 \frac{\sin \xi}{\xi} + a_1 \frac{\cos \xi}{\xi}$$

The boundary condition for a physical polytrope demands that $\theta(\xi) \to 1$ as $\xi \to 0$. It can be possible when $a_0 = 1$ and $a_1 = 0$. Thus, the solution is

$$\theta(\xi) = \frac{\sin \xi}{\xi}$$

(iii) For $n = 5$, after a sequence of substitutions, it can be shown that the Lane-Emden equation has a further solution

$$\theta(\xi) = \frac{1}{\sqrt{1 + \xi^2/3}}$$

This solution is finite in mass but infinite in radial extent, and therefore the complete polytrope does not represent a physical solution.

Bibliography

R.H. Baker and L.W. Fredrick, Astronomy IX Edition (1964) van Nostrand Reinhold Company, New York.

B. Basu, An Introduction to Astrophysics, (1997) Prentice-Hall of India Pvt. Ltd., New Delhi.

J.C. Brandt and S.P. Maran, New Horizon in Astronomy (1972) W.H. Freeman & Company, San Francisco.

B.W. Carroll and D.A. Ostlie, Modern Astrophysics, (1996) Addison-Wesley Publishing Company, Inc., New York.

E. Chaisson, Universe: An Evolutionary Approach to Astronomy (1988) Printice Hall, Englewood Cliffs, New Jersey.

B.T. Draine, Physics of Interstellar and Intergalactic Medium (2011) Princeton University Press, Princeton.

M. Harwit, Astrophysical Concepts, II Edition (1988) Springer-Verlag, New York.

K. Holliday, Introductory Astronomy (1999) John Wiley & Sons Ltd., New York.

W.J. Kaufmann, Universe (1985) W.H. Freeman & Company, New York.

K.F. Kuhn, In Quest of the Universe (1991) West Publishing Company, St. Paul.

C. Payne-Gaposchkin and K. Haramundanis, Introduction to Astronomy, II Ed. (1961) Prentice Hall, Inc., New York.

W.M. Protheroe, E.R. Capriotti and G.H. Newsom, Exploring the universe, III Ed. (1984) Bell and Howell Company, Columbus.

F.H. Shu, The Physical Universe: An Introduction to Astronomy, (1982) University Science Books, Sausalito, California.

T.L. Swihart, Astrophysics and Stellar Astronomy (1968) John Wiley & Sons, Inc., New York.

E.W. Woolard and G.M. Clemence, Spherical Astronomy (1966) Academic Press, New York.

M. Zeilik, Astronomy: The evolving universe, III Ed. (1982) Harper & Row, Publishers, New York.

M. Zeilik and J. Gaustand, Astronomy: The cosmic perspective (1983) Harper & Row, Publishers, New York.

M. Zeilik and S.A. Gregory, Introductory Astronomy and Astrophysics, IV Edition (1998) Saunders College Publishing.

References of articles used

A simple representation of the Gregorian calendar

Suresh Chandra and Rahul:(Bull. I.A.P.T. **5**, 83 (1988).

Tide

Suresh Chandra: Physics Education (UGC, New Delhi sponsored) **17**, 227 (2000).

Pluto is a planet?

Suresh Chandra: Bull. I.A.P.T. **23**, 191 (2006).

Pluto is no more a planet

Suresh Chandra: Bull. I.A.P.T. **23**, 328 (2006).

Earth similarity index with two free parameters

Suresh Chandra, Subas Nepal & Mohit K. Sharma: Phys. Astron. Int. J. **2**, 156 (2018).

Index